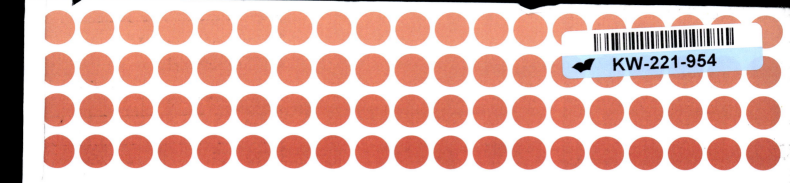

Biology

for CSEC® examination

Joanna George-Johnson and
Lisa Greenstein

with Kaylene Kellman-Holder

HODDER
EDUCATION
AN HACHETTE UK COMPANY

Acknowledgements

The Publishers would like to thank the following for permission to reproduce copyright material.

Text credits

page 102 Mathers C.D., Loncar D., 'Projections of global mortality and budens of disease from 2030' from *PLoS Medicine*, 3(11): 442. Additional information obtained from personal communication with C.D. Mathers. Source of revised HIV/AIDS figures: AIDS epidemic update. Geneva, Joint United Nations Programme on HIV/AIDS (UNAIDS) and World Health Organization (WHO), 2007; **page 278** AVERT (www.avert.org), 'The number of people in Caribbean countries infected with HIV/AIDS' from *UNAIDS/WHO 2008 Report of the global AIDS epidemic'*. **Biology at work (CD material):** **Chapter 6** Shelly-Ann Harris, 'Farmers go organic … reaping success with a natural fertilizer' from *The Jamaica Gleaner* (May 10, 2008); **Chapter 9** Oscar Ramjeet, 'No-smoking law comes in for criticism by Trinidad public' from *Caribbean Net News*; **Chapter 13** 'Living with kidney disease' from *The Sunday Gleaner* (September 11, 2005); **Chapter 15** 'Bone marrow implants save lives' from *Bio-Medicine* (December 22, 2008); **Chapter 19** Guttmacher Institute, 'Abstinence-only programs do not work, new study shows' from *News in Context, www.guttmacher.org/media/inthenews/2007/04/18/index.html* (New York: Guttmacher Institute, 2007); **Chapter 21** Nicholas Agar, 'Designer Babies: Ethical Considerations' from *www.actionbioscience.org/biotech/agar.html*; **Chapter 24** 'Artificial soil' from *www.newscientist.com*.

Photo Credits

p.7 *t all* © Mike van der Wolk (mike@springhigh.co.za), *b l −r* © geogphotos/Alamy, © Bon Appetit/ Alamy, © Denkou Images/Alamy, © Don Bayley/istock photo.com; **p.10** *t* © Steve Hamblin/Alamy, *ml, mr & br* © Mike van der Wolk, *bl* © stephan kerkhofs/istockphoto.com; **p.35** *all* © Mike van der Wolk; **p.59** © Mike van der Wolk; **p.64** © Mike van der Wolk; **p.65** © Mike van der Wolk; **p.70** © Nigel Cattlin /Alamy; **p.72** Lis Burch **p.73** © Mike van der Wolk; **p.76** © Dr P. Marazzi/Science Photo Library; **p.91** *tl* © John Birdsall/Press Association Images, *tr* © Wolfgang Schmidt/Das Fotoarchiv/Still Pictures; **p.101** © Mike van der Wolk; **p.103** Stockdisc/Getty Images; **p.105** *t* © Nancy Nehring/istockphoto.com, *b* © 2008 Nancy Nehring/istockphoto.com **p.117** *all* © Biodisc/Visuals Unlimited/Alamy **p.120** © Dennis Kunkel Microscopy, Inc./Visuals Unlimited/Corbis; **p.121** *t* © blickwinkel/Alamy, *m* © Andrew Syred/ Science Photo Library; **p.123** *l* © Imagestate Media (John Foxx), *c* © Miguel Angelo Silva/istockphoto.com, *r* © istockphoto.com; **p.127** *l* © Dr. George Wilder/Visuals Unlimited/Getty Images, *c* © J.C. Revy/Science Photo Library; **p.128** © Mike van der Wolk; **p.129** *t* © Frans Lanting/Corbis, *b* © Mike van der Wolk; **p.130** *all* © Mike van der Wolk; **p.132** © David T Gomez/iStockphoto.com; **p.134** PhotoAlto; **p.150** © Lehtikuva OY/Rex Features; **p.153** *tl* © Shawn Banton, *tr* © Mike van der Wolk, *ml* © J.C. Revy/Science Photo Library, *mc* © M.I. Walker/NHPA/Photoshot, *mr* © E. R. Degginger/Science Photo Library, *bl* © Bob Gibbons/Science Photo Library, *bc* © Biosphoto/Hazan Muriel/Still Pictures, *br* © Paul Bucknall/Alamy **p.160** © AJ Photo/Hop Americain/Science Photo Library; **p.165** © Western Ophthalmic Hospital/Science Photo Library; **p.169** *l* © Biodisc/Visuals Unlimited/Alamy, *r* © Carolina Biological Supply Company/Phototake/Alamy; **p.170** *t all* Ingram Publishing Ltd, *b* © urosr − Fotolia.com; **p.171** © David Cook/blueshiftstudios/ Alamy **p.182** *l* © PR. G Gimenez-Martin/Science Photo Library, *r* © Eric Grave/Science Photo Library; **p.186** *t* © mypokcik − Fotolia.com, *m* © Stephen Gibson − Fotolia.com, *b* © Tim Gainey/Alamy; **p.187** © Richard Griffin − Fotolia.com; **p.200** *l* Tryphosa Ho/Alamy, *r* Bob Krist/Corbis; **p.202** *l* © Tom Adams/ Visuals Unlimited/Getty Images, *r* © Brandon Cole Marine Photography/Alamy; **p.203** *t* © Michele Campini − Fotolia.com, *m* © Dean Holland /Alamy, *b* © Geoffrey Kidd/Alamy; **p.206** *l* © Brian Hoffman/Alamy, *r* Shawn Banton; **p.209** *l* © Finnbarr Webster/Alamy, *c* © Kevin Schafer/Alamy, *r* © Nick Gordon/ ardea.com; **p.216** © D. Phillips/Science Photo Library; **p.220** *l* © travelpixs/Alamy, *c* © Scott Camazine/Sue Trainor/Science Photo Library, *r* © Ray Ellis/ Science Photo Library; **p.221** *l* © Edd Westmacott /Alamy, *c* © Peter Andrews/Corbis, *r* © Michael Keller/Corbis; **p.222** *l* © Dominique Vernier − Fotolia. com, *r* Courtesy by Bayer; **p.225** © Mary Evans Picture Library/Alamy; **p.231** © Fabrice Coffrini/AFP/Getty Images; **p.236** © 2009 Oxford Scientific/ photolibrary.com; **p.240** © Bob Thomas/Popperfoto/Getty Images; **p.243** *t* © Dennis MacDonald/Index Stock/photolibrary.com, *b* © Arco Images GmbH/ Alamy; **p.248** *t* © Monika Adamczyk − Fotolia.com, *b* © imagebroker/Alamy; **p.257** © The Granger Collection, NYC/TopFoto; **p.270** *l−r*: © Mike van der Wolk, © B.Mason/Alamy, © Stockbyte/Getty Images, © Angela Hampton Picture Library/Alamy, © Steven May/Alamy; **p.274** © Shawn Banton; **p.282** *tl* © Niall Benvie/Corbis, *tr* © 2009 Juniors Bildarchiv/photolibrary.com, *ml* © Tina Manley/Alamy, *mr* © Daniel J Cox/Stone/Getty Images, *bl* © doraemon − Fotolia.com, *br* © Dante Fenolio/Science Photo Library; **p.288** Lis Burch; **p.290** Robert Harding Picture Library Ltd/Alamy; **p.307** *t* © Shawn Banton, *c* © Mike van der Wolk *b* 1996 J. Luke/PhotoLink/Photodisc/Getty Images; **p.308** © Andrew Woodley /Alamy.

Although every effort has been made to ensure that website addresses are correct at time of going to press, Hodder Education cannot be held responsible for the content of any website mentioned in this book. It is sometimes possible to find a relocated web page by typing in the address of the home page for a website in the URL window of your browser.

Hachette UK's policy is to use papers that are natural, renewable and recyclable products and made from wood grown in sustainable forests. The logging and manufacturing processes are expected to conform to the environmental regulations of the country of origin.

Orders: please contact Bookpoint Ltd, 130 Milton Park, Abingdon, Oxon OX14 4SB. Telephone: (44) 01235 827720. Fax: (44) 01235 400454. Lines are open 9.00–5.00, Monday to Saturday, with a 24-hour message answering service. Visit our website at www.hoddereducation.co.uk

First published in 2010 by
Hodder Education, an Hachette UK Company
338 Euston Road
London NW1 3BH

Impression number   5 4 3 2 1
Year   2012 2011 2010

Cover photo © Peter Arnold, Inc./Alamy
Illustrations by Barking Dog Art
Typeset in 11.5pt Bembo by Fakenham Photosetting Ltd, Fakenham, Norfolk
Printed in Dubai
A catalogue record for this title is available from the British Library

ISBN 978 0 340 985427

Contents

Chapter 1   Introduction: What you will learn and do in this course   1

Chapter 2   Life and living: living organisms in the environment   6
2.1   Characteristics of living things   6
2.2   Classifying organisms   7
2.3   Interacting with the environment   9
2.4   Feeding relationships   11
2.5   Special relationships   17
2.6   Nutrient cycling   19

Chapter 3   Cells: basic units of living organisms   22
3.1   An introduction to cells   22
3.2   Functions of parts of cells   27
3.3   Specialization and organization   29

Chapter 4   Moving substances: movement into and out of cells   33
4.1   Transport across membranes   33
4.2   Diffusion   35
4.3   Osmosis   37
4.4   The behaviour of cells in solutions   38

Chapter 5   Biological molecules: building blocks of life   42
5.1   The chemical needs of living organisms   42
5.2   Water, vitamins and minerals   43
5.3   Carbohydrates   44
5.4   Lipids   48
5.5   Proteins   50
5.6   Enzymes   53

Chapter 6   Photosynthesis: how plants make food   58
6.1   Photosynthesis and plant nutrition   58
6.2   The role of the leaf in photosynthesis   65
6.3   Understanding photosynthesis   68
6.4   Plant nutrition and mineral requirements   69

Chapter 7   Heterotrophic nutrition: how humans obtain and use food   72
7.1   Heterotrophic nutrition and digestion   72
7.2   Types and functions of teeth   75
7.3   Digestion in the alimentary canal   77
7.4   Absorption and the fate of digestive products   81
7.5   Health and diet   82

Chapter 8   Cellular respiration: how cells get energy   87
8.1   Respiration and its importance   87
8.2   Aerobic respiration   89
8.3   Anaerobic respiration   90

Chapter 9   Gas exchange and breathing: moving gases in plants and animals   95
9.1   Gas exchange   95
9.2   Breathing in humans   97
9.3   Smoking and health   102

Chapter 10 Transport in animals: blood and circulation   105
    10.1 The role of transport in animals   105
    10.2 Blood vessels   107
    10.3 The human heart   110
    10.4 Blood and its functions   113

Chapter 11 Transport in plants: moving substances around plants   116
    11.1 The role of transport in plants   116
    11.2 Transport of water   117
    11.3 Transport of manufactured foods   127

Chapter 12 Food storage: dealing with excess food   129
    12.1 The importance of storing food   129
    12.2 Food storage in plants   130
    12.3 Food storage in animals   132

Chapter 13 Excretion: getting rid of metabolic wastes   134
    13.1 Plants and animals must excrete wastes   134
    13.2 Excretion in humans   137

Chapter 14 Regulation and control: maintaining an internal balance   143
    14.1 Homeostasis   143
    14.2 Osmoregulation   147
    14.3 Temperature regulation   149

Chapter 15 Sensitivity and coordination: responding to the environment   152
    15.1 What are sensitivity and coordination?   152
    15.2 Nervous coordination   155
    15.3 The brain   160
    15.4 The eye   162
    15.5 Chemical coordination   166

Chapter 16 Support and movement: keeping upright and moving around   169
    16.1 Support systems in plants and animals   169
    16.2 Plant movements   175
    16.3 Animal movements   176

Chapter 17 Growth and development: getting bigger and more complex   180
    17.1 Basic mechanisms of growth   180
    17.2 Mitosis   182
    17.3 Germination and growth in plants   186
    17.4 Measuring growth   190
    17.5 Controlling growth and development   194

Chapter 18 Asexual reproduction and sexual reproduction in plants: continuing life   200
    18.1 Asexual and sexual reproduction   200
    18.2 Sexual reproduction in plants   205

Chapter 19 Sexual reproduction in humans: reproductive success   210
    19.1 Meiosis   210
    19.2 Sexual reproduction in humans   211
    19.3 Contraceptive methods   218

Chapter 20 **Continuity and variation: how characteristics are passed on and changed** 225
  20.1 Mendel and genetics 225
  20.2 Genetic variation 229
  20.3 Patterns of inheritance 231

Chapter 21 **Selection: choosing characteristics for survival** 240
  21.1 The principle of common descent 240
  21.2 Evolution and natural selection 241
  21.3 Artificial selection 247

Chapter 22 **Health and disease: transmission and control of disease** 252
  22.1 Disease – what is it? 252
  22.2 Transmission of diseases 256
  22.3 The human immune system 260
  22.4 Treating and controlling disease 264

Chapter 23 **Drugs and disease: the control and treatment of disease and the impact of drugs** 270
  23.1 Drugs and their uses 270
  23.2 Substance abuse and addiction 272
  23.3 The social impact of disease 277

Chapter 24 **Life and the environment: what organisms need for survival** 281
  24.1 The physical environment 281
  24.2 Investigating an ecosystem 287
  24.3 Human population growth 293

Chapter 25 **Humans and their environment: human activity and its environmental impact** 297
  25.1 Resources and their limits 297
  25.2 Environmental damage 299
  25.3 Environmental solutions 305

Index 311

On the CD
Biology at work – Chapters 1–25
Sample examination Papers 1 and 2
Glossary
Answers to sample examination papers

1 Introduction

What you will learn and do in this course

Welcome to *Biology for CSEC examination*. This course has been designed to help you understand and master all the content and skills you need to cover for the CXC examination in Biology.

This book is organized so that it covers the five main sections in the Biology syllabus:

A – Living organisms in the environment

B – Life processes

C – Continuity and variation

D – Disease and its impact on humans

E – Environment and human activities

This book is divided into clear and manageable chapters, each dealing with a topic from the syllabus. The chapters have many useful features to help you make sense of the topic and to make sure you understand all aspects of it.

How to get the best from your textbook

Your textbook is divided into 25 chapters, with the following features:

An overview and introduction to the topic is provided in clear, everyday English. This helps you to see where the topic you are studying fits into your everyday experience.

A list of objectives helps you see exactly what you should be able to do when you have completed the chapter. These objectives are closely linked to the syllabus so that all material is covered adequately in the book.

Each unit has a clear heading linked to the syllabus.

Chapter

2 Life and living

Living organisms in the environment

No living thing can exist on its own. Think about humans as an example. Human beings need other human beings for company, support and reproduction; they need plants and animals for food and as a supply of raw materials. They also need the natural resources of the Earth (water, soil, minerals and fossil fuels).

In this chapter you are going to study a range of living organisms to find out how they interact with one another and how they interact with and depend on the environment in which they live.

By the end of this chapter you should be able to:
- Classify things as either living or non-living
- Observe similarities and differences between living organisms and use these to group or classify organisms
- Explain how living organisms interact with each other and the environment
- Define the terms environment, habitat, population, community, ecosystem
- Interpret and draw food chains and food webs for given environments and identify organisms at different feeding levels in the food webs
- Construct simple ecological pyramids of number, biomass and energy
- Show that energy transfer through food chains is inefficient
- Discuss commensalism, mutualism and parasitism
- Explain how nutrients such as carbon and nitrogen are constantly cycled

Unit

1 Characteristics of living things

Biology is the study of living organisms. But it is not always clear that something is living or non-living. For example, is a coconut living or non-living?

All living organisms share certain characteristics. Biologists use these seven characteristics to decide whether things are living or non-living.

- **Respiration** – the process by which living cells release energy
- **Nutrition** – plants make their own food by photosynthesis; animals eat plants and/or other animals to supply the food they need to carry out life processes
- **Growth and development** – changes in size and structure as an organism gets bigger and more complex
- **Excretion** – life processes produce waste which the organism needs to get rid of by excretion
- **Reproduction** – plants and animals produce offspring (new plants and animals) either by vegetative or sexual reproduction
- **Sensitivity** (or irritability) – the ways in which living organisms detect changes happening inside or around them and respond to these changes

Key vocabulary is highlighted and the meanings of words are explained in the text. These words are also included in the glossary which you can find on the accompanying CD for easy reference.

Concepts are well illustrated with clear diagrams and photographs. These help to simplify complex biological processes and allow you to visualize what you are learning about.

Throughout the text you will find simple practical activities. These will help you understand and observe processes. They will also help you develop the skills you need to combine and use when you do more formal SBA practicals.

SBA practicals are a requirement of the CSEC examination. They are designed to test your skills in practical work and experimentation. SBA practicals are integrated into the chapters so that you can deal with them in context.

SBA practicals provide an abbreviated list of the specific SBA skills that may be assessed in the activity. These are:

ORR – Observation/recording reporting

MM – Manipulation/measurement

D – Drawing

AI – Analysis and interpretation

PD – Planning and design

More about transpiration

In leaves, water diffuses out of the cells into the air spaces in the spongy mesophyll layer as water vapour. This increases the concentration of water in the air spaces. This increased concentration creates a concentration gradient with the layer of air just outside the leaf. So water diffuses out of the leaves through small pores on the underside of the leaf called **stomata**. Each stomata is surrounded by a pair of **guard cells** which control the opening and closing of the stomata.

Transpiration can also occur through loosely packed cells on the surface of woody stems called **lenticels**. Gases are lost through these openings too.

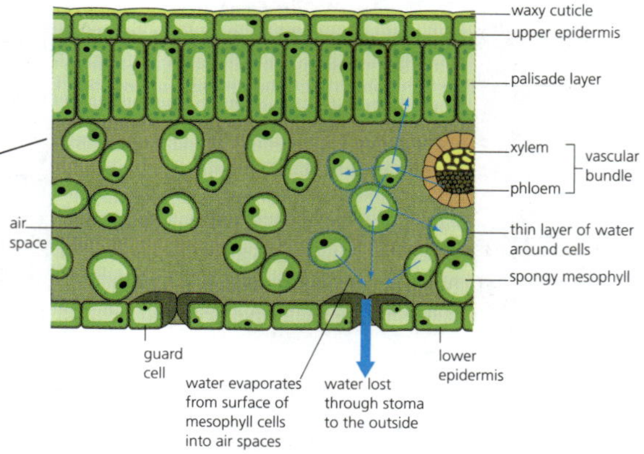

Figure 11.5 The passage of water through a leaf.

120

Cells

Activity 3.1 Observing cells

For this activity your teacher will provide you with prepared slides or show you how to prepare a slide.

1 Examine at least one example of an animal cell and one example of a plant cell under the light microscope.
2 What things do these cells have in common?

Activity 5.1 Testing for reducing sugars

SBA skills
Observation/recording/reporting (ORR)
Manipulation/measurement (MM)
Analysis and interpretation (AI)

Aim: To test for the presence of reducing sugars.

Apparatus and materials
■ reducing sugar sample
■ dropping pipette
■ Benedict's solution
■ test tube
■ hot water bath
■ safety glasses

Method
1 Place about 1 ml of the reducing sugar sample into a clean test tube using the dropping pipette. Note its colour.
2 Add about 1 ml of Benedict's solution to the tube and shake it well. Note any colour changes.
3 Heat the test tube containing the mixture in a hot water bath for 1 to 2 minutes. Make a note of any observations you make. Remove the test tube from the hot water bath after this time.

Observations and results
1 Was there a colour change in the solution in the test tube?
2 How many colours did the solution change to before the last colour?
3 What was the colour change after heating the solution?

A list of apparatus and equipment is often given for an SBA practical. Step-by-step instructions enable you to complete the practical.

Where appropriate, we have included a 'Biology at work' focus to bring what you are learning into context and to relate the Biology to real life situations and applications. These are real case studies or articles drawn from a range of Caribbean and international sources. These can be found on the accompanying CD.

Discussion questions will help you work through the information in the Biology at work sections, make sense of what you have read, and see how it relates to your own studies.

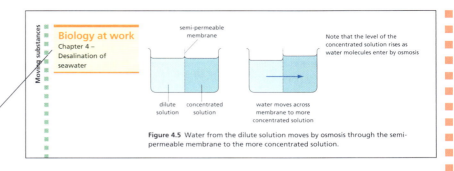

Moving substances

Biology at work
Chapter 4 – Desalination of seawater

semi-permeable membrane

Note that the level of the concentrated solution rises as water molecules enter by osmosis

dilute solution
concentrated solution

water moves across membrane to more concentrated solution

Figure 4.5 Water from the dilute solution moves by osmosis through the semi-permeable membrane to the more concentrated solution.

The skills boxes identify and teach the skills you need to excel at Biology. Once a skill is taught, you will apply it and then reuse it as you work through the course.

Carrying out and recording observations

Observation is a vital science process skill. You can learn about the world around you by using your five senses to observe objects, organisms and events. Two types of observations are used in Biology.

- **Qualitative observations** – using only your senses to observe, for example, that an animal is large, it eats grass and it has long brown hair
- **Quantitative observations** – using numbers or amounts to describe things: for example, the mass of a leaf or the fact that fruits contain ten seeds each.

To make a good observation you should be as detailed as possible. For example, if you are observing a leaf you should describe its colour, shape, size and texture. If you are observing something that is developing or changing, you should describe what happens before, during and after the development or change.

Recording your observations

There are many different ways of recording your observations:

- verbal reports
- written reports
- tables
- sketches with labels
- graphs and charts
- maps and diagrams.

Tables are useful for recording observations in the field because they allow you to record what you have seen and to compare and classify things at the same time. When you use tables you should remember to:

9

Each unit ends with a bank of questions and activities which test your understanding of the work covered. Many of these questions can be answered using information in the unit. Others require more thought and application. All of these activities will help you prepare for examinations. Practice exam papers can be found on the accompanying CD.

Check your knowledge and skills

1 Explain what is meant by classification and why it is useful in scientific terms.
2 Why do scientists use body structure and function rather than size, colour or behaviour to classify animals?
3 Name one difference between a plant and a fungus.
4 Animals can move from place to place, but they do so in different ways.
   a Develop a table that you could use to record animals you might observe in your local area according to the ways in which they move.
   b Is type of movement a useful method of organising your observations? Give a reason for your opinion.
5 Make a key to classify the large cats shown in Figure 2.7. Use the labels to help you. You can decide whether to use a branching key or a word key.

Each chapter ends with:

- a summary of key concepts to help you revise
- a list of skills and/or practical activities that you should be able to do, with page references, so that you can look them up if you need to.

The table below lists the chapters in the book and shows you where SBA practicals, skills and case studies can be found.

Chapter	SBA practicals	Skills	Biology at work
1 Introduction: what you will learn and do in this course			
2 Life and living: living organisms in the environment	Observing organisms in and around a water source Drawing a food web	Carrying out and recording observations Active reading	
3 Cells: basic units of living organisms	Observing plant cells	Drawing diagrams in Biology	Microbiology, African dust and coral reefs
4 Moving substances: movement into and out of cells	The effects of osmosis in plant cells Observing flaccid and turgid cells	Recording and discussing results	Desalination of seawater
5 Biological molecules: building blocks of life	Testing for reducing sugars, non-reducing sugars, starch, lipids and proteins Testing a range of food samples pH and the action of catalase Investigating the properties of enzymes	Drawing line graphs	Can a person take too many vitamins?
6 Photosynthesis: how plants make food	Testing a leaf for starch Photosynthesis and light intensity Showing that carbon dioxide and chlorophyll are necessary for photosynthesis Observing leaf structure Observing and drawing the internal structures of a leaf Investigating whether a water plant gives off oxygen	How to set up and carry out a fair test	Farmers go organic ... reaping success with natural fertilizer
7 Heterotrophic nutrition: how humans obtain and use food	Investigating the effect of pH on the action of pepsin Antacids and digestion		Over-the-counter antacids
8 Cellular respiration: how cells get energy	The effects of exercise on pulse and breathing rates	Understanding test and exam terms	
9 Gas exchange and breathing: moving gases in plants and animals	Drawing a bony fish gill Investigating breathing and carbon dioxide levels		No-smoking law comes into criticism by Trinidad public
10 Transport in animals: blood and circulation	Examining a mammalian heart	How to answer an essay question	Donating blood
11 Transport in plants: moving substances around plants	Observing the movement of coloured water through a stem Investigating the rate of uptake of water by a plant Measuring the uptake of water by a plant using a simple potometer Investigating water loss from leaves	Developing and testing a hypothesis	Eating themselves to death
12 Food storage: dealing with excess food	Drawing storage organs	Answering 'use of knowledge' questions in the exam	Calculating body mass index (BMI)

Chapter	SBA practicals	Skills	Biology at work
13 Excretion: getting rid of metabolic wastes		How to draw a bar graph	Living with kidney disease
14 Regulation and control: maintaining an internal balance		Exam skills – answering questions	Why do you feel hot when you have a cold?
15 Sensitivity and coordination: responding to the environment	Observing responses to stimuli in invertebrates	Concept mapping	Juggle your brain to a new brain in one week!
16 Support and movement: keeping upright and moving around	Studying support in plants, and the vertebral column Creating a model to demonstrate muscle action		Bone marrow implants save lives
17 Growth and development: getting bigger and more complex	Investigating mitosis in plant roots, and germination Evaluating the different methods of measuring growth Measuring growth of a potted plant Investigating the effect of an auxin on root growth in seedlings	Taking accurate measurements	
18 Asexual reproduction and sexual reproduction in plants: continuing life			
19 Sexual reproduction in humans: reproductive success			Does abstinence work as a contraceptive?
20 Continuity and variation: how characteristics are passed on and changed		How to draw a Punnett square	The human genome project (HGP)
21 Selection: choosing characteristics for survival			Designer babies
22 Health and disease: transmission and control of disease			Biological terrorism
23 Drugs and disease: the control and treatment of disease and the impact of drugs			Poison is in the dose Should marijuana be legalized? The papaya ring spot virus
24 Life and the environment: what organisms need for survival	Taking a soil profile Analysing soil composition, moisture and permeability Investigating percentage and density of weeds in a lawn		Artificial soil
25 Humans and their environment: human activity and its environmental impact			

2 Life and living

Living organisms in the environment

No living thing can exist on its own. Think about humans as an example. Human beings need other human beings for company, support and reproduction; they need plants and animals for food and as a supply of raw materials. They also need the natural resources of the Earth (water, soil, minerals and fossil fuels).

In this chapter you are going to study a range of living organisms to find out how they interact with one another and how they interact with and depend on the environment in which they live.

By the end of this chapter you should be able to:

- Classify things as either living or non-living
- Observe similarities and differences between living organisms and use these to group or classify organisms
- Explain how living organisms interact with each other and the environment
- Define the terms environment, habitat, population, community, ecosystem
- Interpret and draw food chains and food webs for given environments and identify organisms at different feeding levels in the food webs
- Construct simple ecological pyramids of number, biomass and energy
- Show that energy transfer through food chains is inefficient
- Discuss commensalism, mutualism and parasitism
- Explain how nutrients such as carbon and nitrogen are constantly cycled

Unit 1 Characteristics of living things

Biology is the study of living organisms. But it is not always clear that something is living or non-living. For example, is a coconut living or non-living?

All living organisms share certain characteristics. Biologists use these seven characteristics to decide whether things are living or non-living.

- **Respiration** – the process by which living cells release energy
- **Nutrition** – plants make their own food by photosynthesis; animals eat plants and/or other animals to supply the food they need to carry out life processes
- **Growth and development** – changes in size and structure as an organism gets bigger and more complex
- **Excretion** – life processes produce waste which the organism needs to get rid of by excretion
- **Reproduction** – plants and animals produce offspring (new plants and animals) either by vegetative or sexual reproduction
- **Sensitivity** (or irritability) – the ways in which living organisms detect changes happening inside or around them and respond to these changes
- **Movement** – living organisms may move parts of themselves in response to stimuli or they may move around from place to place (birds fly to warmer climates during winter).

All of these characteristics must be present for an organism to be alive. If even one is not present, then the object being studied is considered to be non-living.

Figure 2.1 Different types of living organisms all display the seven characteristics of life.

Check your knowledge and skills

1 Three characteristics of living organisms are listed in the table below.
  a Copy the table and complete column 2. Give one reason why each characteristic is essential in living organisms.
  b Referring to Figure 2.1, complete columns 3–5 in the table by naming the structures in each organism that are linked to each characteristic.

Characteristic	Why characteristic is essential to life	Related structures		
		fungus	cattle	banana tree
Nutrition				
Growth				
Reproduction				

2 Leroy says that sugar crystals in the bottom of a beaker are alive because they are growing and developing. Do you agree with him? Give one reason for your answer.
3 Identify and list five things in the classroom that are:
  a living   **b** non-living (but were once alive)   **c** non-living (and were never alive).

Classifying organisms

Unit 2

Scientists use the similarities and differences between organisms to classify organisms into groups. Organisms that share similar characteristics are placed in the same group. For example, the fruits in Figure 2.2 can all be classified as apples.

The science of formally classifying, naming and grouping organisms is called **taxonomy**.

Figure 2.2 Four varieties of apples.

Activity 2.1 Classifying plants by observing their characteristics

Work in pairs. Observe the plants in Figure 2.3 carefully and make a note of the characteristics that you can see (the visible characteristics).

Figure 2.3

1 Place the plants in Figure 2.3 into two groups.
2 Rearrange the plants so that you have four groups.
3 What characteristics did you use to make the two groups?
4 What characteristics did you use to make the four groups?
5 What other visible characteristics could you use to make groups? Try to think of at least three.
6 Are the methods that you used to place the plants in groups scientific enough to eliminate confusion? Give a reason for your answer.

Carrying out and recording observations

Observation is a vital science process skill. You can learn about the world around you by using your five senses to observe objects, organisms and events. Two types of observations are used in Biology.

- Qualitative observations – using only your senses to observe, for example, that an animal is large, it eats grass and it has long brown hair
- Quantitative observations – using numbers or amounts to describe things: for example, the mass of a leaf or the fact that fruits contain ten seeds each.

To make a good observation you should be as detailed as possible. For example, if you are observing a leaf you should describe its colour, shape, size and texture. If you are observing something that is developing or changing, you should describe what happens before, during and after the development or change.

Recording your observations

There are many different ways of recording your observations:

- verbal reports
- written reports
- tables
- sketches with labels
- graphs and charts
- maps and diagrams.

Tables are useful for recording observations in the field because they allow you to record what you have seen and to compare and classify things at the same time.

When you use tables you should remember to:

■ give your table a clear heading
■ plan the table before you start. Decide how many rows and columns you will need to record your observations
■ provide headings for all rows or columns
■ find the quickest and easiest way to record the data – if you are counting species use tally marks; if you are comparing characteristics you can use ticks or crosses.

Activity 2.2 Observing organisms in and around a water source

SBA skills

Observation/recording/reporting (ORR)

Aim: To find out more about the living organisms in and around a pond by observing them.

Method
1 Select a site close to school where you can observe living organisms in and around water. (You can use a rock pool, river, fresh-water pond or swamp area.)
2 Identify and list the living organisms found at the site you have chosen. You can use common names for these organisms.
3 List at least ten non-living components found at the site.
4 Draw up a table to record and compare the plants and animals found in the water and close to the water.

Check your knowledge and skills
1 Explain what is meant by classification and why it is useful in scientific terms.
2 Why do scientists use body structure and function rather than size, colour or behaviour to classify animals?
3 Name one difference between a plant and a fungus.
4 Animals can move from place to place, but they do so in different ways.
   a Develop a table that you could use to record animals you might observe in your local area according to the ways in which they move.
   b Is type of movement a useful method of organizing your observations? Give a reason for your opinion.

Unit 3 Interacting with the environment

Everything around you is part of your **environment**. If you are in a classroom environment then you are probably surrounded by living and non-living things, such as desks, books, your teacher, other students, windows, plants, air and germs.

When your lessons end for the day and you leave the classroom, your environment changes. But you are still surrounded by living and non-living things. We call the living things in the environment the **biotic** components of the environment and the non-living things the physical or **abiotic** components of the environment (see pages 283–4). Organisms are not found just anywhere in the environment. The place that suits an organism and where it can live is called its **habitat**.

Figure 2.4 This aquatic habitat is home to populations of mussels, anemones, seaweeds and crabs.

Living organisms need particular things in order to survive. For example, tropical fish need warm salty water, so they are found in habitats such as the Caribbean Sea where the water is warm and salty. Tropical fish are not found in the cold, icy waters near the poles because those waters do not provide a suitable habitat for them.

Biotic and abiotic components that organisms need in their habitats are:

- food source
- water
- air
- light
- suitable temperature.

Living things do not usually exist on their own. One habitat may be home to several different groups. For example, a rock pool may be the habitat for mussels, anemones, seaweed and crabs. Each of these groups is called a **population**.

Figure 2.5 What conditions would you expect to find in each of these places?

When two or more populations are found living in the same habitat they form a **community**.

A community of living organisms and the physical environment associated with it (the air, water and soil) forms an **ecosystem**. Within an ecosystem, the living organisms (biotic elements) interact with each other and with the abiotic elements around them. There are many different ecosystems on Earth. Land ecosystems are called **terrestrial** ecosystems. Fresh and salt-water ecosystems are called **aquatic** ecosystems. The photographs in Figure 2.5 show you four different examples.

Check your knowledge and skills

1 Write a sentence to explain what the following words mean:
   a environment   b habitat
   c population   d community
   e ecosystem.

2 Study the environments shown in the photographs in Figure 2.5.
   a What conditions would you expect to find in each one?
   b Make a list of the plants and a list of the animals that you would expect to find in each of these environments.

3 Why would you not find exactly the same plants and animals in each place?

Unit 4

Feeding relationships

Feeding relationships are a good example of how living organisms in an ecosystem interact with each other.

Autotrophs or **producers** are organisms that make their own food. In most terrestrial ecosystems the producers are green plants that make food by photosynthesis. In aquatic ecosystems green algae produce food by photosynthesis; some bacteria produce their own food from inorganic compounds.

Heterotrophs or **consumers** are organisms that feed on other organisms. The following are all heterotrophs.

- **Herbivores** – organisms that feed on plants or algae. We call these the primary consumers because they obtain their food directly from the producers.
- **Carnivores** – these feed on other animals. We call these secondary consumers because they eat other consumers.
- **Omnivores** – animals that eat both plants and animals (producers and consumers). Human beings are omnivores.
- **Decomposers** – fungi and bacteria which obtain food by decomposing or breaking down dead organic material and wastes. These organisms play an important role in ecosystems because they break down complex substances and release simple inorganic substances from dead plants so that these can be returned to the environment.

Food chains

We can represent feeding relationships within an ecosystem using a diagram called a **food chain**. The arrows in a food chain show the direction of energy flow from organism to organism in the ecosystem. The different stages or levels of feeding in a food chain are called **trophic levels**. Figure 2.6 shows a simple food chain for a terrestrial ecosystem.

producer   primary   secondary   tertiary   decomposer
             consumer   consumer   consumer

grass   grasshopper   snake   hawk   fungi

Figure 2.6 A simple food chain.

Energy transfer in food chains

The sun is the original source of energy in most food chains. However, only about 1% of the sun's energy actually reaches the leaves of green plants and algae and of this 1%, less than 10% is converted into food. The producers use some of this energy for their own life processes and they release some energy to the environment in the form of heat. When the plant dies, its stored energy is transferred to decomposers (from where it is ultimately lost as heat).

When secondary consumers eat producers, they tend not to eat the whole plant and some plant material (cellulose in particular) is not very digestible, so it does not provide food energy. As a result, less than 10% of the energy contained in the plants is transferred to the next trophic level. Again, these consumers use some of the energy for their own life processes and release some energy to the

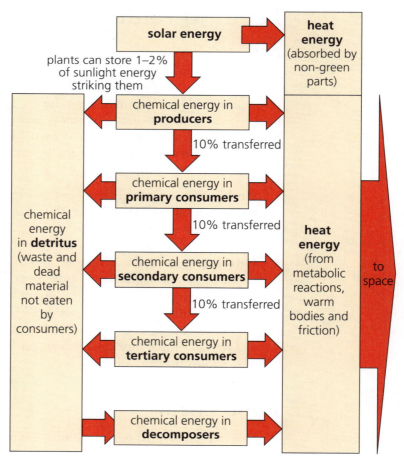

Figure 2.7 Energy gets lost as it flows through a food chain.

environment in the form of heat. When the consumers die, the energy stored in their bodies is also transferred to decomposers and ultimately lost as heat. Figure 2.7 shows how much energy is transferred and lost in a typical food chain.

The diagram shows clearly that the amount of energy transferred from one trophic level to the next in a food chain is reduced at every level.

Level 1	→	Level 2	→	Level 3	→	Level 4	→	Level 5
100%		10%		1%		0.1%		0.01%

The rate of energy transfer through food chains also means that there is less energy to be passed on to higher trophic levels, so there tend to be fewer organisms at each successive level. For example, in the food chain shown in Figure 2.6 there might be 3000 grass plants, 150 grasshoppers, five snakes and only one hawk.

Food webs

The feeding relationships that exist in nature are usually more complicated than the ones shown in a simple food chain. For example, in a rainforest ecosystem (Figure 2.8) there would be many different producers, including fallen leaves, living leaves, grasses, seeds, fruits and flowers. These producers would, in turn, be eaten by many different primary consumers. Some of these primary consumers would themselves be food for more than one type of secondary consumer and so on.

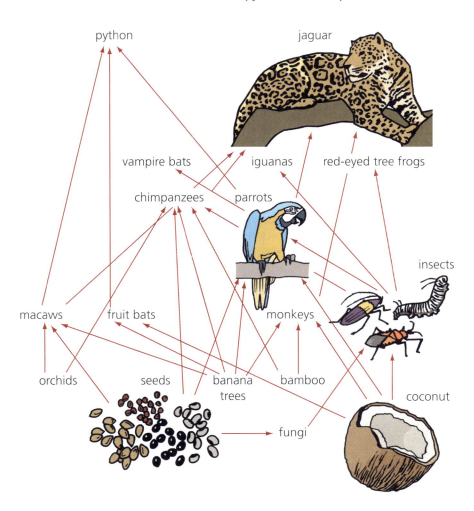

Figure 2.8 Some of the feeding relationships in a rainforest ecosystem.

In order to show complicated feeding relationships we join all the food chains together into a **food web**. A food web outlines the feeding relationships and overlaps in these relationships for a whole ecosystem. Figure 2.9 shows a food web for a coral reef ecosystem. Food chains and food webs for aquatic ecosystems are usually longer and more complex than those for terrestrial ecosystems because there is more sunlight available to producers. Many aquatic food chains begin with very small green plants called phytoplankton or green algae.

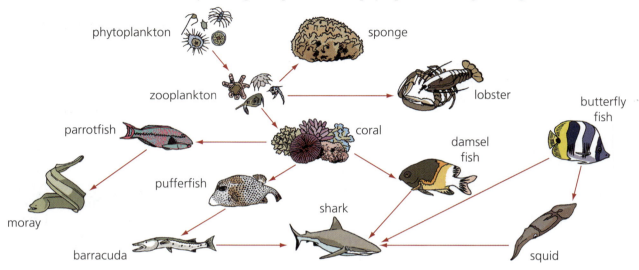

Figure 2.9 A coral reef food web.

SBA skills
Observation/recording/
reporting (ORR)
Analysis and
interpretation (AI)

Activity 2.3 Drawing a food web

Study Figure 2.10 carefully. The organisms can be identified using the key.

Use the information from the diagram to construct a food web to show the feeding relationships in this ecosystem.

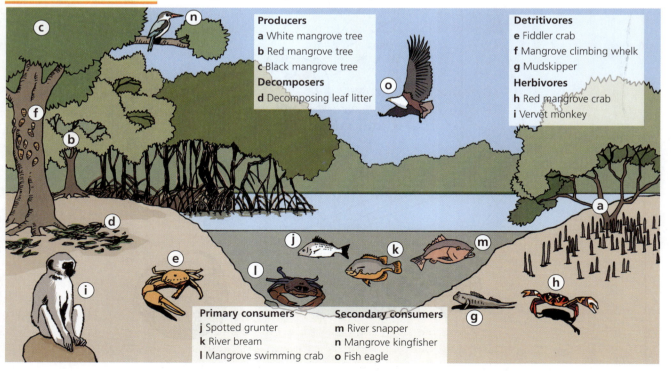

Figure 2.10 A mangrove ecosystem.

Ecological pyramids

Food webs are useful for showing feeding relationships, but they do not tell us about the number of organisms at each trophic level or how energy flows through the ecosystem. In other words, food webs provide qualitative data about ecosystems, but they do not provide quantitative data (look back at page 8 for a description of the terms quantitative and qualitative). Quantitative data about different trophic levels in an ecosystem and the energy flows can be depicted using a mathematical model or graph, called an **ecological pyramid**. There are three main types of ecological pyramids.

Pyramids of number

The actual numbers of organisms at different trophic levels are represented as bars or tiers of the pyramid. In most pyramids of number the producers are the most plentiful, but this is not always the case.

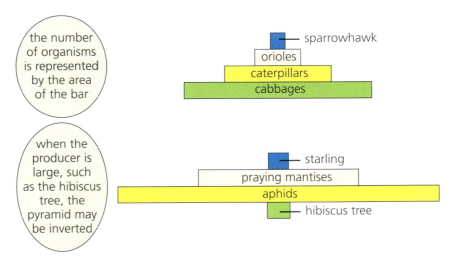

Figure 2.11 Pyramids of number show the number of organisms at each trophic level.

Pyramids of biomass

Biomass is the mass of living organisms (the number of organisms multiplied by the average mass of each organism) at each trophic level at a given time.

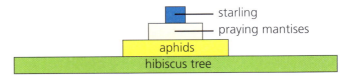

Figure 2.12 Pyramid of biomass.

The mass of all organisms at a level is represented by the area of the bar and given as a mass in kilograms. Pyramids of biomass are easy to draw to scale and they are not inverted (as in the case of a large producer shown in Figure 2.11). However, pyramids of biomass can be misleading because organisms reproduce at different rates and this can impact on the biomass measurements.

Pyramids of energy

Energy pyramids measure the amount of energy flowing through a system over a period of time, normally a year. Values are expressed as units of energy per unit area per unit of time, for example, kJ per m² per annum (kJ/m² pa).

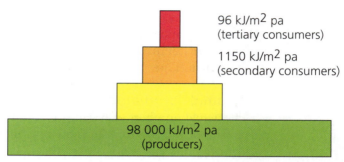

96 kJ/m² pa
(tertiary consumers)

1150 kJ/m² pa
(secondary consumers)

98 000 kJ/m² pa
(producers)

Figure 2.13 Pyramid of energy showing the amount of energy transferred between trophic levels in a fish pond in one year.

Check your knowledge and skills

1 Why do most food chains rely on energy from the sun?
2 What is the difference between a producer and a consumer?
3 What type of food do carnivores eat?
4 How is energy lost to the environment at each level of a food chain?
5 Draw a food web that includes algae, snails, small fish, water beetles, frogs and snakes.
   a On your food web label the producers, primary consumers and secondary consumers.
   b What is likely to happen in the food web if the number of algae-eating snails increases rapidly?
6 Construct a pyramid of biomass based on the following data:
   ■ 30 000 producers, each of average mass 45 g
   ■ 2000 primary consumers, each of average mass 120 g
   ■ 12 secondary consumers of total mass 1800 g
   ■ 3 tertiary consumers of total mass 3200 g

Special relationships

If two different species in an ecosystem rely on each other for survival we say they are interdependent or symbiotic. There are three main types of symbiosis.

Commensalism

This is a symbiotic relationship between two animal or plant species where only one species benefits from the relationship. The other species is not significantly affected by the relationship. Figure 2.14 shows some examples of commensalism.

a remora and shark

b cattle egrets eating insects while cattle graze

c epiphyte on tree

d small animals living in marine worm burrows

e barnacles on whales

f pseudoscorpion on beetle

Figure 2.14 Commensal relationships benefit only one of the species involved in the relationship.

Mutualism

This is a symbiotic relationship between two species in which both species benefit from the interaction and neither is harmed by the interaction. Figure 2.15 shows some examples of mutualism.

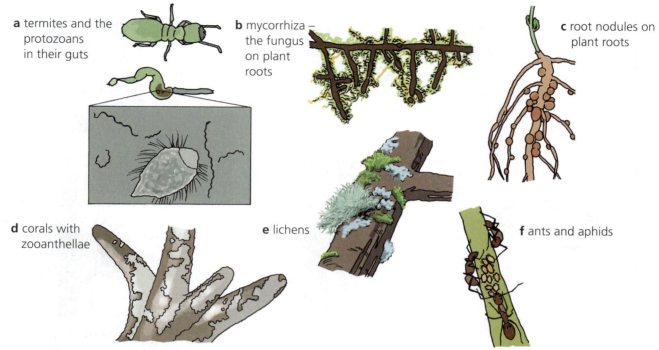

a termites and the protozoans in their guts

b mycorrhiza – the fungus on plant roots

c root nodules on plant roots

d corals with zooanthellae

e lichens

f ants and aphids

Figure 2.15 Mutual relationships benefit both species involved in the relationship.

Parasitism

This is a symbiotic relationship in which one species (the parasite) lives on or in another (the host). The parasite obtains food and/or shelter from the host and it often harms the host in return. Figure 2.16 shows some examples of parasitism.

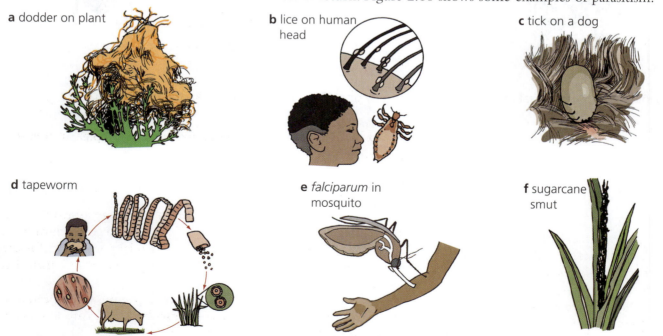

a dodder on plant

b lice on human head

c tick on a dog

d tapeworm

e *falciparum* in mosquito

f sugarcane smut

Figure 2.16 A parasite lives on or in a host and often harms or even kills the host.

Check your knowledge and skills

1 Copy and complete the table below.

Type of symbiosis	Examples	Names of partners	Advantages of relationship	Disadvantages of relationship
Mutualism				
Parasitism				
Commensalism				

2 Explain why an epiphyte growing on a palm tree is not a parasite.

Unit 6 Nutrient cycling

Nutrients are not lost from ecosystems. When an organism dies, the nutrients in its body are broken down and returned to the environment. Figure 2.17 shows how both producers and decomposers play a role in the recycling of nutrients.

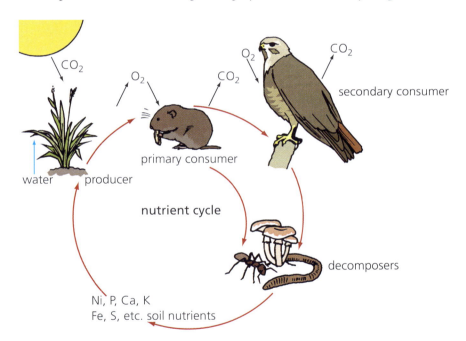

Figure 2.17 Complex organic molecules are broken down into simpler substances that can be absorbed by plants.

The carbon cycle

Carbon is found in all organic molecules. Plants are the only organisms that can manufacture organic molecules by using the carbon in carbon dioxide (during the process of photosynthesis). All other organisms are dependent on plants for the carbon-based compounds they need as a source of energy and as raw materials for cellular growth.

Through photosynthesis, feeding, respiration and decomposition, carbon is constantly recycled in ecosystems. The various processes by which carbon is taken up, used and released are collectively known as the carbon cycle.

Human activities
You will learn more about human activities and how these impact on the oxygen–carbon balance in the atmosphere in Chapter 25.

The nitrogen cycle

Plants need nitrogen because it is built into nitrogen–containing proteins, amino acids and other organic compounds. Nitrogen is the most common gas in our atmosphere but plants cannot use nitrogen in its gas form. They can only use it when it is fixed in organic compounds in the form of nitrates.

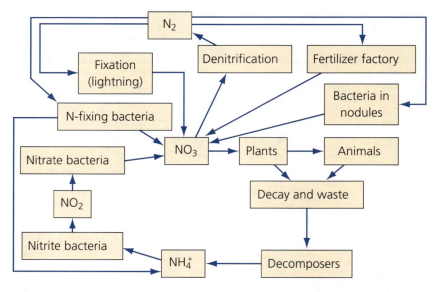

Figure 2.18 Nitrogen is cycled between living organisms and the environment in the nitrogen cycle.

Nitrogen-fixing is a process in which nitrogen is combined with oxygen to form nitrate (NO_3). This process depends on various nitrogen-fixing bacteria that make use of chemical action to convert nitrogen and other nitrogen–containing compounds to nitrates. These bacteria are found in the soil and they are very important in the cycling of nitrogen. Nitrogen can also be fixed in nature by the action of lightning.

Nitrates are taken up by roots of plants. Once the plant dies, the nitrogen compounds in its remains are decomposed and returned to the soil and the cycle continues.

Check your knowledge and skills

1 What is carbon and why is it important for living organisms?
2 Explain how respiration, photosynthesis and the burning of fossil fuels contribute to the carbon cycle.
3 What role do nitrogen-fixing bacteria play in the nitrogen cycle?
4 Explain how nutrient cycling involves both the biotic and abiotic components of an ecosystem.

How to make sense of what you read

The following steps will help you get the most out of your Biology textbook and will give you guidance in how to make links between different topics.

- Read the Chapter introduction to see what objectives and content will be covered.
- Read the list of learning outcomes that you should know once you have read the whole chapter – *do not be put off by this* – use it as a checklist.
- Next, read through the chapter. Think of questions you'd like answered as you read. It helps to read with a pencil and paper beside you to write down questions and make notes. This is called active reading.
- Stop as you read and try to remember what you have read. Some people find it is useful to repeat what they have learnt out loud and to make notes from the textbook next to the notes they have made during lessons.
- As you read take notice of examples, equations or the notes in the margin that will help you to answer the questions at the end of each unit.
- As you read each section, stop and analyse how it fits in with what you already know. For example, gas exchange is linked to respiration because it is how cells get the oxygen they need to be able to respire. Similarly, gas exchange is linked to circulation and transport because the gases that are exchanged need to get to and from the cells in the body. This method of reading, making links and problem-solving is part of a positive cycle of learning which involves asking questions, finding answers and asking more questions. If you get into the habit of this 'active' reading you will find revision easier when it comes to exams.

Chapter summary

Do you know?

If you are unsure of any of the facts in the list, refer to the page number in brackets.

- Living organisms all respire, feed, grow and develop, reproduce, excrete, respond to their environment and move. (page 6)
- Living organisms can be grouped or classified according to their similarities and differences. (page 7)
- Living organisms interact with each other and with the non-living environment around them. (page 9)
- Feeding relationships can be shown as food chains and food webs for given environments. The levels at which organisms feed are known as trophic levels. (page 11)
- Energy transfer through food chains is inefficient and energy is not recycled in an ecosystem so it needs to be replenished constantly (by plants using energy from sunlight during photosynthesis). (page 12)
- A symbiotic relationship is one in which two species interact closely with each other. There are three types of symbiosis: commensalism, mutualism and parasitism. (page 17)

- Nutrients such as carbon and nitrogen are constantly cycled between living organisms and the environment. The processes by which these nutrients are cycled are known as the carbon cycle and the nitrogen cycle. (page 19)

Are you able to?

If you have trouble in doing these things, refer to the page number in brackets.

- Differentiate between living organisms and non-living objects based on the characteristics of living organisms. (page 7)
- Observe and classify organisms using visible characteristics and construct tables to record your observations. (pages 8–9)
- Define the key terms environment, habitat, population, community and ecosystem. (pages 9–11)
- Construct food chains, food webs and simple ecological pyramids. (pages 11–16)
- Identify the partners in symbiotic relationships and state the advantages and disadvantages of the relationship for each species. (pages 17–19)
- Read and interpret flow diagrams to find information and draw conclusions about the carbon cycle and the nitrogen cycle. (pages 19–20)

3 Cells

Basic units of living organisms

You can think of cells as the building blocks of organisms. In this chapter you will study cells to learn more about their structure and how they function. Once you understand that, you will learn how cells group together and perform specialized functions in living organisms.

By the end of this chapter you should be able to:

■ Identify and name the structures found in cells

■ Draw and label diagrams of typical animal and plant cells

■ State the functions of structures found in cells and say why they are important

■ Identify and explain the differences between plant and animal cells

■ Explain why cell specialization is important and give examples of specialization

1 An introduction to cells

Cells are the simplest structures that can live on their own. In other words, single cells exhibit all the characteristics of living things – they respire, grow, reproduce, feed, move, excrete and respond to stimuli.

Organisms that consist of only one cell are called **unicellular** organisms. An example of unicellular organisms is bacteria. Most plant and animal species are **multicellular**. This means they consist of many different cells that work together as a system to keep them alive.

Viruses

Viruses are a special case. Many scientists today believe that viruses are non-living particles and not cells. Viruses consist of two parts: an outer capsule made mostly of protein and an inner core of nucleic acid (RNA or DNA but not both). On their own, viruses do not display the characteristics of living things; in fact, many can be dried, crystallized and frozen without destroying them. Viruses do reproduce and evolve, but they can only do this inside a living cell.

Cells from different organisms

The shape, structure and size of a cell depends on the organism that the cell is part of. Cells from bacteria, fungi, protists, plants and animals are all different.

Figure 3.1 shows you cells from different organisms. Notice that all these cells have structures inside them. These structures are known as **organelles**.

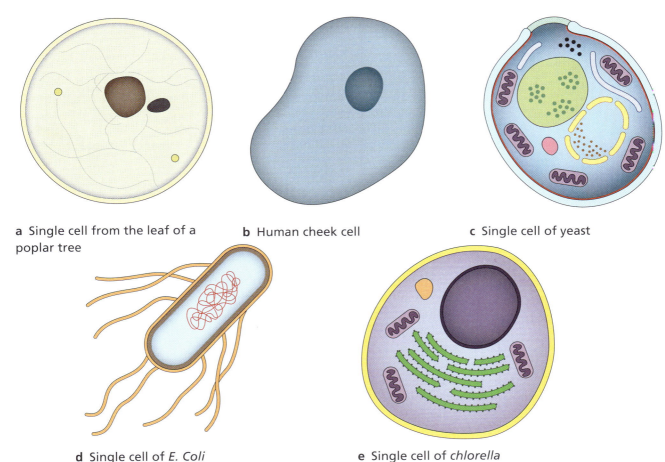

a Single cell from the leaf of a poplar tree

b Human cheek cell

c Single cell of yeast

d Single cell of *E. Coli*

e Single cell of *chlorella*

Figure 3.1 Cells from different organisms show different characteristics.

Studying cells using a microscope

A microscope is an instrument that allows you to view a magnified (enlarged) image of an object. In the laboratory you will use a light microscope to examine and observe cells. A light microscope magnifies objects so they appear between 10 and 400 times larger than they really are.

To examine the smaller organelles found inside cells, you need to use an electron microscope. These powerful microscopes can magnify cells so they appear up to 10 000 times larger than they really are.

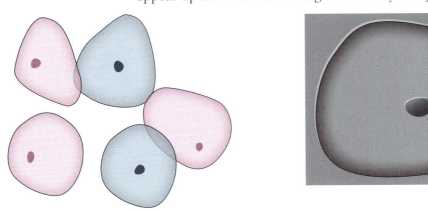

a Cheek cell seen under a light microscope at ×400 magnification

b Cheek cell seen under an electron microscope at ×1500 magnification

Figure 3.2 Human cheek cells seen under a light microscope and an electron microscope.

Activity 3.1 Observing cells

For this activity your teacher will provide you with prepared slides or show you how to prepare a slide.

1 Examine at least one example of an animal cell and one example of a plant cell under the light microscope.
2 What things do these cells have in common?

The structure of cells

Figure 3.3 shows you a typical cell from an animal and a typical cell from a plant.

If you study Figure 3.3, you can see that animal cells and plant cells have some different features. However, both share the following characteristics:

- the cell is surrounded by a **cell membrane**
- the cell contains **cytoplasm** which is the living part of the cell
- the organelles are found in the cytoplasm
- the **nucleus** controls the working and function of the cell, it also contains the chromosomes
- the vacuole is an organelle with a membrane that holds liquid
- the organelles are mostly too small to be seen with the naked eye but they can be seen with an electron microscope.

> ### Did you know?
> Diagrams of cells are two-dimensional so they show cells as flat objects. In reality, cells are three-dimensional. The plant cell shown in Figure 3.3 is actually shaped more like a box than a rectangle.

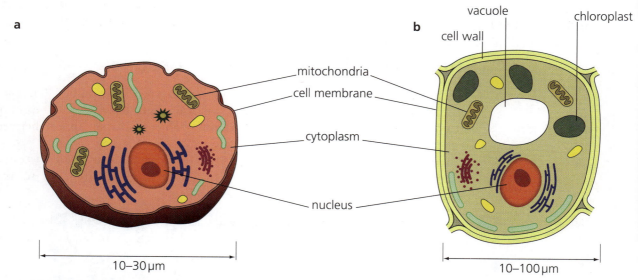

a

b

vacuole
chloroplast
cell wall

mitochondria
cell membrane
cytoplasm
nucleus

10–30 μm

10–100 μm

Figure 3.3 The structures of **a** a typical animal cell and **b** a typical plant cell.

Drawing diagrams in Biology

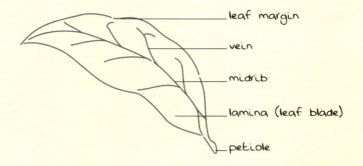

DRAWING OF EXTERNAL VIEW OF
DICOTYLEDONOUS LEAF Mag. ×75

Figure 3.4 A student's diagram of onion cells correctly captioned and labelled.

1 **Drawing lines** – always work in pencil and draw lines that are clean, unbroken and of even thickness (not feathery).

2 **Label lines**
   ■ use a ruler to make sure your label lines are straight
   ■ draw label lines parallel to the bottom of the paper
   ■ do not let label lines cross over
   ■ do not use arrowheads at the ends of your label lines
   ■ make sure your label line ends accurately on the structure you are labelling
   ■ either place lines all to the right of the drawing or distribute them evenly.

3 **Title**
   ■ your title should fully describe the specimen
   ■ place the title at the bottom of your drawing
   ■ write the title in capitals
   ■ state the magnification of your drawing where appropriate.

Biology at work
Chapter 3 –
Microbiology, African dust and coral reefs

Activity 3.2 Drawing cells

1 Look at the diagrams of the animal and plant cells in Figure 3.3. Make a table that lists the structures present in animal cells and those found in plant cells.
2 Refer to the skills box above and follow the guidelines to draw your own correctly labelled drawings of a plant and animal cell.

Comparing animal and plant cells

Plant and animal cells contain the following structures and organelles:

■ cell membrane
■ cytoplasm
■ nucleus
■ mitochondria
■ vacuoles (in animal cells these may be very small or absent)
■ ribosomes (these cannot be seen with the light microscope).

Animal cells	Plant cells
No cell wall	Have a cell wall
No chloroplasts	Many contain chloroplasts
Not rigid, can change shape	Rigid, have a fixed shape
Don't often have vacuoles	Have a central vacuole with cell sap
When present, vacuoles are very small	Vacuole is normally single and large

Table 3.1 A comparison of animal and plant cells.

Activity 3.3 Observing plant cells

Aim: To observe and draw plant cells.

Apparatus and materials
- sample of plant cells (onion epidermis, *Rheo discolor*, or other suitable plant cell)
- clean slides and cover slips
- dropping pipette
- pointer
- forceps and a scalpel or sharp knife
- tissue or filter paper
- light microscope

Method
1 Cut a small thin layer from the plant sample. Do not let the cut piece dry out.
2 Using the dropping pipette, drop some water on to the centre of a clean slide. Carefully place the thin section of plant in the water and, using the pointer, gently spread the section out so it is flat.
3 Cover the section with a cover slip.
4 Use tissue or filter paper to carefully suck up the excess water from the slide.
5 View the slide on a low magnification using the light microscope.
6 Increase the magnification so that you can identify various cell structures.
7 Make a correctly labelled diagram of two or three cells. Remember to include the magnification.

Questions
1 Why do you need to use a thin section when you prepare a slide?
2 Why do you need to use water when you prepare a slide?
3 What is the function of the cover slip?
4 How do you focus the microscope so that you can see the cells clearly?

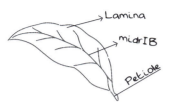

Diagram of Leaf

→ Lamina

→ midrIB

Petiole

Figure 3.5 Maria's drawing.

Check your knowledge and skills

1 Why do scientists need to use microscopes to study cells?
2 a What is an organelle?
  b Where are organelles found in cells?
3 a What do you think would happen to an animal cell if its cell membrane burst?
  b Would the same thing happen to a plant cell? Explain your answer.
4 John observes a cell under a microscope. He decides that it must be a plant cell.
  a What features would he have to identify to decide that it is a plant cell?
  b The real cell is $50\,\mu m$ long. If John views it at a magnification of 400, what length will the image of the cell be?
5 Maria drew the diagram shown in Figure 3.5 of a cell in her practical book.
  a Study Maria's drawing and list the things that are wrong with it.
  b Redraw the sketch correctly using the guidelines for Biology drawing that you learnt on page 25.

Unit 2

Functions of parts of cells

Much like the different parts of the body, the different parts of cells work together as a whole to keep the cell alive. Each structure found in cells has a specific function.

Structure	Function
Cell wall (plant cells only)	Supports and protects the cell, maintains shape of cell
Cell membrane	Contains the cytoplasm within the cell Regulates the movement of substances into and out of the cell
Nucleus	Governs the activities of the cell Contains the genetic material of the cell (chromosomes and genes) Synthesizes DNA and RNA
Ribosome	Synthesizes proteins
Vacuole (usually small if found in animal cells; large in plant cells)	Stores substances, particularly cell sap which contains dissolved sugars and other minerals
Mitochondrion (plural mitochondria)	Important for cellular respiration and the release of energy in the cell
Chloroplast (certain plant cells only)	Absorbs light energy during photosynthesis Contains the green pigment chlorophyll

Table 3.2 Main parts of cells and their functions.

Activity 3.4 Modelling cells

1 Build a model of a plant cell using materials that you can find in your environment. For example, you could use stiff card to represent the cell wall and a thin layer of plastic to represent the cell membrane.

2 Work with a partner. Use your models to identify the main parts of the cell. Take turns to name the parts and to state their functions in the cell.

The nucleus and genetic material

The nucleus is in fact very important in cells, because it contains coded information that governs the activities of cells and the characteristics of different organisms.

All cells with a nucleus contain genetic material. When a cell grows and divides (to reproduce), the nucleus also grows and divides. During cell division, the nucleus changes from a single organelle into individual thread-like structures. These structures are called **chromosomes**.

Chromosomes are made up of a chemical called deoxyribonucleic acid (DNA). DNA carries the coded instructions or genes which determine the characteristics of cells and organisms. Each **gene** governs a particular aspect of cell chemistry.

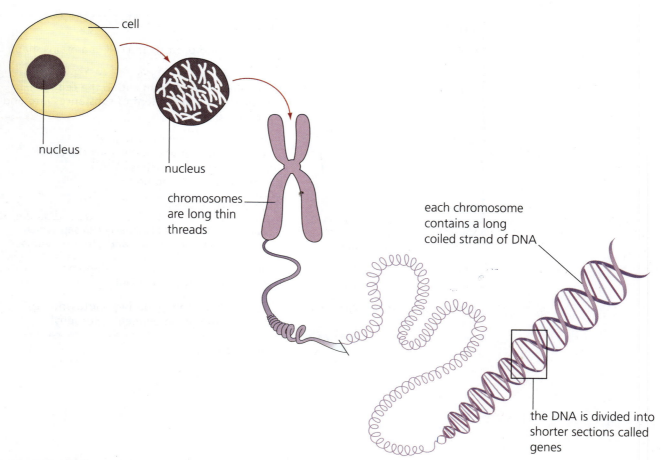

Figure 3.6 All plant and animal cells with a nucleus contain genetic material.

Chromosomes

The number of chromosomes found in the nucleus of a cell is fixed and differs from species to species. Humans have 46 chromosomes, frogs have 26 chromosomes and tomato plants have 24 chromosomes. Each body cell in these different organisms will have the same number of chromosomes.

Chromosomes are always arranged in pairs. The chromosomes in a pair are normally similar in shape and size because one chromosome in each pair comes from the male parent whilst the other comes from the female parent. We call a pair of similar chromosomes a **homologous pair**.

Normal body cells are called diploid cells and the number of chromosomes found in each body cell of a plant or animal is known as the **diploid number**. Because chromosomes are arranged in pairs, the diploid number is always an even number (a multiple of two).

Male and female sex cells are called haploid cells. These cells contain half the number of chromosomes that body cells contain. This number of chromosomes is called the **haploid number**.

You will learn more about these concepts when you deal with mitosis and meiosis on pages 182 and 210.

Check your knowledge and skills

1 Which organelle, found only in plant cells, allows plants to make their own food by photosynthesis?
2 Which organelles are responsible for protein synthesis in cells?
3 How do plant cells maintain their regular shape?
4 What is likely to happen to a cell if its cell membrane bursts?
5 Why is the nucleus important in cells? What is likely to happen to a cell if its nucleus is removed? Why?
6 Explain the terms: **a** chromosome **b** gene.

Specialization and organization

Multicellular plants and animals are made up of many different kinds of cells, each with its own specialized job or function. The function that a **specialized cell** performs in the organism often affects its structure, shape and size.

■ Specialized cells

Table 3.3 overleaf shows you examples of some specialized animal and plant cells and describes how they are adapted for special functions.

Specialized cell	Nature of specialization and functions
Muscle cell	Long fibrous cells which can contract (shorten and thicken) to produce movement
Motor nerve cell	Long fibre called an axon, and a branched ending, allow the cell to control movement by passing messages from the brain to various muscles in the body
Brain cell	Has many connections to pass electrical impulses to many other cells
Red blood cell	Flexible cells which can move through blood vessels, they are filled with haemoglobin which binds with and carries oxygen in the bloodstream
Goblet cells	Cells which are shaped like cups or glasses to increase their surface area, they produce mucus to lubricate and protect the internal surfaces of the body (intestines, stomach, trachea)
Sex cells	Haploid cells which combine to produce the diploid number of chromosomes; the tail (flagellum) on a sperm cell allows it to 'swim' towards the female egg cell
Root hair cell	Extends outwards from root into hair-like structure to increase cell surface area and allow for absorption of minerals and water from the soil
Guard cells (around stomata)	Able to expand and contract to open or close stomata (pores) on leaf surface, and control the amount of water and gas that leaves the plant via the stomata
Xylem cells	Hollow cells (contain no cytoplasm) which are joined together end to end to form tubes allowing water to be transported through the plant

Table 3.3 Cell specialization in animals and plants. (The animal and plant cells shown are not to scale.)

Tissues and organs

Cells which have a similar structure and function are grouped together in multi-cellular organisms to form **tissue**. Tissue cells all work together to perform the same function. For example, in mammals millions of epithelial (skin) cells combine to form epithelial tissue which lines inner body surfaces and covers the outside of the body (with skin). In plants, millions of xylem cells combine to form vascular tissue (veins or tubes) that transport water around the plant.

The four main tissue types found in animals are:

- epithelial tissue (epithelium) – lines surfaces inside the body (such as the intestine) and covers the outside of the body (as skin)
- connective tissue – joins and strengthens other tissues, examples include blood, lymph, cartilage and bone
- muscle tissue – sheets or bundles of cells which are capable of producing movement
- nerve tissue – connects the central nervous system with other tissues and organs in the body, found in motor neurones and sensory neurones.

The following tissue types are found in plants:

- epidermal tissue – covers the outer layer of plants and prevents loss of water from plant surfaces
- mesophyll – found inside green leaves and allows photosynthesis to take place
- parenchyma – tissue which fills spaces between other tissues and helps to support the plant; also has a role to play in food storage, for example in carrots
- vascular tissue – system of tubes or veins in the roots, stems and leaves of plants that allows movement of food, minerals and water throughout the plant
- supportive tissue – thick fibrous tissue which supports the plant and allows the parts of the plant to move in the wind and in response to other stimuli.

Tissues combine to form organs. An organ is a group of tissues that works together to perform a specific function. For example, the lungs are organs that are made up of muscle tissue and epithelial tissue, nerves and blood vessels. Plant tissue also combines to form three main organs – the roots, stems and leaves. In most living organisms, the organs work together in a system to perform a particular function. Complex organisms are made up of many different systems.

Check your knowledge and skills

1 List five examples of specialized animal cells and their functions.
2 How are root hair cells adapted to suit their function in plants?
3 Explain the difference between an organelle and an organ.
4 Copy the table below and place the terms below the table in the correct columns.

Organelle	Cell	Tissue	Organ	System	Organism

liver   nucleus   bird   potato   palm tree   chloroplast   stomach
phloem   sperm   brain   heart, veins and blood   mitochondria
the inner lining of the intestine   motor neurone   lizard   skin   blood

Chapter summary

Do you know?

If you are unsure of any of the facts in the list, refer to the page number given in brackets.

- The cell is the simplest unit of living organisms. Organisms may be unicellular, but most plants and animals are multicellular. (page 22)
- Cells are too small to be seen with the naked eye. They can be viewed with a light microscope and viewed in more detail with an electron microscope. (page 23)
- All cells have a cell membrane, cytoplasm and organelles. All cells have a nucleus (except red blood cells), containing chromosomes. Some organelles, such as chloroplasts, are found only in plant cells. (page 24)
- The cytoplasm is the living part of the cell. Cytoplasm is a watery solution containing dissolved substances and organelles. (page 24)
- Plant cells which are involved in photosynthesis contain chloroplasts. These are organelles which contain the chemical chlorophyll. Light energy is absorbed by chlorophyll and used by plants to make food during photosynthesis. (page 27)
- The nucleus of the cell contains the chromosomes. Chromosomes contain the genes which are made up of DNA. (page 28)
- Cells group together to perform specialized functions in multicellular organisms. Groups of cells are called tissues. Groups of tissues are called organs. Groups of organs which work together to perform a specific function are called organ systems. (page 31)

Are you able to?

If you have trouble in doing these things, refer to the page number in brackets.

- Use a light microscope to observe and examine different plant and animal cells. (page 24)
- Apply your knowledge of cells to differentiate between plant and animal cells. (page 24)
- Draw and label a simple diagram of an unspecialized animal cell and an unspecialized plant cell. (page 25)
- State the function and importance of structures found in cells. (page 27)
- Use local materials to construct a model of a cell. (page 28)
- Give examples of different types of tissues found in animal and plant cells. (page 31)

4 Moving substances

Movement into and out of cells

Living things have to transport water, nutrients, wastes and gases into and out of cells and between the cells in their bodies. In this chapter you will learn how small particles are able to pass through cell membranes as necessary by the processes of diffusion, osmosis and active transport.

By the end of this chapter you should be able to:

- ▪ Explain why there is a need for transport into and out of cells
- ▪ Name substances that are transported into and out of cells
- ▪ Carry out experiments to explain osmosis
- ▪ Discuss the importance of diffusion and osmosis in living organisms
- ▪ Describe active transport and give examples to show where it takes place

Transport across membranes

One of the functions of the cell membrane is to separate the cell's internal environment from its external environment. The cell membrane forms a barrier between the substances inside the cell and everything outside the cell. This separation is important because it prevents substances in the cell from flowing away and it stops foreign materials from entering cells and doing damage. However, in living organisms, some substances do move into and out of cells and from one cell to another. This is possible because the cell membrane is both **partially** and **semi-permeable** – it allows some substances to pass through it but not others.

Cell membranes and permeability

Materials that are permeable allow substances to pass freely through them. Materials that are impermeable allow nothing to pass through them. Partially permeable (or semi-permeable) materials, such as cell membranes, allow small molecules of water and certain solutes in solution to pass through them but they block the passage of large solute molecules. Because cell membranes allow some small particles to pass through them easily, whilst other small particles cannot pass through, they are also selectively permeable (the membrane 'selects' which substances it will allow through it).

The fact that certain substances can move across a partially permeable cell membrane allows the cell to:

- ▪ obtain nutrients
- ▪ excrete wastes
- ▪ secrete useful substances such as hormones
- ▪ maintain optimum concentrations of ions
- ▪ maintain a pH suitable for the action of enzymes.

How are substances transported across the cell membrane?

Substances will only move across the cell membrane when there is a difference between the concentration of the substance inside and outside the cell. For example, when there is a high concentration of waste materials inside the cell and a low concentration of wastes outside the cell, the waste material will be transported from the area of high concentration to the area of low concentration. The difference in levels of concentration is called the **concentration gradient**.

Substances can be transported across the cell membrane by one of two mechanisms: **passive transport** or **active transport**.

Passive transport

Passive transport happens on its own without needing energy. In this type of transport substances move down the concentration gradient from a region where they are in high concentration to one where they are in low concentration. There are two types of passive transport: **diffusion** and **osmosis**.

Figure 4.1 The concentration gradient works in the same way as a slope.

Active transport

Active transport is the opposite of passive transport. The particles move up the concentration gradient from a region where they are in low concentration to a region where they are in high concentration.

This process requires energy in the form of **adenosine triphosphate (ATP)**. It is possible for osmosis and diffusion to occur outside a living cell but active transport can only take place in respiring cells. This is because respiration is the process that produces ATP.

Active transport takes place in the following cases:

- ion carriers in the cell membranes of red blood cells actively pump sodium and potassium out of and into the cells respectively
- companion cells in plants draw sucrose into the phloem tubes
- root hair cells absorb nitrate ions from the soil around them.

Check your knowledge and skills

1 Give two reasons why it is important for substances to be transported across cell membranes.
2 List three ways in which substances can be transported across a cell membrane.
3 Why does passive transport require less energy than active transport?

Unit 2

Diffusion

Diffusion is the movement of particles (molecules and ions) down a concentration gradient, from an area of high concentration to an area of lower concentration.

Diffusion

Diffusion can be observed in gases and liquids in everyday life. For example, if you spray perfume at the back of a room, the particles move through the area of the room, from the area of high concentration (where you sprayed) towards the areas of lower concentration. Or, if you add milk to a cup of tea, the milk particles are concentrated near the top of the cup where you poured them, but they spread through the liquid until they are evenly distributed.

Figure 4.2 demonstrates the tendency for like particles to move away from each other. The particles of dye move away from each other and become dispersed in the water. This is an example of diffusion.

a When the dye is placed in the water it is concentrated in one area.

b The dye dissolves in the water and the particles move from the area where the dye is concentrated to the rest of the water.

c After some time the dye particles are evenly distributed throughout the water and diffusion stops because there is no longer a concentration gradient.

Figure 4.2 The process of diffusion can be observed when dye is dispersed in water.

In living things diffusion takes place across cell membranes. Because the cell membrane is selectively permeable only certain small molecules are able to enter and leave the cell by diffusion. Substances which dissolve in lipids diffuse easily across the cell membrane, so do gases such as oxygen and carbon dioxide.

Diffusion is important in living systems

Plants and animals rely on diffusion to get many of the substances they need and to get rid of wastes.

During photosynthesis, carbon dioxide must be taken into the leaf through the stomata. The atmosphere has a higher concentration of carbon dioxide than the cells inside the leaf; therefore the carbon dioxide diffuses into the leaf through the stomata. Oxygen produced during photosynthesis diffuses out of the leaf down the concentration gradient as the concentration of oxygen outside the leaf is lower than the concentration in the leaf cells.

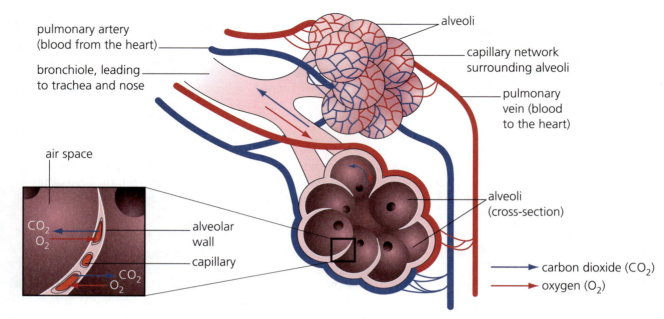

Figure 4.3 Oxygen diffuses from the alveoli of the lungs into capillaries. Carbon dioxide diffuses from capillaries into the lungs.

In all animals that breathe air, oxygen diffuses from the alveoli (small grape–like sacs) in the lungs into the bloodstream through the blood capillaries (see Figure 4.3). Carbon dioxide diffuses out of the bloodstream into the capillaries in the same way. This happens because the concentration of carbon dioxide is higher in the blood than it is inside the lung alveoli.

During digestion in animals, food is broken down into glucose and amino acids in the intestine. The products of digestion diffuse from the intestine into the blood through the capillaries.

Cell membranes are adapted to maximize diffusion

Diffusion can be a fairly slow process because no energy is involved. However, cell membranes are adapted to ensure that diffusion takes place at its maximum rate.

- The membranes are extremely thin so that particles have to travel a very short distance to get through the membrane.
- The surface area of membranes may be increased to maximize the area available for diffusion. The greater the area, the faster diffusion will take place.

Check your knowledge and skills

1 Figure 4.4 shows how molecules enter a cell by diffusion. Copy the diagram and label it clearly to explain what happens during the process of diffusion.

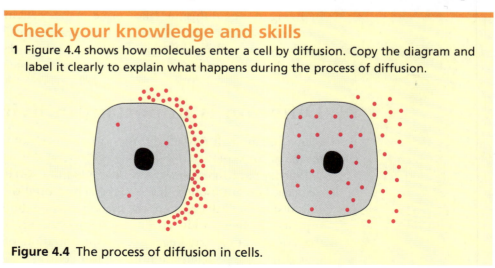

Figure 4.4 The process of diffusion in cells.

2 Explain how the following factors may affect the rate at which diffusion occurs:
  a the thickness of the cell membrane
  b the difference in concentration of a substance inside and outside of a cell
  c the surface area of the cell membrane.

3 A helium-filled balloon will gradually deflate if it is left in a room.
  a What happens to the helium molecules in the balloon?
  b What does this suggest about the material that the balloon is made from?

Osmosis

Osmosis is the process by which water molecules move from a region of high concentration to a region of low concentration through a partially permeable membrane. In other words, osmosis is the name given to diffusion of water molecules.

Solutions

In order to understand the concept of osmosis you need to understand the terms used when talking about solutions. Remember that:

solute + solvent = solution

- The **solute** is the substance which is dissolved.
- The **solvent** is the substance (usually a liquid) in which substances are dissolved.
- The **solution** is the substance formed when a solute is dissolved in a solvent.

For example, in a solution of salt and water, salt is the solute and water is the solvent.

Depending on the ratio of solute to solvent in a solution, the solution is either **dilute** or **concentrated**. If the solution contains a high ratio of solvent to solute then the solution is dilute. If the solution contains a high ratio of solute to solvent then the solution is concentrated (see Table 4.1).

Solution	Solute	Solvent
Concentrated	High solute concentration	Low solvent concentration
Dilute	Low solute concentration	High solvent concentration

Table 4.1 Summary of dilute and concentrated solutions.

Describing osmosis

Osmosis takes place across a partially permeable membrane, as shown in Figure 4.5 overleaf. The water always moves from the dilute solution to the concentrated solution, i.e. down its concentration gradient. Thus osmosis may also be defined as the movement of water molecules from a dilute solution to a concentrated solution across a partially permeable membrane.

Osmosis

Osmosis may take place through a membrane that is not selectively permeable as long as the membrane has pores which are large enough for the water molecules to pass through. Visking tubing is an example of a membrane that allows water molecules to pass through.

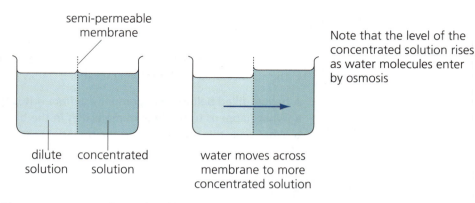

semi-permeable
membrane

Note that the level of the
concentrated solution rises
as water molecules enter
by osmosis

dilute   concentrated
solution   solution

water moves across
membrane to more
concentrated solution

Figure 4.5 Water from the dilute solution moves by osmosis through the semi-permeable membrane to the more concentrated solution.

Unit 4 The behaviour of cells in solutions

Osmosis takes place all the time in living organisms. For example, water from digested food in the large intestine moves back into the cells of the body by osmosis. You need to understand how cells behave in solution so that you understand how osmosis works in different situations and conditions.

The way that plant cells and animal cells behave when they are placed in a solution depends on the difference in concentration of the solutions inside the cells and outside the cells (see Figure 4.6).

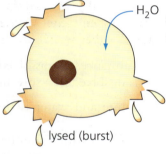

cell placed in a hypotonic
solution (less concentrated)

H_2O

lysed (burst)

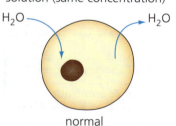

cell placed in an isotonic
solution (same concentration)

H_2O H_2O

normal
cell placed in a
hypertonic solution
(more concentrated)

H_2O

shrivelled

Figure 4.6 What happens when an animal cell is placed in a dilute solution?

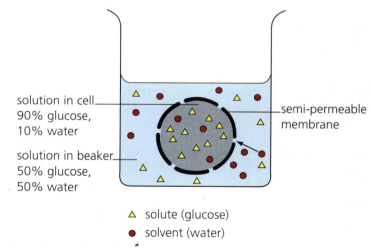

solution in cell
90% glucose,
10% water

semi-permeable
membrane

solution in beaker
50% glucose,
50% water

△ solute (glucose)
● solvent (water)

Figure 4.7 The effects of putting an animal cell in a concentrated solution.

Osmosis in animal cells

When an animal cell is placed in a dilute solution as in Figure 4.6, water moves into the cell by osmosis. This is because the solution inside the cell is more concentrated than the solution outside the cell. As water enters, the cell swells up. If osmosis continues the cell membrane may rupture, causing cellular death.

If the animal cell is placed in a more concentrated solution, water will move out of the cell (see Figure 4.7). The more concentrated the solution outside the cell, the more water will leave the cell. If osmosis continues, the cell will crinkle and shrivel, as the cell becomes dehydrated. This also causes cellular death.

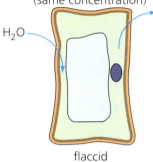

cell placed in a hypotonic solution (less concentrated)

H_2O

turgid (normal)

cell placed in an isotonic solution (same concentration)

H_2O

H_2O

flaccid

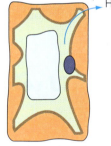

cell placed in a hypertonic solution (more concentrated)

H_2O

plasmolysed

Figure 4.8 The effects of putting a plant cell in a dilute or a concentrated solution.

Osmosis in plant cells

When plant cells are placed in a solution that is more dilute than the solution inside the cell, water will move into the cell. Plant cells are strengthened by their cellulose cell wall and this prevents the plant cell from bursting. However, the cell still swells up as water enters and it becomes swollen and very firm in texture. A plant cell that is filled with water is said to be **turgid**.

Solutions

■ A solution that is more dilute than the solution inside a cell is called a hypotonic solution.

■ A solution that is more concentrated than the solution inside a cell is called a hypertonic solution.

■ A solution that has a concentration equal to the one inside a cell is called an isotonic solution.

When plant cells are placed in a concentrated solution, water moves out of the cell into the surrounding solution. As water leaves the cell, the cell shrinks and begins to lose its firmness. A cell that has lost its firmness is said to be **flaccid**. It is possible for plant cells to lose so much water that the cytoplasm shrinks and the cell membrane pulls apart from the cell wall. This process is called **plasmolysis**.

When a cell is placed in a solution that has the same concentration as the solution inside the cell there is no net movement of water. In other words, water will neither enter nor leave the cell.

Recording and discussing results

In many practical activities you will be asked to record and discuss your results.

Recording results

■ Your results include what you found out or discovered, your careful observations and your measurements.

■ Results are best recorded in a table and/or a graph.

■ Labelled diagrams can also be used to record results.

Discussing results

Your discussion should summarize and outline what the results of your experiment show you. Consider the following questions in your discussion:

■ Did I get the results I expected?

■ Is my hypothesis supported or disproved?

■ What sources of error were there?

■ Could data or measurements be collected more accurately or in a different way?

■ What problems were encountered?

■ How could the experiment design be improved?

Activity 4.1 The effects of osmosis in plant cells

Aim: To observe osmosis in plant cells and measure its effects.

Apparatus and materials
- 6 potato cylinders (about 5 cm in length)
- unknown solutions labelled A, B and C (your teacher will supply these)
- 3 Petri dishes
- marker pen
- ruler and scalpel blade (be careful when using the scalpel)
- paper towel

Method
1 Cut the six potato cylinders to a length of 3–4 cm, making sure they are all the same length.
2 Record the length of the potato cylinders.
3 Label the Petri dishes A, B and C and add solutions A, B and C respectively.
4 Place two potato cylinders into each Petri dish.
5 Let the potato cylinders stand in the solutions for 30 minutes.
6 Remove the potato cylinders and gently blot the excess solution with dry paper towel.
7 Measure the length of each potato cylinder. Calculate the change in length for each piece.
8 Observe the texture of each potato cylinder. Compare it with untreated potato.
9 Rank the solutions in order of concentration from lowest to highest.

Observations and results
Draw up a table of results and observations.

Discuss
Discuss all your results using the following terms: osmosis, concentration gradient, water and solute concentration.

Activity 4.2 Observing flaccid and turgid cells

Aim: To observe the effects of hypotonic and hypertonic solutions on plant cells.

Apparatus and materials
- 2 small pieces of onion leaves
- forceps
- ruler and scalpel blade
- marker pen
- water
- sucrose solution
- 2 microscope slides
- teat pipettes

Method
1 Working in pairs, use forceps to remove the inner layer from a fleshy onion leaf.
2 Cut the inner layer into a 1 cm square.

3 Label the slides A and B and mount one piece of onion skin on each. Use the water to mount the onion skin on slide A and the sugar solution to mount the onion skin on slide B.
4 Observe the slides under a microscope.
5 Draw two cells from each slide, paying close attention to the position of the cell membrane in each case.

Questions
1 Discuss what you observed on each slide.
2 Suggest a different solution that you could use to improve this activity.

Check your knowledge and skills

1 Kalia sliced pineapple into a bowl and sprinkled some sugar on it. She left it for 15 minutes. When she came back there was a large amount of liquid in the bowl with the pineapple. Use your knowledge of osmosis to explain fully what happened.
2 When lettuce becomes limp you can put it into cold water for about 20 minutes to plump it up again. How does this work?
3 Copy and complete the table to summarize what you have learnt about transport across the cell membrane.

Process	Description of process	Requirements	Examples in plant and animal cells
Diffusion			
Osmosis			
Active transport			

Chapter summary

Do you know?

If you are unsure of any of the facts in the list, refer to the page number in brackets.

- Living systems must be able to transport materials into and out of the cells and between cells. (page 33)
- Cells are surrounded by a cell membrane which is partially or semi-permeable. Transport into and out of the cells takes place across the cell membrane. (page 33)
- Diffusion and osmosis are passive forms of transport. Passive transport does not require energy (in the form of ATP). (page 34)
- Active transport takes place against the concentration gradient. It requires energy in the form

of ATP so it can only take place in living, respiring cells. (page 34)
- Osmosis is the passive diffusion of water only across a partially permeable membrane. (page 37)

Are you able to?

If you have trouble in doing these things, refer to the page number in brackets.

- Give examples of kinds of active and passive transport in living systems. (page 34)
- Set up and carry out an experiment to observe and measure osmosis. (page 40)
- Correctly measure the length of materials in the laboratory. (page 40)
- Make and record observations. (page 40)
- Report the results of an experiment. (page 40)

5 Biological molecules

Building blocks of life

All matter, including living organisms, is made up of a combination of elements in the form of molecules and other compounds. Although living organisms are very different from each other, they are all built from similar basic compounds.

In this chapter you will learn more about chemical substances and how they combine to form biological compounds.

By the end of this chapter you should be able to:

- Distinguish between organic and inorganic compounds
- Identify and describe carbohydrates, lipids and proteins
- State the chemical and physical properties of carbohydrates, lipids and proteins
- Carry out a range of tests to distinguish between food substances
- Explain the role and importance of enzymes in living organisms
- Investigate the effects of changing conditions on enzyme action

Unit **1**

The chemical needs of living organisms

All living organisms are made up of a combination of chemical substances. Some of these substances are complex compounds of carbon, hydrogen and oxygen. Compounds with carbon and hydrogen are called **organic** substances and they are an important component of all living things. Other chemical compounds are simpler and they lack the element carbon, these are called **inorganic** compounds.

◼ Organic molecules

Organic molecules are those which are produced in living organisms and which contain carbon–hydrogen bonds. An organic molecule always contains bonded carbon and hydrogen, and the organic molecules found in living organisms usually contain oxygen as well. Carbohydrates, proteins, lipids and nucleic acids are all examples of organic molecules.

◼ Inorganic compounds

Inorganic compounds are those which do not contain a carbon–hydrogen bond. Some inorganic molecules contain carbon or hydrogen, but they never contain both. Water, minerals and salts are all examples of inorganic compounds.

Hydrogen, carbon and oxygen

Hydrogen, carbon, oxygen and nitrogen are very important elements; they make up about 99% of the mass found in all living organisms.

Check your knowledge and skills

1 A molecule of water (H_2O) contains hydrogen and oxygen.
  a Is water an organic compound? Give a reason for your answer.
  b Why is water an important component of all living cells?
  c Apart from water, human cells contain about 1% inorganic compounds. Give two examples of such compounds.
  d How do humans and other living organisms get the raw materials they need to form organic compounds?

Unit 2

Water, vitamins and minerals

All living organisms contain water and water makes up almost 70% of the human body. Humans and other animals take in water when they drink and in their food. Plants take in water from the soil through their roots by the process of osmosis.

Why is water so important?

Animals and plants need water to survive. Cellular processes cannot take place without water. When cells do not have enough water they become dehydrated, cellular processes stop and eventually the cell dies.

Water is an important component of living organisms because:

■ it is involved in all metabolic reactions in the cells of plants and animals
■ blood systems in animals transport substances dissolved in water. Water in the blood also helps to maintain blood at the correct consistency
■ water can absorb heat, so it helps with the maintenance of body temperature
■ waste materials are passed out of the body dissolved in water
■ substances are transported around plants dissolved in water.

■ Vitamins

Vitamins are complex organic substances that are needed in small amounts by the body to maintain health.

Vitamin	Use in the body	Deficiency diseases and their symptoms
A	Keeps cells along the respiratory system healthy; Used to make the pigment in the rods of the eye needed for seeing in dim light	Respiratory infections and night blindness (xerophthalmia)
B (there are about 12 different kinds)	Involved in many metabolic reactions	Beriberi – weak and painful muscles, depression, irritability
C	Keeps tissues healthy	Scurvy – bleeding gums, wounds taking longer to heal, heart failure
D	Controls calcium and phosphorous absorption; Important in bone and teeth formation	Rickets – lack of calcium in bones causes 'bow legs' in children and 'knock knees'

Table 5.1 Some of the vitamins needed by humans.

Biology at work
Chapter 5 – Can a person take too many vitamins?

■ Minerals

Minerals are inorganic substances that are required in very small amounts by animal cells. Minerals are usually found in the form of salts. Both plants and animals need minerals for healthy functioning (see Tables 5.2 and 5.3 overleaf). Plants can make their own vitamins, but animals have to obtain them in their diet.

Mineral and its chemical symbol	Use in the body	Deficiency diseases and their symptoms
Calcium (Ca) and phosphorus (P)	For strong bones and teeth	Brittle bones and teeth
Iron (Fe)	Formation of haemoglobin	Anaemia – 'weak blood', tiredness, reduced energy due to a lack of red blood cells
Iodine (I)	Formation of the hormone thyroxine	Goitre (in adults) – reduced metabolic rate, swelling of the thyroid gland in the neck Cretinism (in children) – physical and mental retardation

Table 5.2 Some of the minerals needed by humans.

Mineral and its chemical symbol	Use in plant	Deficiency symptoms
Nitrogen (N)	To make amino acids, proteins, chlorophyll and DNA	Yellow leaves; stunted growth
Magnesium (Mg)	Contained in chlorophyll molecules	Yellow leaves
Phosphorus (P)	Part of cell membranes	Poor root growth; dull green leaves
Potassium (K)	Aids enzymes involved in respiration and photosynthesis	Yellow leaves

Table 5.3 Some of the minerals needed by plants.

Check your knowledge and skills

1 Why is water important for all living organisms?

2 Why do we need iron in our diets?

3 What symptoms would you see in a person suffering from a vitamin C deficiency?

Carbohydrates

Carbohydrates are organic molecules which contain carbon, hydrogen and oxygen. Examples of carbohydrates are sugar, starch and cellulose (fibre).

The simplest type of carbohydrate is the **monosaccharide**. Monosaccharides are simple, soluble sugars which are able to join together easily to form more complex carbohydrates. The best-known monosaccharide is glucose, but fructose (fruit sugar) is also a monosaccharide. The basic structure of glucose is $C_6H_{12}O_6$, which means it has 6 carbon atoms, 12 hydrogen atoms and 6 oxygen atoms. Monosaccharides consist of a single sugar molecule. These can be represented either in a linear form or in the hemiacetal form. The transformation from one form into the other is shown in Figures 5.1 and 5.2 for the monosaccharide glucose. When glucose and fructose molecules join up they form a **disaccharide** called sucrose. Sucrose is the sugar that is used to sweeten certain drinks and snacks.

The structure of carbohydrates

When many monosaccharides join together they form larger more complex carbohydrates. These large carbohydrates are called **polysaccharides** and they are less soluble than simple carbohydrates. As carbohydrates get larger they lose their sweet taste.

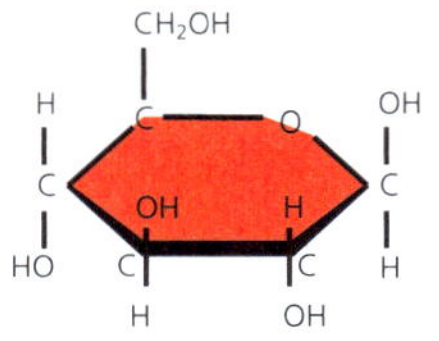

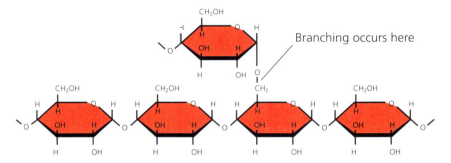

Branching occurs here

Figure 5.1 The formula for glucose.

Figure 5.2 The structure of a carbohydrate.

Monosaccharides

'Mono' means one and 'saccharide' means sugar. A monosaccharide can also be called a single or simple sugar.

Important carbohydrates

- **Starch** – a very important polysaccharide found in plants. It is made of lots of glucose molecules joined together.
- **Glycogen** – is also a very important carbohydrate found in animal cells. Glycogen is very similar to starch. It is made of many glucose molecules joined together in long chains.
- **Cellulose** – the fibre found in plants. Cellulose is a large carbohydrate made from glucose molecules joined together in long straight chains. The chains in cellulose molecules are longer than those in glycogen molecules. Animals do not produce cellulose.

The glucose molecules that make up cellulose have a different structure to the glucose molecules in starch and glycogen. The glucose molecules in cellulose are called beta glucose and those in starch and glycogen are called alpha glucose. The two different forms of the glucose molecule have slightly different shapes, which gives the substances their different properties.

The functions of carbohydrates

The main function of carbohydrates is to provide energy for organisms. Dietary sources of carbohydrates for humans include cakes, breads, yam, potato and any other starchy foods. When carbohydrates break down in organisms they release energy in the form of ATP. You will learn more about this on page 88 when you study respiration.

In cellulose, the carbohydrate performs a structural function in plants. Plant cell walls are made from cellulose which is responsible for the rigidity of plant cells.

Simple and complex sugars have different physical and chemical properties (see Table 5.4, overleaf).

Type of carbohydrate	Physical characteristics	Chemical characteristics
Monosaccharides	Simple sugars such as glucose and galactose Sweet crystalline substances Soluble in water	Reduce Cu^{2+} to Cu^+ ions – they are called 'reducing sugars'
Disaccharides	Double sugars formed by condensation (joining of two simple sugars by the removal of water) Sucrose is made up of glucose and fructose Maltose (or malt sugar) is made up of glucose and glucose Lactose is made up of glucose and galactose Crystalline and sweet Soluble in water	Except for maltose they are usually 'non-reducing sugars'
Polysaccharides	Insoluble in water White powdery substances Not sweet – may even taste bitter	'Non-reducing sugars', but can be broken down into reducing sugars

Table 5.4 The physical and chemical properties of different types of carbohydrates.

Testing for carbohydrates

Carbohydrates react in the presence of certain substances, called **reagents**. You need to know which reagents to use and how to test for simple sugars, complex sugars and starch.

Activity 5.1 Testing for reducing sugars

Aim: To test for the presence of reducing sugars.

Apparatus and materials
■ reducing sugar sample
■ dropping pipette
■ Benedict's solution
■ test tube
■ hot water bath
■ safety glasses

Method
1 Place about 1 ml of the reducing sugar sample into a clean test tube using the dropping pipette. Note its colour.
2 Add about 1 ml of Benedict's solution to the tube and shake it well. Note any colour changes.
3 Heat the test tube containing the mixture in a hot water bath for 1 to 2 minutes. Make a note of any observations you make. Remove the test tube from the hot water bath after this time.

Observations and results
1 Was there a colour change in the solution in the test tube?
2 How many colours did the solution change to before the last colour?
3 What was the colour change after heating the solution?

Activity 5.2 Testing for non-reducing sugars

SBA skills
Observation/recording/
reporting (ORR)
Manipulation/
measurement (MM)
Analysis and
interpretation (AI)

Aim: To test for non-reducing sugars.

Apparatus and materials
- non-reducing sugar sample
- dropping pipette
- Benedict's solution
- test tube
- dilute hydrochloric acid
- sodium hydrogen carbonate solution
- hot water bath
- tongs
- safety glasses

Before carrying out the test for a non-reducing sugar you should test a small amount of the sample for reducing sugar. If the test is negative (the solution does not turn orange/brick red) then you can proceed.

Method
1 Place about 1 ml of the non-reducing sugar sample into a clean test tube using the dropping pipette.
2 Add 1 ml of hydrochloric acid to the sample and shake well.
3 Heat the test tube in a hot water bath for 2 to 3 minutes.
4 Remove the tube from the hot water bath using the tongs and cool it under a running tap.
5 Add sodium hydrogen carbonate solution to the tube a little at a time until the fizzing stops.
6 Test the solution for reducing sugar.

Observations and results
1 Why was the test for a reducing sugar negative?
2 Why did you have to add sodium hydrogen carbonate to the mixture of acid and sugar in the test tube?
3 What was the colour change of the solution after heating?
4 What can you conclude from this test?

Activity 5.3 Testing for starch

SBA skills
Observation/recording/
reporting (ORR)
Manipulation/
measurement (MM)
Analysis and
interpretation (AI)

Aim: To test for the presence of starch.

Apparatus and materials
- starch sample
- iodine solution
- test tube
- 2 dropping pipettes

Method
1 Place about 1 ml of the starch sample into a clean test tube using the dropping pipette.
2 Add three drops of iodine solution to the contents of the test tube using a clean dropping pipette and shake it gently.

Observations and results
What colour change was observed?

Table 5.5 summarizes the tests that you have carried out in the activities. The tests are used to find out what types of carbohydrates foods contain.

Food test for carbohydrate	Method	Results
Test for starch, using iodine	Add a few drops of iodine solution to the food.	A blue-black colour indicates starch is present. Brown indicates no starch is present.
Test for simple reducing sugars, using Benedict's test	Mash food and mix it with water. Put mixture into a clean test tube and add equal amount of Benedict's solution. Heat the mixture in a water bath.	Orange-red, yellow or green means reducing sugar is present. Mixture will be blue if no reducing sugar is present.
Test for complex non-reducing sugars (break down into reducing sugars by hydrolysis using hydrochloric acid), then use Benedict's test	First test for reducing sugar using Benedict's test. This will be negative. Then break the complex sugars into reducing sugars by mixing the solution with a few drops of hydrochloric acid. Heat the mixture in a water bath. Neutralize the acid mixture by adding sodium hydrogen carbonate. Test for reducing sugars using Benedict's test.	Orange-red, yellow or green means reducing sugar is present.

Table 5.5 Tests for carbohydrates and expected results.

Check your knowledge and skills

1 What tests would you do on a piece of food to see if:
   a it contained starch
   b it contained reducing sugar?
2 Kayla tests an unknown substance for reducing sugar. The Benedict's solution remains blue.
   a Can Kayla be sure the substance doesn't contain sugar?
   b What could she do to make sure?

Unit 4 Lipids

Lipids are a diverse group of organic compounds found in all living organisms. Fats and oils and substances made from fats and oils are lipids. Lipids, such as butter, which are solid at room temperature, are called **fats**. Lipids, such as sunflower oil, which are liquid at room temperature, are called **oils**.

Foods that contain oils and fats such as fatty meats, cooking oils, butter, margarine and lard are a good source of lipids.

The structure of lipids

Like carbohydrates, lipids are made from the elements carbon, hydrogen and oxygen. However, in lipids these elements are found in a different ratio and they are arranged differently.

Structure of fats and oils

The fatty acid chains of a triglyceride are straight when they contain only single bonds. However, sometimes the chains have kinks or bends in them because they contain double bonds.

Triglycerides that have straight-chain fatty acids tend to make fats, while those with kinks in the chains tend to make oils.

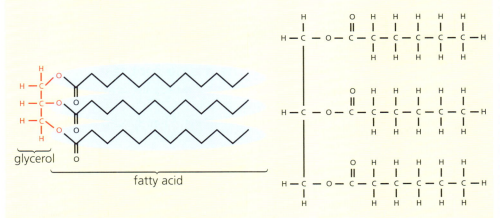

Figure 5.3 Structure of triglycerides.

Lipids contain one molecule of glycerol ($C_3H_8O_3$) and three molecules of fatty acid. Fatty acids are organic acids such as stearic acid, oleic acid and palmitic acid, which all have a long hydrocarbon chain (–COOH). The combination of glycerol and three fatty acid chains is also known as a triglyceride (see Figure 5.3).

Different lipids are distinguished by the types of fatty acids which combine with the glycerol. You have probably heard of **saturated** and **unsaturated fats**. Saturated fatty acids result in solid fats like butter and the thick layer of blubber found in seals and whales. Unsaturated fatty acids tend to result in liquid oils such as canola and olive oil.

The functions of lipids

- Energy storage – this is the main function of lipids because they contain more energy per gram than carbohydrates. Many plants and animals convert excess food into fats and oils. These can be converted back to energy if needed at a later stage.
- Helps to maintain constant body temperature – in mammals, the layer of fat under the skin helps to insulate the body.
- Cushions and protects – most vital organs are surrounded by a layer of fat that helps to cushion and protect them from mechanical damage.
- Cell membranes – lipids make up a major part of cell membranes and droplets of lipid may be stored in the cytoplasm of cells to provide energy.

Testing for lipids

Lipids do not dissolve in water, but they do dissolve in organic substances. This property is used to test for lipids.

Activity 5.4 Testing for lipids

Aim: To test for the presence of lipids.

Apparatus and materials
- lipid sample
- paper
- paper towel
- ethanol
- distilled water
- test tube
- dropping pipette

Method 1 The grease-spot test
1 Place a drop or smear of the lipid sample on a piece of paper. Blot any excess using paper towel.
2 Leave the spot to dry. Then hold the paper up to the light and look for a translucent spot.

Method 2 The emulsion test
1 Place about 1 ml of the lipid sample into a clean test tube using the dropping pipette.
2 Add 1 ml of ethanol to the sample and shake vigorously.
3 Add 2 ml of distilled water to the mixture and observe what happens.

Observations and results
1 What did you observe when you held the paper up to the light?
2 What did you observe when you added ethanol to the fat sample and shook the solution?

Lipids leave a translucent, greasy mark on paper. When lipids are dissolved in ethanol and then mixed with clean water, a cloudy emulsion results. This is a sign that lipids are present in the mixture.

Check your knowledge and skills
1 What is the difference between saturated and unsaturated fats?
2 What are the functions of lipids in the body?
3 Christopher wants to test an item of food for the presence of lipids. Suggest a simple way he could do this.

Unit 5 Proteins

After water, **proteins** are the most abundant molecules in most organisms. Proteins are especially abundant in animals where they perform many different functions.

Proteins are found in animal flesh (beef, chicken or fish), milk, eggs and peas and beans.

■ The structure of proteins

There are many different kinds of proteins, but they all contain carbon, hydrogen, oxygen and nitrogen. Proteins are made up of sub-units called **amino acids**. Figure 5.4 shows the structure of a typical amino acid. There are about 20 amino acids that are found naturally in the food we eat.

Amino acids are able to join together in long chains called polypeptides. Polypeptide chains bend and form specific shapes which result in particular protein molecules. Each molecule has its own specific shape. Its properties are determined by its structure.

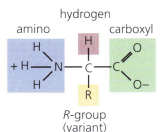

Figure 5.4 An amino acid is the smallest unit of a protein.

The functions of proteins

Proteins carry out many important functions in organisms. For example, proteins are needed to:

- produce muscle and tissues
- speed up reactions in the body (enzymes)
- provide structure and elasticity in animal cells (collagen and cartilage)
- carry oxygen (haemoglobin in red blood cells)
- build hair, nails and horns in animals (keratin)
- produce hormones which control and regulate body functions
- transport substances across cell membranes.

Testing for proteins

You can test for the presence of protein using the Biuret test. This involves using solutions of potassium hydroxide and copper sulphate. The test can also be carried out using sodium hydroxide but this should be used with care as it is harmful if it comes into contact with your skin.

Activity 5.5 Testing for proteins

Aim: To test for the presence of proteins.

Apparatus and materials
- protein sample
- copper sulphate solution
- potassium hydroxide
- test tube
- 3 dropping pipettes

Method
1 Place 1 ml of the protein sample into a clean test tube using the dropping pipette.
2 Add 1 ml of potassium hydroxide to the test tube and shake the mixture.
3 Add a few drops of copper sulphate solution and shake the test tube again.

Observations and results
What colour change did you observe?

If the mixture turns violet or purple when you add copper sulphate then protein is present in the substance being tested.

SBA skills
Observation/recording/
reporting (ORR)
Manipulation/
measurement (MM)
Analysis and
interpretation (AI)

Activity 5.6 Testing a range of food samples

SBA skills

Observation/recording/
reporting (ORR)
Manipulation/
measurement (MM)
Analysis and
interpretation (AI)

Aim: To test for the presence of carbohydrates, lipids and proteins in different foods.

Apparatus and materials
- crushed food samples (tomato, cucumber, onion, potato, beans/peas, lime juice, egg white)
- Benedict's solution
- hydrochloric acid
- copper sulphate
- sodium hydroxide
- iodine solution
- test tubes
- dropping pipettes
- tongs
- hot water bath
- safety glasses

Method
1 Test each food substance for carbohydrates, lipids and/or proteins.
2 Use a table to record your results.

Living organisms need a balance of inorganic molecules and organic molecules for healthy functioning. Table 5.6 summarizes what you need to remember about the sources and functions of carbohydrates, lipids and proteins.

Nutrient	Sources	Functions
Carbohydrate	Plantain, yam, sugar, cereals, bread	Primary supply of energy Assists in the utilization of fats Structural purposes, e.g. cellulose
Lipid	Cheese, margarine, cream, lard, milk, vegetable oil, avocado pear, butter	Secondary supply of energy after carbohydrates have been used up Aids in the absorption of fat-soluble vitamins, e.g. vitamins A and D Protects soft organs by acting as a shock absorber Provides insulation
Protein	Beef, liver, fish, lamb, eggs, poultry, legumes, nuts	Builds and repairs body tissues Builds antibodies – the parts of the blood which fight infections Builds enzymes

Table 5.6 Sources and functions of nutrients.

Check your knowledge and skills
Write a short paragraph explaining why it is important to have carbohydrates, lipids and proteins in a balanced diet.

Unit **6** Enzymes

An enzyme is a biological molecule that speeds up the rate of chemical reactions in a living organism. Reactions occurring in an organism's cells are known as **metabolism**. In general, enzymes are found in cells; however there are cases in which they are released outside the cells and then their effects are felt outside. An enzyme is a type of protein, and therefore has all the properties of a protein.

■ How do enzymes work?

An enzyme works on a substance called a **substrate**. A substrate is a reacting molecule that binds to the enzyme. An enzyme combines with a substrate forming an enzyme–substrate complex. In the breakdown of starch to maltose, the substrate would be starch. The enzyme that it binds to is amylase. There is a specific area on the enzyme that this substrate binds to, called the **active site**. At the active site, the enzyme binds to the reactants (substrates) and in doing so, it reduces the amount of energy needed to start the reaction; this speeds up the reaction.

It is important to note that enzymes themselves do not undergo change during a reaction. Enzymes **catalyse** a reaction. Only the substrates are changed during a reaction. Figure 5.5 shows an enzyme-controlled reaction.

The way in which enzymes work is often described as 'the lock and key hypothesis' (see Figure 5.6, overleaf).

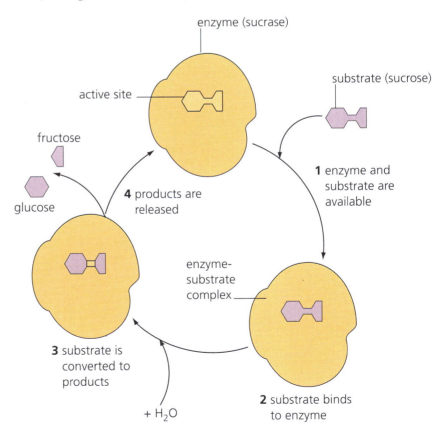

Figure 5.5 How an enzyme works.

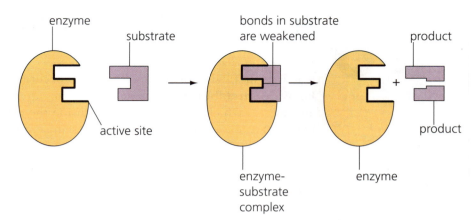

Figure 5.6 An enzyme has an active site which a specific substrate fits into, like a key into a lock. Products are formed at the active site during the reaction.

The properties of enzymes

- Enzymes are specific – this property is essential in the working of enzymes. Enzymes are usually responsible for catalysing one type of reaction. This means that the enzyme maltase, which breaks down maltose into glucose, cannot break down sucrose. This can be explained in terms of a 'lock and key' structure. The lock would behave like an enzyme which the key (the substrate) is trying to open. Therefore, the active site of the enzyme must fit with the substrate binding to it in order for the enzyme to work.

- Enzymes work best at a specific pH – a particular enzyme reaction will not take place efficiently if the pH is greater or less than the pH required for the enzyme reaction to occur. Different enzymes works best at different pHs. For example, amylase works at a pH of 8 and pepsin works at a pH of about 2. The pH at which an enzyme catalyses a reaction most efficiently is known as its optimum pH. The graph shown in Figure 5.7 shows how pH affects the action of three different enzymes.

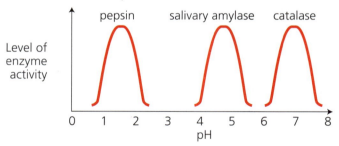

Figure 5.7 The effect of pH on an enzyme-catalysed reaction.

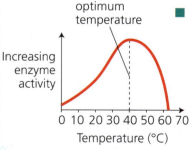

Figure 5.8 The effect of temperature on an enzyme-catalysed reaction.

- **Enzymes work at a specific temperature** – the optimum temperature at which most enzymes work is at body temperature. An increase in the temperature increases the kinetic energy associated with the movement of the enzymes and substrates. An increase in energy increases the chances of the enzymes colliding, thus increasing the chance of reactions. There comes a point, however, where enzyme activity reduces because the temperature is too high. At this temperature the enzymes is **denatured**. When an enzyme is denatured its active site is destroyed and it becomes inactive. When the active site is destroyed the substrate can no longer fit into the active site of the enzyme. At low temperatures, enzyme controlled reactions occur very slowly. The graph in Figure 5.8 demonstrates how temperature influences the rate of an enzyme-controlled reaction.

Activity 5.7 pH and the action of catalase

Apparatus and materials

- liver extracts of pH 1, 4, 7, 10, 14
- dropping pipette
- test tube rack
- hydrogen peroxide (H_2O_2)
- test tubes
- marker pen
- ruler
- stopclock
- safety glasses
- water

Method

1 Label the test tubes with the pH value of the liver extracts.
2 Place two drops of a liver extract and 1 ml of water in each labelled test tube.
3 Add 1 ml of H_2O_2 to the first test tube.
4 Measure the height of the foam that is formed after 10 seconds.
5 Record your result in an appropriate table (see Table 5.7).
6 Repeat steps **2–5** with the other tubes.
7 Write-up your activity, using the questions and table to help you.

Questions

1 Write a hypothesis for the experiment.
2 Write an aim for the experiment.
3 What is the independent variable?
4 What is the dependent variable?
5 Do the results support your hypothesis?
6 State a conclusion from your results.
7 Draw a graph of your results.

Tube	pH	Height of foam (mm) after 10 seconds
1		
2		
3		
4		
5		

Table 5.7 Results table.

Drawing line graphs

You may often have to display results in the form of graphs. A graph can be used to display information that gives a 'picture' of an occurrence. In these types of graphs you are plotting data pairs of two physical quantities in a co-ordinate system with horizontal and vertical axes. It is important to make the graph as clear as possible. You should draw your graphs using the following guidelines.

- Draw your graph on standard graph paper. The axes must be clearly labelled.
- Always give your graph a title.
- The **independent variable** is plotted on the x-axis; label the axis with the variable and its units (if appropriate).
- The **dependent variable** is plotted on the y-axis; label the axis with the variable and its units (if appropriate) (e.g. temperature, °C).
- Choose an appropriate scale for plotting your graph and label the axes accordingly.
- Use a sharp pencil to draw your curve.
- Label your curves if there are more than one on your graph.

Activity 5.8 Investigating the properties of enzymes

SBA skills

Observation/recording/reporting (ORR)

Manipulation/measurement (MM)

Analysis and interpretation (AI)

Apparatus and materials

- Petri dish filled with agar jelly containing starch
- 10% amylase solution (agar well 1)
- boiled amylase solution (agar well 2)
- amylase solution at pH 2 (agar well 3)
- drinking straw
- marker pen
- stopclock
- droppers
- iodine solution
- water

Method

1 As far as possible make four wells in the agar gel using the drinking straw.
2 Using the marker pen label the holes 1, 2, 3 or 4 on the bottom of the dish.
3 To each well add the correct amylase solution, and to the last well add water.
4 Be careful not to spill any of the solutions on the agar gel.
5 Leave the Petri dish for 40 minutes.
6 After 40 minutes pour in enough iodine solution to cover the agar gel.
7 Record your observations in a suitable table.

Explain your observations at each well. Explain the purpose of the well with only water.

Check your knowledge and skills

1 The enzyme catalase liberates oxygen from hydrogen peroxide. In an experiment carried out by Stacy, the rate of the reaction was half that of her friend Patrick's. This confused Stacy because both of them used similar amounts of hydrogen peroxide and catalase. When she checked the pH of the solution it was 7, which is the pH required for the reaction to occur. The pH of Patrick's reaction was also 7. Stacy's experiment was carried out at room temperature. Using the information given, deduce what Patrick may have done differently for his reaction rate to have doubled. State how this affected the rate of his reaction.

2 Equal samples of a food sample in solution were placed in two separate test tubes. The test tubes were labelled 'A' and 'B'. After carrying out food tests on test tube 'A', it tested positive for starch, negative for reducing sugar, and positive for protein. Solution 'A' was then added to test tube 'B'. The food was then tested again after 30 minutes. The results showed a negative test for starch, a positive test for reducing sugar and positive again for protein. Using your knowledge of enzymes explain the results of the experiment.

3 Explain why there was very little reaction occurring after a student placed a small piece of fat into an aqueous solution of lipase for 24 hours in a beaker that was left in the refrigerator.

4 What is the main function of enzymes?

5 State three reasons why enzymes are important.

6 Give two sources of carbohydrates in our diet.

7 State the type of molecules in:
   a water   b carbohydrate.

8 In what metabolic process is the energy from carbohydrates released?

9 Give two sources of lipids in our diet.

Chapter summary

Do you know?

If you are unsure of any of the facts in the list, refer to the page number in brackets.

- All matter is made up of elements. (page 42)
- Organic compounds contain carbon and hydrogen. (page 42)
- Inorganic compounds do not have carbon and hydrogen at the same time. (page 42)
- Water is a small inorganic molecule that makes up about two-thirds of the total number of molecules in our bodies. (page 43)
- The simplest units in carbohydrates are called monosaccharides. (page 44)
- Triglycerides are the molecules found in lipids. (page 49)
- Proteins are made up of amino acids. (page 50)
- Enzymes are proteins that catalyse (speed up) metabolic reactions. (page 53)
- Enzymes are specific. (page 53)
- Enzymes are affected by temperature and pH. (page 54)

Are you able to?

If you have trouble in doing these things, refer to the page number in brackets.

- Test for reducing sugars. (page 46)
- Test for non-reducing sugars. (page 47)
- Test for starch. (page 47)
- Test for lipids. (page 50)
- Test for proteins. (page 51)
- Construct a line graph from given data. (page 56)
- Investigate the properties of enymes. (page 56)

Chapter 6

Photosynthesis

How plants make food

All living things depend on energy from the sun. The sun gives out energy in the form of light. Plants convert sunlight into food (chemical energy) in a process called **photosynthesis**. Plants are able to store the food they make. Humans and animals eat plants and then convert the stored food in plants into energy for their own movement, respiration, growth and repair.

In this chapter you will learn how plants make their food by photosynthesis and about some of the factors that affect this important process.

By the end of this chapter you should be able to:
- Describe the process of photosynthesis
- Draw and label the external and internal structures of a leaf
- Describe and explain how the structure of a leaf is adapted for photosynthesis
- Demonstrate that light and chlorophyll are needed for photosynthesis
- Explain why plants need minerals

Unit **1**

Photosynthesis and plant nutrition

In Chapter 5 you tested parts of food plants (onions, tomatoes, potatoes) for sugar and starch and obtained positive results. You can also test the leaves of non-food green plants such as hibiscus or geranium to show that the leaves of all green plants contain starch. In order to do this you have to soften the leaf and remove its green colour so that you can see the results of the iodine test.

Activity 6.1 Testing a leaf for starch

Aim: To test the hypothesis that all green plants contain starch.

Apparatus and materials
- green leaf, freshly picked from a plant growing in full sunlight
- Bunsen burner
- tripod
- gauze
- glass jar or beaker half-filled with water
- test tube
- test-tube rack
- ethanol
- iodine solution
- dropping pipette
- protective mat
- white tile
- tongs

Method
1. Heat the beaker of water over a Bunsen burner.
2. Holding the leaf in the tongs dip it in the water for 20 seconds which is just long enough to soften it.
3. Turn off the Bunsen burner.
4. Pour some ethanol into a test tube and put the leaf in the ethanol.
5. Stand the test tube in a beaker of warm water for 10 minutes. This removes the green colour from the leaf.

SBA skills
Observation/recording/reporting (ORR)
Manipulation/measurement (MM)
Analysis and interpretation (AI)

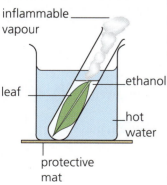

Figure 6.1 Decolourizing a leaf.

58

6 Rinse the leaf well in running water. Your leaf is now ready for the starch test.
7 Place the leaf on a white tile. Using the dropping pipette place a drop of iodine solution on to the leaf. What happens to the iodine solution on the leaf?

We know that plants contain sugar and starch because we can test for these substances and get a positive result. We can also show that plants don't absorb sugar and starch from the soil because we can test soil to show that it doesn't contain sugar and starch. These facts suggest that plants must be able to produce sugar and starch from other materials.

Photosynthesis

Plants make their own food. The ability to produce food places plants in a group of organisms known as **autotrophs** ('auto' means 'self' and 'troph' relates to feeding). **Autotrophic nutrition** describes the ability of an organism to make its own food from simple **inorganic** compounds. The form of autotrophic nutrition carried out by plants is called photosynthesis.

'Photo' literally means 'light' and 'synthesis' is the manufacture (making) of something complex by building up simple substances. Photosynthesis therefore refers to a chemical reaction in which plants use light energy from the sun, together with simple inorganic substances (the reactants), to manufacture complex food substances such as carbohydrates (the products).

The glucose produced in photosynthesis contains energy that was originally contained in the sunlight which reached the plant. This is one of the main reasons why we say all life on Earth depends on energy from the sun.

In order for photosynthesis to occur plants need:

- light
- chlorophyll
- carbon dioxide
- water.

If any of the above factors is unavailable to the plant, photosynthesis will not occur.

The reactants in photosynthesis are carbon dioxide and water. Minerals also assist in making the range of food materials that are required by the plant. The reactants are combined using the energy from sunlight that is trapped by a green pigment (hence plants are green) called **chlorophyll** found in special structures called **chloroplasts**. It is in chloroplasts that glucose is made. Chloroplasts are plant organelles found inside some plant cells and they are sometimes described as acting like small 'solar converters'. During the process of photosynthesis, oxygen is given off as a waste product.

The general formula for the process of photosynthesis is:

$$\text{carbon dioxide} + \text{water} \xrightarrow[\text{chlorophyll}]{\text{sunlight}} \text{glucose} + \text{oxygen}$$

$$6CO_2 + 6H_2O \xrightarrow[\text{chlorophyll}]{\text{sunlight}} C_6H_{12}O_6 + 6O_2$$

Figure 6.2 These sugar cane plants produce sugar by converting sunlight, carbon dioxide and water into carbohydrates.

Photosynthesis

The equation for photosynthesis shows glucose is the food compound produced in the reaction. It is important to realize that plants also make additional food products including other carbohydrates, lipids, proteins and vitamins.

Table 6.1 shows where the factors needed for photosynthesis come from.

Factor	Route of entry into plant	Use in photosynthesis
Carbon dioxide	Carbon dioxide gas in the atmosphere diffuses into the leaf through the stomata (small pores on the underside of leaves)	Combines with hydrogen in water to make glucose
Water	Water from the soil enters the plant's roots by osmosis (see page 118). Water then travels up the plant stem to the leaves through specialized cells called xylem	Provides a source of hydrogen for the production of glucose
Sunlight	Sunlight penetrates through the transparent epidermis of the leaf into the chloroplasts in the mesophyll cells (light sources other than sunlight can also be used in photosynthesis as long as they have the correct wavelength)	Provides the energy for photosynthesis

Table 6.1 Where the factors needed for photosynthesis come from and how they are used in the reaction.

Obtaining the factors for photosynthesis

- Carbon dioxide – the concentration of carbon dioxide in the air is extremely low (approximately 0.04%). Leaves therefore must be very efficient at absorbing it. However, the cells which need carbon dioxide are in a layer in the middle of the leaf (the mesophyll layer) (see Figure 6.10, page 67). As carbon dioxide is being used up in the leaf it will be at a lower concentration in the mesophyll layer than in the atmosphere. Therefore, carbon dioxide moves down its concentration gradient from an area of higher concentration (outside the leaf) to an area of lower concentration (inside the leaf) by diffusion. This occurs through the stomata. Once inside the air spaces, carbon dioxide diffuses through the cell walls, cell membranes and chloroplast membranes into the chloroplast where it is utilized.
- Water – water in the soil is at a higher concentration than in the cells inside root hairs. Water therefore enters the root hair cells by diffusion. Water is continually used up in the mesophyll layer (in photosynthesis) and lost through the stomata in **transpiration**. Transpiration is the word used to describe the concentration gradient that exists between the atmosphere and the soil – water travels up the xylem vessels in the stem and into the leaves where most of it is lost through the stomata (by evaporation) but some of it reaches the mesophyll cells and the chloroplasts inside photosynthetic cells by osmosis. Transpiration therefore serves to supply the mesophyll cells with water, cool the plant and provide a solution in which to transport important minerals.
- Sunlight – the upper epidermal cells are transparent, therefore sunlight easily enters the leaf. The thinness of the leaf also means that the sunlight radiates easily to the mesophyll cells where it is needed for photosynthesis.

What affects the rate of photosynthesis?

By simply looking at the chemical equation (look back at page 59), you will see that when there are more of the substances on the left (the reactants) more of the substances on the right (the products) will be produced.

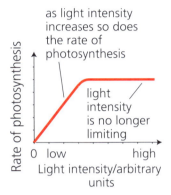

as light intensity increases so does the rate of photosynthesis

light intensity is no longer limiting

Rate of photosynthesis

0 low   high
Light intensity/arbitrary units

Figure 6.3 Light is a limiting factor in the rate of photosynthesis.

The following **variables** affect the rate of photosynthesis.

- Light intensity – when no light is present, photosynthesis does not occur. However, as light intensity increases, so does the rate of photosynthesis. Light intensity therefore is a **limiting factor**. As light intensity increases above an optimum level, the rate of photosynthesis levels off. This is because other factors such as temperature and carbon dioxide concentration become limiting factors (see Figure 6.3).
- Carbon dioxide concentration – the same principle applies to carbon dioxide concentration. An increase in carbon dioxide increases the rate of photosynthesis up to a point. After this point an increase in carbon dioxide concentration has no effect on the rate of photosynthesis.
- Temperature – for a chemical reaction to take place particles need to collide with enough energy for them to react. Increasing the temperature increases the rate and energy with which the reactants of photosynthesis collide, thus increasing the rate of photosynthesis. Above a certain temperature the rate may stop increasing, or even decrease! This is because stomata close at very high temperatures to prevent water loss from the leaf. The closed stomata mean that less carbon dioxide can enter the leaf from the atmosphere. Can you think of another reason why the rate of photosynthesis stops increasing at very high temperatures? (*Hint:* enzymes.)
- Wind speed – wind removes the layer of air saturated with water from around the surface of the leaves and therefore the stomata. This makes the diffusion gradient between the air spaces inside the leaf and the atmosphere steeper leading to an increase in the transpiration rate. The mesophyll layer therefore gets more water and the rate of photosynthesis increases.

Experiments to investigate photosynthesis

De-starching a plant

Before carrying out experiments to show that plants make starch in photosynthesis, you need to start with leaves that have no starch! A plant is placed in a dark box or cupboard for 48 hours. During this time the plant uses up all the starch in its leaves. This is called 'de-starching'. After this period, a starch test should be carried out on a leaf from the plant to confirm a negative result.

Figure 6.4 overleaf shows an experiment done by two students to investigate whether photosynthesis occurs in the absence of light.

When the students tested the leaf for starch, they observed that:

- parts of the leaf that were covered by foil stayed brown in the iodine test
- parts that were not covered by foil turned blue-black in the iodine test.

They concluded that light is needed for photosynthesis because only the parts of the leaf exposed to sunlight produced starch. Parts of the leaf that were in the dark (covered with foil) did not produce starch.

a Start with a de-starched pot plant.

b Use foil to cover part of a leaf.

c Put the plant in a sunny place for a day.

d Remove the foil from the leaf and test the leaf for starch.

Figure 6.4 An experiment to show that plants need light for photosynthesis.

Photosynthesis cannot take place without light. However, the intensity and frequency of light affect the rate at which photosynthesis takes place. In general, photosynthesis takes place at a faster rate in bright light. The brighter, or more intense the light, the faster is the rate of photosynthesis, until the optimum light conditions are exceeded (see Figure 6.3, page 61).

Chlorophyll is most effective at absorbing light when the plant is exposed to frequencies of visible light (red, orange, yellow, green, blue, indigo and violet) so photosynthesis occurs at a faster rate when the light contains these colours. When sunlight contains high levels of ultraviolet light, photosynthesis actually slows as the chloroplasts cannot absorb ultraviolet light.

Activity 6.2 Photosynthesis and light intensity

Aim: To investigate the effect of light intensity on the rate of photosynthesis.

Apparatus and materials
- large beaker with water
- boiling tube
- piece of pondweed (*Elodea*)
- stopclock
- metre rule
- thermometer
- lamp with 60 watt incandescent bulb

Method
1 Set up the apparatus as shown in Figure 6.5.
2 Place the lamp at the 100 cm mark on the ruler and start the stopclock.
3 After five minutes, count the number of bubbles given off from the plant in one minute. Record your result in a table.
4 Move the lamp closer to the beaker in increments of 20 cm, wait five minutes, then count the number of bubbles given off in one minute (ensure that the temperature of the water in the beaker remains the same). Take five sets of readings and record the results in your table.
5 Plot your data on a graph. Plot bubble rate on the *y*-axis and distance on the *x*-axis.

SBA skills
Observation/recording/reporting (ORR)
Manipulation/measurement (MM)
Analysis and interpretation (AI)

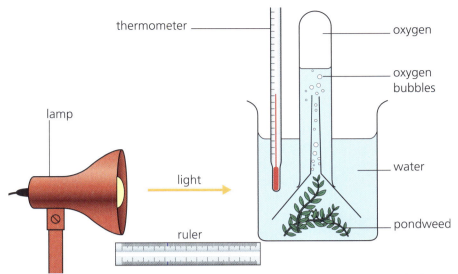

Figure 6.5 Apparatus used to measure the rate of photosynthesis in *Elodea* pondweed.

Questions

1 It takes a lot of energy to change the temperature of water.
   a What do you think was the purpose of the water in the beaker?
   b Why was the water needed?
2 What are the possible sources of error in this experiment?
3 What was the purpose of waiting for five minutes each time the lamp was moved before counting started?
4 Did the number of bubbles increase as the lamp got closer? Explain your answer.
5 Can you think of a way to investigate the effect of temperature on the rate of photosynthesis? (*Hint:* You can use the same apparatus as shown above.)

We cannot really test to show that plants need water for photosynthesis because removing the water kills the plant. However, scientists have done experiments with radioactive isotopes to show that water is used during photosynthesis. We can therefore assume that plants can make food as long as they have enough water.

Water for photosynthesis is absorbed through the roots and transported to the leaves through the xylem vessels. When water is plentiful, photosynthesis takes place at an increased rate; in dry conditions, the rate decreases.

How to set up and carry out a fair test

When you carry out an experiment in Biology you aim to test whether your hypothesis is true or false. For example, you may want to test the hypothesis 'photosynthesis cannot take place without carbon dioxide'. In this case, you want to test whether an absence of carbon dioxide prevents photosynthesis from taking place. You need to do a **fair test** in which you observe or measure the effects of changing *one factor* only (the carbon dioxide) whilst keeping all other conditions the same.

In order to make a test fair, you normally carry out your test with a **control**. In the experiment in Activity 6.3 the control would be a plant which is kept in exactly the same conditions as the experimental plant *except* that it has a supply of carbon dioxide. So, you keep all variables the same for the experiment and control except for the variable whose effect you wish to investigate.

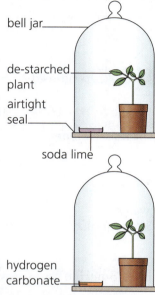

bell jar

de-starched plant

airtight seal

soda lime

hydrogen carbonate

Figure 6.6 Apparatus used to investigate whether photosynthesis occurs when carbon dioxide is removed from around a plant.

Figure 6.7 Variegated leaves.

Activity 6.3 Showing that carbon dioxide is necessary for photosynthesis

Aim: To investigate whether photosynthesis occurs when carbon dioxide is removed from the air around a plant.

Apparatus and materials
- 2 de-starched pot plants
- soda lime
- 2 Petri dishes
- hydrogen carbonate
- 2 bell jars (or 2 large clear plastic bags)
- elastic bands or similar, to secure plastic bag around pot
- starch test equipment (see Activity 6.1 on page 58)

Method
1 In the pot of one de-starched pot plant place a Petri dish of soda lime; and in the other place a Petri dish of hydrogen carbonate.
2 Place a bell jar or clear plastic bag over each plant. Make sure the plastic bags are fastened securely around the pot (Figure 6.6).
3 Place each plant in a brightly lit room for 30 minutes.
4 Test a leaf from each plant for starch (see Activity 6.1 on page 58). Record your observations.

Questions
1 Which plant contains starch? Was this plant the experiment or the control?
2 What can you conclude from this experiment?

As with water, it isn't possible to remove the chlorophyll from a leaf without killing the leaf. However, we can do experiments to show that chlorophyll is needed for photosynthesis because some plants, such as the one shown in Figure 6.7, have variegated leaves. In a variegated leaf, the green parts contain chlorophyll, the light parts do not.

Activity 6.4 Showing that chlorophyll is necessary for photosynthesis

Aim: To show that chlorophyll is necessary for photosynthesis.

Apparatus and materials
- de-starched variegated plant
- starch test equipment (see Activity 6.1 on page 58)

Method
1 Work in pairs. Discuss how you could use the equipment listed above to show that plants need chlorophyll for photosynthesis.
2 Write your hypothesis.
3 Develop a plan for your experiment. Check your plan with your teacher.
4 Carry out your experiment and record your results.
5 What can you conclude from your experiment?

Check your knowledge and skills

1 Where do plants get the energy to convert carbon dioxide and water into sugar?
2 Why does an equation for photosynthesis show chlorophyll and sunlight above the arrow?
3 How many molecules of carbon dioxide and water are needed to produce one molecule of glucose?
4 Why are plants sometimes called the Earth's lungs?
5 Where do plants get the inorganic substances they need for photosynthesis?
6 List the conditions that must be present in order for photosynthesis to take place.

Unit 2 The role of the leaf in photosynthesis

Most photosynthesis takes place in the leaves of plants. Leaves are plant organs that are specially adapted for the process of photosynthesis. They have several features, or adaptations, that allow them to carry out photosynthesis efficiently.

Activity 6.5 Observing leaf structure

SBA skills
Observation/recording/reporting (ORR)
Manipulation/measurement (MM)
Analysis and interpretation (AI)

Aim: To observe and record the characteristics of the outsides of leaves.

Apparatus and materials

- leaves from different plants – a selection of different shapes, thicknesses, surface textures, rough-edged, smooth-edged, etc.

Method

1 Examine the different leaves carefully.
2 Practise the drawing skills you learnt on page 25 by making a drawing of the external structure of a leaf of your choice.

Questions

1 What do all the leaves have in common?
2 What differences exist between the leaves?
3 What do you think are the purposes of the different structures that you have observed in the leaves?

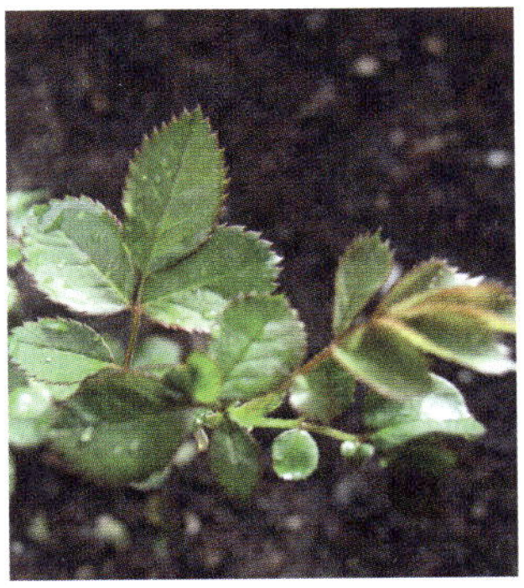

Figure 6.8 The leaves on plants are arranged so that all leaves are exposed to sunlight.

If you look at how leaves grow on a plant you will see that they are arranged so that there is very little overlap – this exposes all the leaves to sunlight. Many plants also have leaves with jagged edges or small leaflets. This design provides space between leaves and allows sunlight to reach the lower parts of the plant. You can see how smaller leaves with jagged edges are arranged in a rose bush in Figure 6.8.

The external structure of the leaf

Leaves are totally green or at least green in part (see Figure 6.7). You should also notice that a leaf has a broad, flat surface. This is called the **lamina**. The large surface area of a leaf allows for the maximum absorption of sunlight and carbon dioxide. On the underside of a leaf are tiny holes in its surface visible through a hand lens. These 'holes' are special pores that open and close. They are called

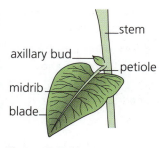

axillary bud

stem

petiole

midrib

blade

Figure 6.9 The external features of a leaf.

CO₂ in the air

The atmospheric concentration of CO_2 on Earth is rising. The original levels of CO_2 in our atmosphere came from gases released by volcanoes. However, today, human activities are responsible for most CO_2 released into the atmosphere. The levels of CO_2 are gradually rising each year. In 1960, CO_2 formed about 0.03% of the atmosphere, but today the level is closer to 0.04%. This increase is contributing to the greenhouse effect on Earth (see Chapter 25.)

stomata. Stomata are surrounded by cells called **guard cells** – these cells pull the stomata open or push them shut. Stomata are important because they:

- allow carbon dioxide to diffuse into the plant
- allow oxygen produced by photosynthesis to diffuse out of the plant
- can open and close to regulate water loss from the plant.

Leaves also have a network of veins which allow substances to be transported to and from the leaf from other parts of the plant. The external features of a typical leaf are shown in Figure 6.9.

Adaptations of the leaf

As a Biology student you will notice that at all levels of organization, form follows function. This means that all biological systems are designed in such a way that their jobs are done as efficiently as possible. Plant leaves must be adapted to ensure that they make best use of limited resources. For example, atmospheric concentration of carbon dioxide can be as low as 0.03% in some places. Carbon dioxide is a requirement for photosynthesis, thus its absorption must be extremely efficient.

Adaptation	Function
Large surface area, due to wide lamina, supported by petiole	Exposes the leaf to as much air and sunlight as possible
Thin in cross section	Allows efficient diffusion of carbon dioxide into the leaf, and of oxygen out of the cells Allows sunlight to penetrate to all photosynthetic cells
Stomata in lower epidermis	Allows diffusion of gases between the leaf and the atmosphere Prevents excessive water loss by transpiration
Transparent epidermis (no chloroplasts)	Allows maximum sunlight to radiate to the mesophyll layer
Vascular bundles (containing xylem and phloem vessels) within short distance of all palisade cells	Ensures all cells are well supplied with water Quickly removes the glucose produced
Air spaces in spongy mesophyll layer	Allows diffusion of carbon dioxide into and oxygen out of all mesophyll cells
Chloroplasts arranged broadside on, especially in dim light	Exposes as much chlorophyll to sunlight as possible
Chlorophyll on membranes within the chloroplast	Increases the surface area, thus exposing more chlorophyll molecules to sunlight

Table 6.2 How a leaf is adapted to its function.

The internal structure of the leaf

You've seen that the external structure of leaves helps maximize exposure to sunlight and carbon dioxide. However, photosynthesis actually takes place inside the leaf, so the internal structure is also well adapted for photosynthesis.

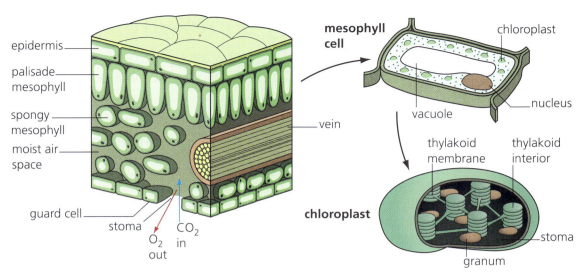

Labels on figure:
epidermis
palisade mesophyll
spongy mesophyll
moist air space
guard cell
stoma
O₂ out
CO₂ in
vein

mesophyll cell
chloroplast
nucleus
vacuole

thylakoid membrane
thylakoid interior
chloroplast
stoma
granum

Figure 6.10 The internal structure of a leaf.

The diagram in Figure 6.10 shows what a thin section of leaf looks like under a microscope. Read the information around the diagram carefully. It will help you to understand how leaves are adapted for photosynthesis.

Activity 6.6 Observing and drawing the internal structures of a leaf

Aim: To observe and draw the internal structures of a leaf.

Apparatus and materials
- thin section of leaf
- tweezers
- slide
- coverslip
- water
- microscope

Method
1 Using the tweezers carefully mount the section of leaf on a slide.
2 Examine the section under a microscope.
3 Try to identify the structures shown in Figure 6.10.
4 Draw a labelled sketch to show what you could see under the microscope.

Check your knowledge and skills
1 List three external features of a leaf which help it get as much sunlight as possible.
2 Why is the underside of leaves often a paler colour of green than the upperside?
3 a In which cells are the most chloroplasts found in a leaf?
   b Why are most chloroplasts found in these cells?
4 State the role that each part of a leaf plays in photosynthesis:
   a air spaces in the spongy mesophyll layer
   b xylem
   c stomata
   d chloroplasts.

Understanding photosynthesis

We have so far shown photosynthesis as one simple chemical equation – the process is actually a lot more complex. It involves several different reactions, most of which require enzymes, and can be divided into two main stages.

The light–dependent stage of photosynthesis

As its name suggests, the **light-dependent stage** requires, or depends on, light. Light energy is trapped by chlorophyll and is used to split water. The process of using light energy to split water is known as **photolysis** (literally 'light splitting' since 'photo' means 'light' and 'lysis' is splitting). When water is split its hydrogen and oxygen are produced. The hydrogen is used in the light-independent stage and oxygen is released as a waste product. These equations outline photolysis:

$$H_2O \xrightarrow[\text{chlorophyll}]{\text{light}} H_2 + \tfrac{1}{2}O_2$$
$$\text{water} \qquad\qquad \text{hydrogen} \quad \text{oxygen}$$

or: $$2H_2O \longrightarrow 2H_2 + O_2$$

You can carry out a simple test using water plants to show that plants produce oxygen during photosynthesis.

Activity 6.7 Investigating whether a water plant gives off oxygen

Aim: To find out if oxygen is produced during photosynthesis.

Apparatus and materials
- *Elodea* (Canadian pondweed) or Hydrilla
- 2 beakers of water
- 2 funnels
- 2 test tubes
- splint
- matches
- safety glasses (for testing gas)

Method
1 Set up each plant as shown in Figure 6.11.
2 Place one plant in a bright, sunny place and the other in a darkened place (a cupboard or similar).
3 Leave the plants for two days, then compare the gas produced in each experiment.
4 Test the gas that has collected for oxygen with a glowing splint. Be careful with the lighted splint. Always wear safety goggles and keep your hair tied back.

Questions
1 Was this a fair test? Give a reason for your answer.
2 Did both plants produce gas?
3 Was the gas produced by the plant which was left in sunlight oxygen? How could you tell?

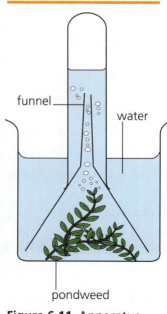

Figure 6.11 Apparatus used to demonstrate that plants produce oxygen during photosynthesis.

■ The light-independent stage of photosynthesis

The **light-independent stage** does not require light or chlorophyll. (It can, however, occur while it is light, so the term 'dark stage' is not quite accurate because the reactions do not occur only in the dark.) In this stage, the hydrogen generated in the light-dependent stage is used to convert carbon dioxide to sugars. This is a **reduction reaction** because it involves addition of hydrogen.

Light-independent reactions (the Calvin Cycle) convert carbon dioxide into sugar, the basic food source for all plants and animals. These light-independent reactions take place in the stroma. The light-dependent reactions occur in the thylakoid membranes.

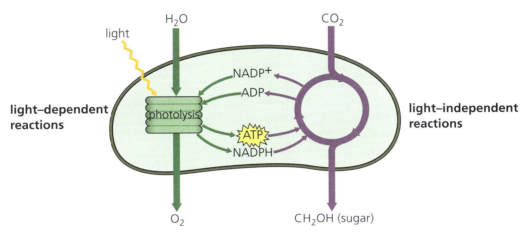

Figure 6.12 Summary of the reactions that take place in the two main stages of photosynthesis.

Check your knowledge and skills

1 Does photosynthesis stop immediately if you remove a plant from sunlight and place it in total darkness? Explain your answer.

2 What are the products of the reactions in the light-dependent stage of photosynthesis?

3 What happens to these products during the light-independent stage of photosynthesis?

Unit 4 — Plant nutrition and mineral requirements

You have learnt that most photosynthesis takes place in the leaves. This means that plants make most of their food in their leaves. But the cells in every part of the plant need food to supply the energy for their life processes. This means that the food has to be moved or transported around the plant.

Plants have a transport system consisting of veins made up of xylem and phloem tubes. Xylem tubes are used to transport water from the roots to the rest of the plant. Phloem tubes are used to transport food substances around the plant. You will learn more about transport in plants in Chapter 11.

Most of the glucose produced by photosynthesis is quickly converted into starch and stored in certain parts of the plant. The starch can then be broken down into soluble sugars when the plant needs the energy.

Figure 6.13 The stringy fibres in the stalk of this celery plant are made of cellulose. Cellulose cannot be turned back into sugar.

Storing glucose as starch

A storage molecule should have the following characteristics; it should be insoluble, compact, and relatively unreactive. Glucose does not fit these criteria for a number of reasons.

- It is highly reactive. This means it gets involved in chemical reactions, interfering with metabolic processes.
- It dissolves easily. This makes glucose 'osmotically active'; through osmosis it can alter the concentration of the solution in the cell. This may lead to water entering the cell, and can eventually cause the cell to burst.
- It is a large molecule, so it takes up too much space to store energy easily.

Therefore glucose needs to be converted to a different kind of molecule in order to be stored effectively by living organisms. Humans mostly convert glucose to starch, which is a more efficient food storage molecule.

Sugars are soluble in water. This means that they dissolve in water. When plants need energy, dissolved sugars are moved along the veins (in the phloem tubes). The dissolved sugars can move to all parts of the plant to provide energy for growth and other cellular processes.

In plants glucose cannot be transported directly. It must therefore be converted to a less reactive, but just as soluble substance. This carbohydrate is sucrose.

Some sugar is converted into **cellulose**. Cellulose is the stringy fibres that are found in the walls of plant cells (see Figure 6.13). Cellulose cannot be converted back into sugar if the plant needs energy.

What happens to the glucose produced in photosynthesis?

To summarize what you have learnt so far, four main things happen to the glucose produced in photosynthesis.

- Glucose is oxidized during respiration to provide energy to drive other reactions in the plant.
- Glucose is involved in polymerization reactions to produce molecules of starch, which can be stored in various parts of the plant.
- Glucose is used to produce cellulose which strengthens cell walls; this provides structural support for the plant as it grows.
- Glucose is involved in synthesis reactions with other chemicals (mineral salts) to produce amino acids, which are used to build the proteins needed for cellular processes in the plant.

Minerals and how plants use them

In Chapter 5 you learnt that plants need minerals and that they grow badly and show signs of deficiency when certain minerals are in short supply. But how do plants obtain and use these minerals?

Plants use chemical elements found in the soil in the form of **mineral** salts (ions) which are soluble (will dissolve) in water, and which are absorbed by root hair cells in the plant's roots.

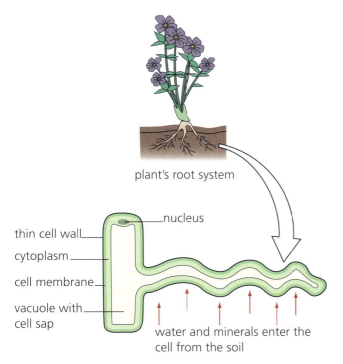

plant's root system

nucleus

thin cell wall

cytoplasm

cell membrane

vacuole with
cell sap

water and minerals enter the
cell from the soil

Figure 6.14 Uptake of mineral salts and water by a root hair cell.

- Nitrogen – plants take this up from the soil in the form of nitrates. The plant builds amino acids by combining nitrogen (from absorbed nitrates) with glucose (from photosynthesis). Amino acids are the building blocks of proteins which are essential components of cell cytoplasm, and which are used to form enzymes in the plant.
- Sulphur – this is an important component of some plant proteins. Plants obtain this by taking it up from the soil in the form of sulphates.
- Phosphorus – plants take this up from the soil in the form of phosphates. It is needed in certain chemical reactions that take place in the plant.
- Zinc, boron, manganese, copper, calcium, molybdenum and iron – a plant also needs these minerals, but only in tiny amounts; because of this they are called trace elements (or micronutrients). They are necessary for healthy plant growth. They often act as catalysts in chemical reactions in the plant.

Check your knowledge and skills

Draw a flow diagram illustrating how glucose is used in a plant.

Chapter summary

Do you know?

If you are unsure of any of the facts in the list, refer to the page number in brackets.

- Plants are autotrophs – this means they make their own food by photosynthesis. (page 59)
- Photosynthesis is a process in which the chlorophyll in green plants absorbs sunlight and combines it with carbon dioxide and water to form glucose and oxygen. (page 59) Photosynthesis can be represented by the chemical equation:

$$6CO_2 + 6H_2O \rightarrow C_6H_{12}O_6 + 6O_2$$

- Photosynthesis relies on the presence of light, water, chlorophyll and carbon dioxide. If any of these factors are missing, photosynthesis doesn't take place. (page 59)
- Plant leaves are adapted to maximize exposure to sunlight and to make sure that photosynthesis takes place as efficiently as possible. They have a large surface area, a thin cross-section and lots of chloroplasts. (page 65)
- Photosynthesis takes place in two stages: the light-dependent stage which produces oxygen

from water and a light-independent stage in which the plant produces carbohydrates in a reduction reaction. (page 68)
- Plants can make all the food they need by photosynthesis but they need a supply of mineral salts to build the sugars into more complex food molecules. (page 69)

Are you able to?

If you have trouble in doing these things, refer to the page number in brackets.

- Test parts of a plant for starch (the products of photosynthesis). (page 58)
- Carry out controlled experiments to show that light, carbon dioxide and chlorophyll are necessary for photosynthesis. (pages 62 and 64)
- Set up and carry out a fair test in which only one variable is changed. (page 63)
- Draw and label the external and internal structures of a leaf. (pages 65 and 67)
- Prepare a slide mount with a thin section through a leaf. (page 67)
- Carry out an experiment with water plants to show that oxygen is produced during photosynthesis. (page 68)

How humans obtain and use food

When we think of food, we usually think of eating; tasting, smelling and enjoying the foods we like. However, have you ever considered what happens to your food once you have swallowed it?

In this chapter, you will learn about the processes by which we use food for energy and as a source of the materials for building our body's cells. Everything you eat must be broken down so that any important nutrients can be absorbed into your bloodstream and become part of your cells.

By the end of this chapter you should be able to:

- Differentiate between heterotrophic and autotrophic nutrition
- Identify the various types of heterotrophic nutrition
- Identify the parts of the alimentary canal
- Describe the processes that take place in the alimentary canal
- Discuss the role of teeth in the process of digestion
- Draw and label a diagram showing the internal structure of a tooth
- Define a balanced diet with reference to age, gender and physical activity
- Discuss the advantages and disadvantages of a vegetarian diet
- Give examples of the results of a surplus or a deficiency of essential nutrients

Figure 7.1 Very tasty Caribbean foods!

Unit **1** Heterotrophic nutrition and digestion

In Chapter 6 you learnt that plants are autotrophs – they can synthesize their own food by photosynthesis. Animals (and some other organisms) are not able to synthesize their own food so they need to feed on other organisms to obtain food. Feeding on organic substances produced by other organisms is called **heterotrophic nutrition** and organisms that obtain nutrition in this way are called **heterotrophs**. Animals, including humans, and fungi are heterotrophs.

■ Types of heterotrophic nutrition

All heterotrophs obtain their food from other organisms but they do this in different ways. We can place heterotrophs into different categories based on the ways in which they feed.

- Saprophytes – these are organisms that feed on the dead remains of other organisms and carry out **extracellular digestion**. This means they secrete enzymes into their food to break it down into soluble substances. Then they absorb the soluble substances. Saphrophytes serve a very useful purpose in ecosystems as they are responsible for decomposition and decay. The fungus in Figure 7.2 is a saprophyte which obtains nutrients from the tree trunk.
- Parasites – these are organisms, such as the tapeworm (*Taenia solium*), that live inside, or on, a host organism that acts as a food source. Parasites usually do not

Figure 7.2 Fungi are saprophytes which means they feed on the dead remains of other organisms.

Is it egestion or excretion?

It is important to note that the release of faeces is *not* a process of excretion – it is part of egestion.

have an alimentary canal but they do have suckers or some structure which attaches them to the host and allows them to obtain nutrients.

- Holozoic – these organisms feed by taking in complex organic substances which are subsequently broken down by an internal digestive system for absorption. Humans and other mammals are holozoic organisms. They eat chunks of plant and animal matter and then break it down to release the nutrients their bodies need.

Holozoic nutrition in mammals

Holozoic nutrition involves five major processes. These are **ingestion**, **digestion**, **absorption**, **assimilation** and **egestion**.

- Ingestion – this involves taking in pieces of food. Organisms normally bite or tear off pieces of food and then chew it before swallowing.
- Digestion – the process by which food is broken down and dissolved so that it can be used by cells in the body. Ingested food is broken down mechanically and chemically into simpler, smaller substances that can be absorbed by the cells of the body.
- Absorption – the movement of dissolved substances (nutrients) across the cell membrane of the cells lining the alimentary canal into the bloodstream where they are carried to the cells in the body.
- Assimilation – when the products of digestion are used in the cells. An example of assimilation is glucose being used in the process of respiration.
- Egestion – the removal of undigested substances from the alimentary canal.

The alimentary canal in humans

The human digestive system consists of an alimentary canal which extends from the mouth to the anus, together with various associated organs and glands. In order to understand the processes of digestion you need to know the location of the main organs of the alimentary canal as shown in Figure 7.3.

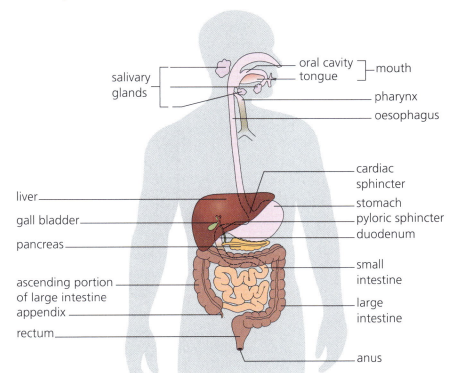

Figure 7.3 The structure of the human alimentary canal.

Types of digestion

During digestion food is broken down by mechanical and chemical means.

Mechanical digestion

Mechanical or **physical digestion** involves breaking food down into smaller pieces. There are no chemicals involved in this process and no chemical changes or reactions take place in the food. Biting into a piece of chicken, breaking off a piece and then chewing it, is a form of mechanical digestion.

The main purpose of mechanical digestion is to increase the surface area of the food so that chemical digestion can take place more efficiently. Mechanical digestion takes places in several areas of the alimentary canal (see Table 7.1). As you read, refer to Figure 7.3 for the location of each part of the alimentary canal.

Part of the alimentary canal	Description of mechanical digestion
Mouth	Food is chewed and broken down by the action of teeth and shaped into a ball (called a bolus) by the tongue
Oesophagus	Muscles contract and relax (called **peristalsis**) to transport food to the stomach
Stomach	The churning action of the walls of the stomach helps to grind and crush food into smaller pieces
Small intestine (duodenum)	Bile is released and this causes the **emulsification** of fats. This process doesn't change the fat, but it breaks down large globules of fats into extremely small droplets of fats

Table 7.1 Mechanical digestion in the alimentary canal.

Chemical digestion

During **chemical digestion** food material is reduced to its simplest components by chemical reactions. This process is important because food material can only be absorbed into the walls of the gut once it has been reduced to simple substances (the undigested molecules are simply too large to cross the gut wall cell membranes).

Chemical digestion involves the use of **enzymes**. Enzymes facilitate (help) break down complex molecules, which are too large to pass through the walls of the alimentary canal, into smaller molecules which are able to pass through.

Enzymes are specific in their action. This means that different enzymes are needed to digest different carbohydrates, proteins and lipids. Other nutrients are small enough to pass directly into the walls of the alimentary canal.

The main differences between chemical and mechanical digestion are:

- In chemical digestion food substances are broken down into molecules, whereas in mechanical digestion the food substances are broken down into smaller pieces.
- No enzymes are involved in mechanical digestion but enzymes are always involved in chemical digestion.

Check your knowledge and skills

1 Define the following terms:
   a heterotrophic nutrition   b digestion   c chemical digestion.
2 Why does the food eaten by holozoic organisms need to be digested?
3 Name the digestive process that relies on enzyme action.

Types and functions of teeth

In humans and other mammals, digestion starts in the mouth. Teeth play a role in both the ingestion of food and in mechanical digestion. Once food is ingested, the teeth act to cut it up, chew it and grind it into smaller pieces.

■ Teeth in humans

Humans are born without any visible teeth, however by about the age of four to five years their first set of 20 teeth are in place. This first set of teeth is called the temporary, deciduous or **milk teeth**.

Milk teeth are lost from the ages of about five to six. As each milk tooth drops out it is replaced by a new one. This second set of teeth is called the **permanent teeth**. Most people have their permanent teeth by their mid to late teens.

A full set of permanent teeth consists of 32 teeth. There are 16 in the upper jaw and 16 in the lower jaw. They are categorized as follows:

- eight incisors
- four canines
- eight premolars/bicuspids
- twelve molars.

The molars at the back of the mouth in the upper and lower jaw are usually the last to grow and are called wisdom teeth.

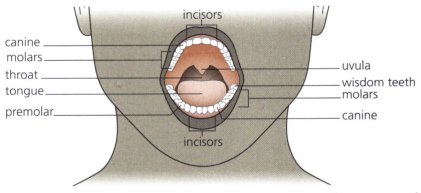

Figure 7.4 The positions and shapes of teeth in the human mouth.

Each category of tooth is shaped differently to suit its functions. Table 7.2 summarizes the shapes of the four main types of teeth and their functions.

Category of tooth	Shape	Functions
Incisors	Chisel shaped	Biting off pieces of food
Canines	Pointed	Biting off and tearing into food
Premolars	Two triangular ridges, quite flat surface	Tearing and grinding food
Molars	Few triangular ridges, largely flat surface	Grinding and chewing food

Table 7.2 How tooth shape is related to its function.

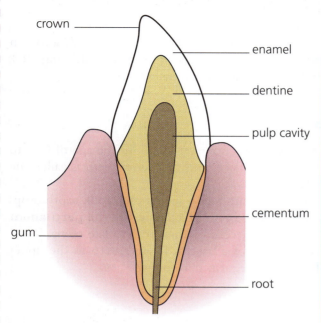

crown

enamel

dentine

pulp cavity

cementum

gum

root

Figure 7.5 The internal structure of a typical tooth.

Internal structure of a tooth

Figure 7.5 shows the main structures in a tooth and Table 7.3 shows how a tooth is adapted to its function.

Structure	Function
Enamel	Extremely hard outer surface of the tooth. Provides the hard surface for chewing and grinding
Dentine	Hard, porous material below the enamel. Supports the enamel. Harder than bone
Pulp cavity	Soft centre of the tooth. Contains blood vessels and nerves. Nourishes the dentine
Crown	Part of the tooth that shows in the mouth beyond the gum
Root	Part of the tooth that is hidden from view below the gum
Cementum	Tough, bone-like material. Helps to hold the tooth in its socket

Table 7.3 Parts of a tooth and their functions.

Did you know?

Tooth enamel is the hardest substance found in living organisms.

Activity 7.1 Looking at your teeth

Apparatus
- mirror

Method
1 Using a mirror, carefully observe your teeth.
   a Count the total number of teeth you have.
   b Count the number of teeth in your upper and in your lower jaw.
   c Identify the four main types of teeth, and describe them and their positions in your mouth.
   d Count the number of teeth in each category in your mouth.

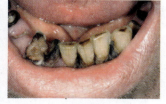

Figure 7.6 Tooth decay is a common but mostly preventable health problem.

Caring for your teeth

Teeth play an essential role in nutrition so it is important to care for them in order to prevent tooth decay and gum disease. Once permanent teeth are lost, they do not grow back.

Tooth decay is largely caused by the action of bacteria. Bacteria live inside the mouth and on the surfaces of the teeth. Some bacteria combine with food deposits to form a sticky layer called **plaque** that covers the teeth. The bacteria in plaque feed on sugars which produces acid. This acid damages the tooth enamel, eventually destroying it, causing holes or cavities in the surface of the tooth. Acid then leaks into the tooth, destroying the dentine layer underneath. When the decay reaches the pulp cavity it causes infection and pain (as there are nerves in the pulp). In some cases, an abscess may develop under the tooth and this can cause serious illness and pain.

You can take steps to care for you teeth and prevent decay. These include:

- Brushing and flossing – done regularly, this removes any food that may be left on your teeth that could become food for bacteria.
- Using a fluoride toothpaste – fluoride forms a protective coating over teeth helping to prevent acids from causing decay.
- Using an alkaline mouthwash – the acids are neutralized by the alkaline nature of the mouthwash.
- Regular checks with your dentist – your dentist will be able to detect any problems early on and address them before they become a major problem.

Check your knowledge and skills
1 Draw a labelled diagram to show the internal structure of a human molar.
2 Explain how the hard outer layer of a tooth can be destroyed by the action of bacteria on the surface of the tooth.
3 Suggest two methods of preventing damage to teeth caused by bacteria.

Unit 3 Digestion in the alimentary canal

In Section 7.1 on page 74 you learnt that food is broken down into smaller pieces and moved through the alimentary canal during the process of mechanical digestion.

Peristalsis is an important element of mechanical digestion. Figure 7.7 shows you how lumps of food are moved through the alimentary canal by peristalsis. To make this process easier, food is mixed with mucus secreted by the mucosa in the alimentary canal. The mucus helps to make the food slippery so it will travel easily along the digestive tract.

As food is mechanically digested, vitamins, minerals and water are released. These important nutrients can be absorbed directly into the bloodstream, mostly in the stomach because their molecules are small enough to be transported across cell membranes.

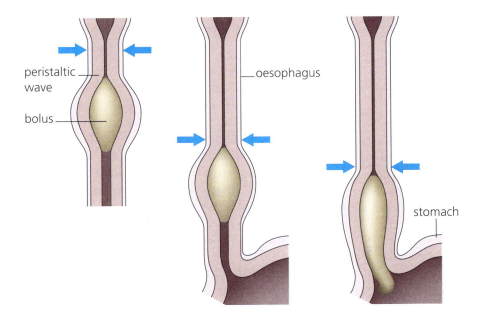

Figure 7.7 Peristalsis is a means of moving food along the alimentary canal.

Breaking down macromolecules by chemical digestion

Although vitamins, minerals and water can be absorbed directly into the bloodstream, other nutrients have to be chemically digested before they can be absorbed into the blood. In humans, proteins, carbohydrates (starch) and lipids (fats) have to be chemically digested. These are large organic macromolecules that cannot pass through the walls of the alimentary canal unless they are broken down into their smaller, constituent units.

■ Starch is digested to form smaller molecules of glucose.
■ Protein is digested to form smaller molecules called amino acids.
■ Lipids (fats) are digested to form smaller molecules of fatty acids and glycerol.

Enzymes needed in chemical digestion

Chemical digestion cannot happen without enzymes. Each nutrient requires the action of a different enzyme for digestion and these enzymes each have specific conditions in which they work best. The names, sources, sites of action and function of the main digestive enzymes are shown in Table 7.4.

Enzymes	Where the enzyme is made	Where it works	Function of enzyme
Amylase	Salivary gland	Mouth	Starch → maltose
Pepsin	Stomach wall	Stomach	Protein → polypeptides
Amylase	Pancreas	Small intestine	Starch → maltose
Trypsin		Small intestine	Protein → polypeptides
Lipase		Small intestine	Fats → fatty acid and glycerol
Maltase	Wall of small intestine	Small intestine	Maltose → glucose
Peptidase		Small intestine	Polypeptides → amino acids

Table 7.4 Enzymes needed in digestion.

Journeying along the digestive tract

The mouth

When food is ingested, the teeth and tongue break it down mechanically and shape it into lumps (a bolus) of food that can be swallowed.

At the same time, chemical digestion also takes place when food is mixed with saliva. Saliva contains the enzyme **amylase**. The temperature and pH of the mouth provide ideal conditions for amylase to work. Amylase in the mouth begins the process of digesting the macromolecule starch by breaking it down into the smaller disaccharide, maltose.

However, disaccharides such as maltose are still too large to be absorbed into the bloodstream, so maltose has to be digested further later. This does not happen in the mouth because amylase does not act on maltose, it acts only on starch – this is an example of an enzyme being specific.

The oesophagus

The oesophagus (throat) is a muscular tube that extends from the mouth to the stomach. When a bolus of food is swallowed, it is moved along the oesophagus by peristalsis. No enzymes are secreted in the oesophagus so no further chemical

Maltose

The disaccharide maltose is a reducing sugar. So the presence of maltose can be tested for using the Benedict's solution test (see page 46).

digestion takes place. However, the amylase from the mouth is mixed with the bolus of food and continues to break down starch as it moves down the oesophagus and into the stomach.

The stomach

The stomach is a muscular sac shaped like a mango. When food enters the stomach it remains there for a few hours. The muscular walls of the stomach constantly contract and relax. This action churns the food and mixes it with gastric juices and mucus to form a soup-like mixture called **chyme**.

Gastric juice is a fluid secreted from special cells, called gastric pits, in the wall of the stomach. Gastric juice contains a mixture of substances including mucus, hydrochloric acid and the enzyme, pepsin (see Table 7.4).

- Mucus is important because it lines the wall of the stomach and protects it from the actions of acid and pepsin.
- Hydrochloric acid is important because it creates a suitable environment for the action of pepsin. Pepsin is one of the few enzymes in the body that acts best in an acidic medium. The hydrochloric acid also destroys any microorganisms that may be found in the bolus entering the stomach.
- Pepsin is an enzyme that catalyses the breakdown of protein into short polypeptides. However, like disaccharides, polypeptides are still too large to be absorbed into the bloodstream.

The stomach has a strong sphincter muscle at its base. After a few hours, the food is let out of the stomach, a little at a time, into the first part of the small intestine.

Activity 7.2 Investigating the effect of pH on the action of pepsin

Aim: To investigate the conditions needed for protein to be digested.

Equipment and apparatus
- 5 test tubes
- 5 cubes of boiled egg white
- 0.5% sodium hydrogen carbonate solution
- 0.8% hydrochloric acid solution
- 1% pepsin solution
- 35 °C water bath
- measuring cylinder
- water
- marker pen

Procedure
1 Label the test tubes 1 to 5.
2 To test tube 1 add 4 ml sodium hydrogen carbonate solution and 1 ml pepsin solution.
3 To test tube 2 add 4 ml hydrochloric acid and 1 ml pepsin solution.
4 To test tube 3 add 4 ml water and 1 m pepsin solution.
5 To test tube 4 add 5 ml hydrochloric acid solution.
6 To test tube 5 add 5 ml water.
7 Place all the test tubes into the water bath.
8 Carefully place one boiled egg-white cube in each tube.
9 Observe the tubes over the next 30–45 mins.

Questions

1 List two precautions which you could take to ensure that this experiment is fair.
2 Give an appropriate hypothesis for this investigation.
3 What is the purpose of test tube 5?

Activity 7.3 Antacids and digestion

Aim: To plan and design an investigation to determine the effect of different antacids on the digestion of protein.

You are going to plan and design an experiment to see what effect different **antacids** have on the digestion of protein. You are not going to carry out your experiment. The skill is the ability to plan and design a fair experiment.
   Here are some tips to help you plan and design your experiment.

■ Be sure that you understand what your experiment is about. Think about and identify the scientific process that can be applied to the experiment.
■ Read background information on the topic before you decide how to plan your experiment.
■ Create a suitable hypothesis. You must be able to test your hypothesis and find evidence to support or disprove it.
■ Aim to design a simple but effective experiment. Keep in mind that the larger the sample size you use the more reliable the results of the experiment will be. Do not use irrelevant materials in your design.
■ Make certain that your experiment is fair, and where possible always use a control. (A control is an experimental setup that is the same as the 'test' setup but instead of changing one factor, *all* the factors remain the same for the duration of the experiment.) A control is necessary so that any change in the results in the test setup can be attributed to changing the one factor.
■ Make sure to list your limitations and any possible sources of errors.
■ Identify the variables.
■ Show how you would present the results of the experiment.

The small intestine

The small intestine is a tubular section of the alimentary canal which is several metres long in an adult human. The small intestine is divided into three parts: the **duodenum**, jejunum and ileum. The duodenum is the shortest section and the ileum is the longest section of the small intestine.
   Small amounts of food are released bit by bit from the stomach into the duodenum. Two glands also empty their contents into the duodenum. The **pancreas** is a gland that makes and secretes pancreatic juice into the duodenum. Pancreatic juice contains several enzymes, including amylase, trypsin, chemotrypsin and lipase, as well as sodium hydrogen carbonate. The sodium hydrogen carbonate neutralizes the acidic chyme from the stomach and stops the action of pepsin. The neutralized medium allows enzymes in pancreatic fluid to work.

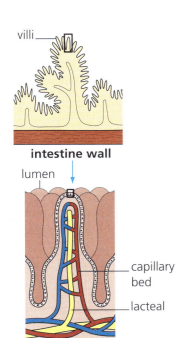

villi

intestine wall

lumen

capillary bed

lacteal

villus

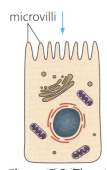

microvilli

Figure 7.8 The structure of the inner wall of the small intestine and a single villus, showing how it is adapted for absorption.

Enzyme action in the small intestine

Amylase has the same substrate as amylase in the mouth. So it is responsible for catalysing the breakdown of any starch not already digested in the mouth.

Trypsin catalyses the breakdown of protein. Therefore any proteins that were not digested in the stomach are now broken down with the help of trypsin.

The enzyme lipase works on lipids (fats). It is responsible for the breakdown of lipids into fatty acids and glycerol. The small intestine is the only place where lipids are chemically broken down, so to ensure that they are efficiently broken down a substance known as **bile** is secreted into the duodenum. Bile is made in the liver and stored in the gall bladder, from where it is released into the duodenum.

Bile causes the **emulsification** of lipids. This means that it causes lipids to form extremely small droplets. This increases the surface area for the enzyme lipase to work on thus increasing the efficiency of the enzyme. There is no enzyme in bile; therefore the emulsification process of bile is a form of mechanical digestion.

Once it leaves the duodenum, the soupy mixture enters into the other parts of the small intestine. On the inner walls of the intestine there are thousands of finger-like projections called **villi**. You can think of these villi giving the surface of the small intestine a texture a bit like a bath towel. The cells that make up the villi are covered in turn with smaller structures called microvilli. Both types of villi serve to increase the surface area of the intestine to facilitate (help) the absorption of nutrients into the blood. Cells found in the walls of the intestine also make and secrete enzymes that are responsible for the final digestive process. Enzymes, such as maltase that catalyse the digestion of maltose and peptidases that are responsible for the final digestion of polypeptides, are among the enzymes made by the special cells in the walls of the small intestine. The jejunum and ileum are responsible for the final digestion and all of the absorption of nutrients.

So, the final results of digestion are:

- starch is broken down to glucose molecules
- proteins are broken down to amino acids
- lipids are broken down to fatty acids and glycerol.

These units are now small enough to pass through the walls of the small intestine.

Unit 4 Absorption and the fate of digestive products

Key fact

Do not confuse the hepatic portal vein that carries nutrient-rich blood to the liver with the hepatic vein which carries deoxygenated blood from the liver to the heart.

All the nutrients, except for the components of lipids, enter the walls of the small intestine by active transport and then continue into the blood vessels by diffusion. This is called absorption.

Nutrients leaving the alimentary canal are carried away by a special blood vessel called the **hepatic portal vein**. This vessel carries the nutrients to the liver where they are processed. The components of lipids (fatty acid and glycerol) diffuse into the walls of the villi where they are taken away by special vessels called **lacteals** which are part of the **lymphatic system**. The fatty acids and glycerol molecules then enter the bloodstream later on via another major blood vessel called the **vena cava**; they do not go directly to the liver like other nutrients.

Once the digested nutrients have entered the bloodstream they are carried throughout the body. These dissolved nutrients enter cells to be used in cellular processes. The use of absorbed food in the cells is called assimilation.

Large intestine

This part of the alimentary canal is shorter than the small intestine but its lumen is wider. It consists of several parts: the caecum, colon, appendix and rectum. The human alimentary canal does not have the enzymes to facilitate the break down of some material that may be in food. One excellent example of this is plant fibre. Plant fibre is mainly cellulose from the cell walls of plant cells. This undigested material and some water spends up to three days in the large intestine; this facilitates the efficient absorption of water to form semi-solid faeces.

Faeces are made up of mainly cellulose, cholesterol, bile, mucus, bacteria and water. They are released through the anus.

Check your knowledge and skills

1 Compare the different types of heterotrophic nutrition.
2 The enamel of the teeth needs to be cared for to prevent it from decaying. Explain fully how a tooth becomes decayed.
3 Explain why the enzyme pepsin works only in the stomach of the alimentary canal.
4 There is a mechanical component to digestion. Why is it preferable to break food items into smaller pieces?
5 For each of the following 'juices' state where they are secreted, name their major components, and give the function of each component in the digestive process.
   a saliva   b gastric juice   c pancreatic juice.
6 Explain the role of bile in digestion.

Unit 5 Health and diet

When people use the word 'diet' they are often talking about an eating plan to help them lose weight. But in Biology the term **diet** refers to all the food you eat on a regular basis.

Diet is an important component of health. Your diet supplies the energy and nutrients that your body uses. If your body is to work properly and stay healthy it needs a balanced diet. A balanced diet is one in which a person eats the correct amount and types of food.

The amount of food that a person needs varies and depends on how active the person is, how old the person is, and whether the person is male or female. In addition to eating the correct amount of food, a balanced diet must contain the correct proportions of food from the different food groups as well as vitamins, minerals and water. A balanced diet provides the body with all the energy it needs and the necessary chemicals of life (see Chapter 5).

The energy in food

You already know that energy is released from food by respiration and you have learnt that respiration is a similar process to combustion (an oxidation reaction). If we assume that respiration and combustion release similar amounts of energy, then we can work out the energy content of food substances by measuring the amount of energy released when food is burned in air (combusted). The energy released by burning a sample of food is used to heat a known volume of water; the increase in temperature of the water is measured and used to calculate the energy value of food in kilojoules per gram (kJ/g).

Carbohydrates, fats and proteins all have different energy values.

- Carbohydrates contain 16 kJ/g
- Fats contain 37 kJ/g
- Proteins contain 21 kJ/g.

How much energy do I need?

The amount of energy your body needs depends on how many metabolic processes are taking place. Your **metabolism** is the total of all the chemical reactions taking place in your body. Your metabolism is affected by:

- Age and gender – in general, females have a lower metabolic rate than males and younger people have a higher metabolic rate than older people. Metabolic rate increases in pregnant and breastfeeding women.
- Activity level – the amount of physical work or exercise you do affects how much energy you need. A person who does hard physical activity requires more energy than a person who sits at a desk all day.

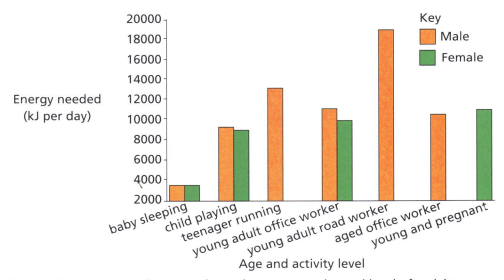

Figure 7.9 Energy requirements depend on age, gender and level of activity.

What is a balanced diet?

A balanced diet contains a mixture of food groups. No single food group contains all the nutrients that the body needs. This means a balanced diet must contain carbohydrates, proteins, fats, vitamins, minerals and water.

In general, a balanced diet is one which contains a variety of foods that are low in sugar and fats and high in fibre. An example of a balanced diet would be:

- Wholegrain cereals, fruits and vegetables – these provide energy, vitamins and minerals and fibre and they should make up about two-thirds of the diet.
- Lean meat, chicken, fish and dairy products – these provide protein and should make up most of the other third of the diet.
- Fats – these should be limited to approximately 15–30 g/day.
- Refined foods high in sugars – these should be avoided, except for the occasional treat.
- Water – approximately 6–8 glasses of fresh water should be drunk each day.

Vegetarian diets

Some people have restricted diets. Reasons for this include age, religion (such as Rastafarianism, Seventh Day Adventists), diseases such as heart disease and hypertension. One type of restricted diet is the **vegetarian diet**.

A true vegetarian diet is one which leaves out all foods derived from animals. There are sub-groups of vegetarian diets that range from strict vegetarian to the lacto-ovo vegetarian.

- Vegans – these people are very strict vegetarians and eat only plant materials. They do not eat butter, milk or lard.
- Lacto-vegetarians – as the name suggests, these people include milk and dairy foods in their otherwise vegetarian diet.
- Lacto-ovo-vegetarians – these people add eggs to the foods of the lacto-vegetarians.
- Semi-vegetarians – these people include fish and poultry, but no red meats, in their otherwise vegetarian diet.

Since protein usually comes from animal sources in the diet, a vegetarian must find other sources of protein. These include peas, beans legumes and nuts.

Malnutrition

Malnutrition is often associated with not getting enough food. However, malnutrition actually means 'bad diet' or incorrect nutrition. It therefore includes conditions where one eats too much or too little food, as well as conditions where food is eaten in the wrong proportions.

Eating too much food

If you consume more energy (in food) than your body needs, the excess is stored either as glycogen or as fat. When a person stores a great deal of fat they become heavier than they should be and they are said to be overweight. Figure 7.10 shows healthy weights for people of different heights. A range of weights is shown for each height because a person's ideal weight depends on their age, gender and also their bone structure.

When a person's weight is 20% or higher than the average for their height, they are said to be **obese**. Obesity can lead to many other health problems. These include coronary heart disease (CHD), hypertension and diabetes. The problem of obesity is a significant problem in the Caribbean. According to the Pan-American Health Organization the problem has increased by about 400% over the last decade. It is now the most important underlying cause of death in the region.

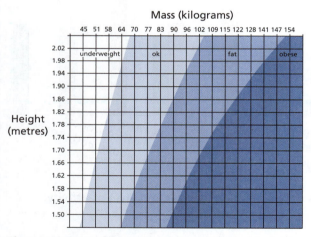

Figure 7.10 Healthy weights for different heights.

- Hypertension is the prolonged or chronic elevation of a person's blood pressure, usually due to blockages in arteries.
- Diabetes is when a person has abnormally high blood glucose levels continually.

Protein energy malnutrition (PEM)

PEM is a condition that is usually prevalent in very poor countries, usually those stricken with famine and wars. Children are usually worse affected. They are usually tired, under-developed for their age, and have stunted growth. This is because they lack the proper nutrients in their diets, so they do not have the necessary raw materials for growth and repair of muscle cells and for energy.

Deficiency diseases

A lack of specific nutrients can cause certain illnesses known as **deficiency diseases**.

Deficiency disease and symptoms	Nutrient/s that is deficient
Kwashiokor (often in children) Muscles develop slowly and limbs have stick-like appearance Abdomen and liver swell	Protein
Marasmus (often in children) Body tissues waste away and child becomes very thin Skin is dry and wrinkled	Protein and general lack of nutrients
Scurvy Bleeding from gums and internal organs Wounds heal slowly Painful joints	Vitamin C
Xerophthalmia Night blindness, poor vision, reduced resistance to infections	Vitamin A
Rickets Soft bones and teeth in children Brittle bones and teeth in adults	Calcium and vitamin D
Anaemia Reduced number of red blood cells which reduces the oxygen-carrying capacity of blood Tiredness and lack of energy Dizziness	Iron and vitamin B_6
Goitre Swelling of thyroid glands Reduced metabolic rate In children may lead to mental retardation	Iodine

Table 7.5 Some deficiency diseases and their symptoms.

Check your knowledge and skills

1 a Make a list of all the food you ate yesterday.
  b Draw up a table like the one below. Place a tick or a cross in each column to show what is contained in each of the foods. Sweet potato has been filled in here as an example.

Food eaten	Carbohydrate	Protein	Fat/oil	Minerals	Vitamins	Fibre
Sweet potato	✓	✓		✓	✓	✓

2 Use your completed table to analyse your diet. Consider the following.
  a Is your diet healthy and balanced?
  b Are you eating too much of any nutrient(s)?
  c Are there any nutrients lacking in your diet?
  d What changes could you make to your diet to make it healthier?

Chapter summary

Do you know?

If you are unsure of any of the facts in the list, refer to the page number in brackets.

- The purpose of digestion is to break down food into small particles that can be absorbed by the body. (page 72)
- The human digestive system consists of the alimentary canal and related organs. (page 73)
- There are two types of digestion – mechanical and chemical. (page 74)
- Mechanical digestion is a physical process that does not involve enzymes. (page 74)
- Chemical digestion involves enzymes and results in changes to the molecular structure of the reactants. (page 74)
- Mechanical digestion starts in the mouth with the action of the teeth. (page 75)
- Humans have 32 permanent teeth. There are four different types of teeth – incisors, canines, molars and premolars. Each type of tooth is adapted to suit its function. (page 75)
- Tooth decay is caused by the action of bacteria in the mouth. Decay can be reduced by good tooth care and regular visits to the dentist. (page 76)
- In the mouth, food is mixed with saliva which contains the enzyme amylase. This enzyme acts on starch to break it down. (page 78)
- Chewed food, mixed with saliva, is shaped into a bolus and swallowed. The oesophagus carries the food to the stomach. (page 78)
- Food is stored in the stomach for several hours. Chemical digestion of protein by the enzyme pepsin begins in the acid medium of the stomach. (page 79)
- The bulk of digestion takes place in the shortest section of the alimentary canal – the duodenum. (page 80)
- Pancreatic juice flows into the duodenum to facilitate the breakdown of nutrients. Bile is also released in the duodenum to break fats down into smaller particles. (page 80)
- Digestion is completed in the small intestine. (page 81)
- Digested food is absorbed into the body through the walls of the small intestine. The surface of the small intestine is covered with villi and microvilli which increase the surface area and facilitate absorption. (page 81)
- Most nutrients are transported to the liver where they are further broken down or stored. (page 81)
- Dissolved nutrients are transported to the cells of the body by the blood. They are then absorbed into cells and used for metabolic processes. (page 82)
- Water is removed from undigested food in the large intestine and faeces are formed. These are removed from the body in the process of egestion. (page 82)
- Human beings need a balanced diet. A balanced diet is a combination of the right nutrients in the correct proportions. (page 82)
- Dietary needs are determined by a person's age, gender and level of activity. (page 83)

Are you able to?

If you have trouble in doing these things, refer to the page number in brackets.

- Draw and label the human alimentary canal. (page 73)
- Draw and label the internal structure of a tooth. (page 76)
- Carry out a test to observe the effects of pH on the enzyme pepsin. (page 79)
- Formulate a proper hypothesis. (page 80)
- Plan and design an experiment. (page 80)
- Carry out a controlled experiment. (page 80)
- Identify limitations and sources of error in an experiment. (page 80)
- Identify the healthy and unhealthy components of a diet based on given information. (page 85)

8 Cellular respiration

How cells get energy

All living things need energy for growth and repair, reproduction, movement and to build up complex molecules. In addition, humans and other mammals need energy to produce heat to maintain a constant body temperature. Plants need energy so that they can get substances into their roots by active transport.

Plants make their own food for energy but animals have to eat food to get the energy for these processes. Once they have digested and assimilated the food, they need to somehow release the energy in a useful form. In this chapter you will learn about the process that releases this energy called **cellular respiration**.

By the end of this chapter you should be able to:

- Define respiration
- Distinguish between respiration and breathing
- Differentiate between aerobic and anaerobic respiration
- Understand the function of ATP
- Carry out experiments to investigate the products and reactants of respiration

Unit 1 Respiration and its importance

Energy is the ability to do work. The body works to sustain all life processes – growing, reproducing, moving, cell division and active transport – all these processes involve some form of work; therefore body processes require energy.

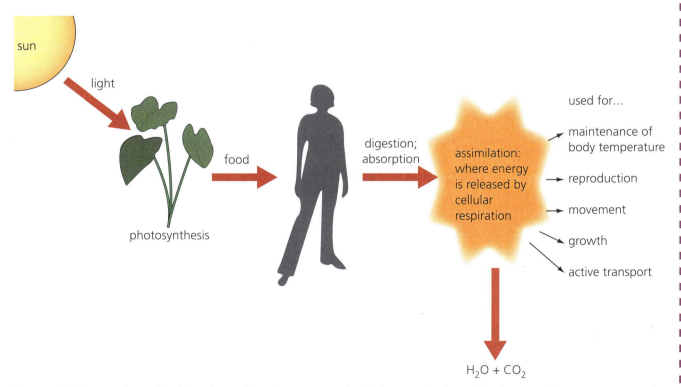

Figure 8.1 Energy from food is released by the process of cellular respiration. It is then used to carry out work.

Energy has to be converted into forms that organisms can use. For example, during photosynthesis plants convert light energy from the sun into chemical energy in organic molecules. The stored energy then has to be released before it can be used. The energy conversion which releases this energy is called cellular respiration.

What is respiration?

Respiration takes place in all the cells of the body. It is a metabolic process in which organic substances are broken down to simpler products with a release of energy. This energy is incorporated into special energy-carrying molecules and is then used for other metabolic processes.

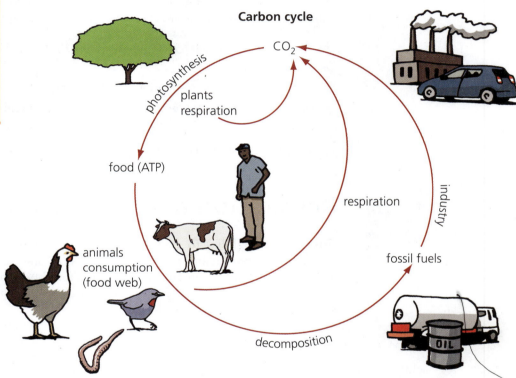

Carbon cycle

CO_2

photosynthesis

plants respiration

food (ATP)

respiration

industry

animals consumption (food web)

fossil fuels

decomposition

Figure 8.2 Respiration takes place in all cells to keep an organism alive. Plants and animals respire all the time to provide the energy they need for cellular functions.

Cells need a constant supply of energy. However, if all the energy released by respiration was made available at the same time, there would be too much energy and cells would become damaged. To avoid this, energy is released more slowly in the form of energy-carrying molecules. These molecules store the energy for a short time and allow it to be used as it is needed by cells. The most common energy-carrying molecule is called **adenosine triphosphate** which is often shortened to **ATP**. ATP is the form in which energy is carried in cells and made available when needed, so it is sometimes called the energy 'currency' of living things. ATP cannot be stored, so a constant supply of it is needed at all times.

In most organisms, glucose is the starting point for respiration but other substances are used in other organisms. In the human body, for example, excess glucose is converted into **glycogen** and stored in the liver and muscles. Glycogen is converted back to glucose for use in respiration.

Respiration can take place with or without oxygen being present. Respiration which needs oxygen is called **aerobic respiration** and respiration that doesn't need oxygen is called **anaerobic respiration**.

Unit 2

Aerobic respiration

Aerobic respiration requires the use of oxygen. It is an oxidation reaction because oxygen is added to glucose. The reaction products are carbon dioxide, water and energy. Respiration involves many different reactions, each catalysed by enzymes, but it can be summarized by the following equation:

glucose   + oxygen   → carbon dioxide + water   + energy (2900 kJ)

$$C_6H_{12}O_6 \;+\; 6O_2 \;\rightarrow\; 6CO_2 \;+\; 6H_2O \;+\; \text{energy (2900 kJ)}$$

Respiration takes place in the **mitochondria** of cells. Cells that need large amounts of energy have large amounts of mitochondria, for example muscle cells.

Although respiration releases large amounts of energy, cells do not over-heat and burn up. This is because the reactions in respiration take place in small steps and the energy is released in stages as small 'packets' of ATP.

In humans and other mammals, most respiration takes place aerobically. This is because our respiratory system and blood circulation allow large amounts of oxygen (the raw material for respiration) to be transported to the cells of the body and for large amounts of carbon dioxide (a waste product of respiration) to be removed from the body.

Aerobic respiration is ideal because it releases more energy than other types of respiration; this is because the glucose is completely broken down during the oxidation process. Notice from the equation that all six carbons in the glucose molecule can be found in the carbon dioxide. For this reason it is the most preferred type of respiration.

Activity 8.1 Measuring the rate of oxygen uptake during respiration

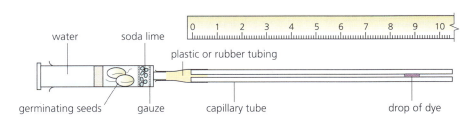

Figure 8.3 A simple respirometer.

Key fact

The process of oxidation can be observed when you burn a substance in air. For example, if you burn sugar in the laboratory it bursts into flames (combustion) and releases carbon dioxide, water and energy. This is similar to what happens in respiration, except that the reactions in respiration are smaller and more controlled.

Equipment and apparatus

- simple respirometer
- 6 to 8 germinating seeds
- dye
- soda lime
- ruler
- stopclock
- marker pen

Procedure

1 Assemble the respirometer as shown in Figure 8.3.
2 Pull up a small amount of dye into the tube.
3 Mark the position of the dye.
4 Start the stopclock and after 1 minute measure the distance that the dye travels.
5 Repeat the steps several times to get at least four readings.
6 Given the diameter of the capillary tube, determine the average volume of oxygen used up by the germinating seeds.

Questions

1 In what direction did the dye move?
2 Explain why the dye moved in that direction.
3 Explain why you can assume that the movement of the dye was due to the uptake of oxygen by the seedlings.

Check your knowledge and skills

1 Write an equation for respiration.
2 Why is respiration important to all living cells?
3 Glucose is oxidized during aerobic respiration. What does this mean?
4 Explain why oxidation does not cause damage to cells.

Unit 3

Anaerobic respiration

Respiration can also take place without oxygen. You learnt earlier that this type of cellular respiration is called anaerobic respiration.

During anaerobic respiration, organic molecules such as glucose, are broken down to form simpler molecules with the release of energy. But, anaerobic respiration releases much *less* energy than aerobic respiration. This is because respiration without oxygen does not break down the glucose completely. Some forms of anaerobic respiration only break the glucose molecule into two, others break it into four and only a small amount of carbon dioxide is released.

■ Anaerobic respiration in yeast

Yeast is a single-celled organism that uses anaerobic respiration to break down sugar to make alcohol. Anaerobic respiration in yeast cells is called **fermentation** and it can be summarized by the following equation.

$$C_6H_{12}O_6 \rightarrow 2C_2H_5OH + 2CO_2 + \text{a little energy}$$

C_2H_5OH is the formula for **ethanol**. Ethanol is an important substance in industry. It is found in all alcoholic drinks and it can be used as a fuel. Because of the release of carbon dioxide during the process, yeast is also important for the baking industry, because it causes the dough to rise. Therefore yeast fermentation is widely used in several industries.

Figure 8.4 Anaerobic respiration in yeast has many commercial uses – alcohol production (left) and bread-making (right) are shown here.

Activity 8.2 Investigating anaerobic respiration in yeast (by making real ginger ale)

Equipment and apparatus

- clean, 2 L plastic soft–drink bottle with top
- funnel
- fine-cut grater
- measuring cup
- teaspoon and tablespoon
- 1 cup sugar
- 2 tablespoons freshly grated ginger root
- juice of 1 lemon in a container
- clean stirrer
- fresh granular baker's yeast (¼ teaspoon)
- cold, fresh pure water (bottled water is fine)

Procedure

1. Add 1 cup of sugar to the 2 L bottle using the dry funnel. (Leave the funnel in place until you are ready to put the top on the bottle.)
2. Measure ¼ teaspoon of fresh granular active baker's yeast.
3. Add the yeast through the funnel into the bottle and shake to disperse the yeast grains into the sugar granules.
4. Grate the ginger root using the grater to produce ½ tablespoon of grated root.
5. Place the grated ginger in the cup measure.
6. Add the lemon juice to the grated ginger.
7. Stir the lemon juice and grated ginger to form a slurry.
8. Add the slurry of lemon juice and grated ginger to the bottle.
9. Fill the bottle to the neck with fresh, cool, clean water, leaving about 4 cm of head space, and securely screw the top down to seal. Invert repeatedly to thoroughly dissolve the sugar.
10. Place the bottle in a warm place for 24 to 48 hours. Or until the bottle feels hard or stiff.
11. Once the bottle feels hard to a forceful squeeze, it is ready.
12. Open the bottle. (Be careful not to point at anyone while opening because the contents of the bottle may spurt out violently.) Have a taste!

Questions

1. Why did the bottle become 'stiff'?
2. Describe the reaction that took place in the bottle.
3. What are the main products of the reaction?

■ Anaerobic respiration in bacteria

Many bacteria are able to survive without oxygen. They respire anaerobically – mostly by fermenting sugars. The end products of anaerobic respiration in bacteria differ from bacteria to bacteria; some produce methane, some produce lactic acid and some produce alcohol.

Anaerobic respiration in some bacteria releases methane. Methane is the chief ingredient in **biogas**. Biogas is an alternative fuel source that is considered to be more environmentally friendly than the more commonly used fossil fuels.

The equation for anaerobic respiration taking place in these bacteria is:

$$C_6H_{12}O_6 \rightarrow 3CO_2 + 3CH_4$$
$$\text{glucose} \qquad \text{carbon dioxide} \quad \text{methane}$$

Bacteria which produce lactic acid are useful for making butter and yoghurt.

■ Anaerobic respiration in humans

There are times when the demand for oxygen in cells is higher than the supply. For example, when a person exercises strenuously, the muscles need more ATP. The muscles do not get enough oxygen to respire aerobically so the cells begin to respire anaerobically. Lactic acid is produced and no carbon dioxide is given off.

$$C_6H_{12}O_{66} \rightarrow 2C_3H_6O_3 + \text{little energy}$$
$$\text{glucose} \qquad \text{lactic acid}$$

As lactic acid builds up in the muscles it can become toxic to the body. Fatigue sets in and the muscles begin to cramp, and if the muscle continues to work it will eventually collapse. By the time the muscle stops working the cells would have built up what is known as an **oxygen debt**. This must be repaid immediately after the exercise. This is why we breathe so heavily after very strenuous exercise. This allows for increased oxygen to get into our bodies so that the lactic acid can be broken down aerobically into carbon dioxide and water.

Once all the lactic acid has been broken down, the oxygen debt is repaid and the heart rate and breathing return to normal.

Aerobic respiration	Anaerobic respiration in muscle cells	Anaerobic respiration in yeast
Breaks down glucose	Breaks down glucose	Breaks down glucose
Uses oxygen	Does not use oxygen	Does not use oxygen
Produces carbon dioxide and water	Produces lactic acid	Produces carbon dioxide and ethanol
Releases large amount of energy (ATP)	Produces small amount of energy (ATP)	Produces small amount of energy (ATP)

Table 8.1 A comparison of aerobic and anaerobic respiration.

Activity 8.3 Feeling the effect of lactic acid build-up on muscles

Procedure

1 Raise one hand straight above your head and leave the other pointing to the floor by your side.
2 Clench and release your fists on both hands repeatedly until you cannot bear it any more. Place both hands on the table.
3 Record and explain your observations.

Activity 8.4 The effects of exercise on pulse and breathing rates

Equipment and apparatus

- stopclock
- notebook
- access to stairs or steps

Procedure

1 Working in pairs, decide who will count and who will take the exercise! Record the number of breaths and pulses of the pair who is going to take the exercise (at rest for 1 minute).
2 She or he then needs to run up and down the stairs or steps 10 times.
3 *Immediately* after the exercise, start the stopclock and take their pulse and breath rates.
4 Take their pulse and breath rates for 1 minute every 2 minutes until both rates return to the resting rate.
5 Draw up a table of the data and present your results on a graph.

Questions

1 Describe the breathing rate and pulse rates immediately after exercise.
2 Explain why the rates of the pulse and breathing went up during exercise.
3 Explain why the rates of the pulse and breathing did not fall to their resting rate immediately after the exercise was finished.

Understanding test and exam terms

When you answer questions in a test or exam you need to know what the examiner expects from you so that you can give the correct type of answer. The instruction words used in the questions give you important information about the type of answer you are expected to give.

Read the information in the table below carefully to see what type of answer you need to give when you see instruction words in questions.

Instruction	Type of answer you need to give
Define	A clear statement of what something means
Describe	Give the main details but you do not need to explain or give reasons why
Discuss	Give your own point of view or present advantages and disadvantages
Explain	Supply reasons for something and the theory behind it
List	Give points with no explanation or reasons, can be just single words
State	Give a short answer with no supporting facts or reasons
Predict	You are not expected to know the answer but you are expected to apply your knowledge to say what you think is likely to happen
Suggest	This means there is more than one answer or that you are expected to answer based on your knowledge and skills
What is meant by …	Give a definition and some reasons or explanation

Check your knowledge and skills

1 Define cellular respiration.
2 Compare aerobic respiration and anaerobic respiration.
3 Using the list of statements below compare aerobic respiration and combustion by writing each statement in the correct column in a table like the one shown. Some of the statements apply to both respiration and burning.

- Releases energy
- Happens slowly
- Happens quickly
- Rate controlled by enzymes
- Rate cannot be controlled
- Uses oxygen
- Produces carbon dioxide
- A one-step process
- A series of many reactions

Aerobic respiration	Combustion

Chapter summary

Do you know?

If you are unsure of any of the facts in the list, refer to the page number given in brackets.

- All living things must respire so that they can have the required energy to carry out their metabolic reactions. (page 87)
- Respiration is a metabolic process that takes place in cells to convert organic substances into usable energy. (page 88)
- ATP is the energy 'currency' for all living cells. (page 88)
- Aerobic respiration uses oxygen to release large amounts of ATP from the breaking down of glucose. (page 89)
- Anaerobic respiration does not require oxygen and releases very little energy. (page 90)

Are you able to?

If you have trouble doing this, refer to the page numbers in brackets.

- Set up a respirometer and use it to investigate oxygen uptake in small organisms. (page 89)

Gas exchange and breathing

Moving gases in plants and animals

Your body needs energy to be able to function. Energy from digested food is released and made available to cells by respiration. This process needs a supply of oxygen, which is supplied by the air you breathe in. Respiration also produces a waste gas, carbon dioxide, which we get rid of by breathing it out.

In this chapter you will learn more about gas exchange (how oxygen is taken in and carbon dioxide is removed from the bodies of plants and animals).

By the end of this chapter you should be able to:

■ Define gaseous exchange and explain why it is important in all organisms

■ Identify features that are shared by all gaseous exchange surfaces

■ Explain how gas exchange surfaces are adapted to increase their surface area

■ Describe and explain the importance of breathing in humans

■ Name and locate the main parts of the respiratory system in humans

■ Discuss the effects of smoking cigarettes on the body

Gas exchange

You have seen that cellular respiration uses oxygen to oxidize glucose and release molecules of ATP that supply the energy that cells need to stay alive. One of the products of respiration is the gas carbon dioxide.

In order for respiration to take place, living organisms have to obtain a regular supply of oxygen from their surroundings. They must also be able to get rid of the excess carbon dioxide by returning it to the surroundings. Taking in oxygen and getting rid of carbon dioxide is called **gas exchange**.

Gas-exchange surfaces

The part of an organism where oxygen and carbon dioxide are exchanged is called a **gas-exchange surface**. The surface is adapted to allow cells to obtain the oxygen they need and get rid of the carbon dioxide they produce. Figure 9.1 overleaf shows a typical gas exchange surface.

Different organisms carry out gas exchange in different ways.

■ Humans and other mammals have lungs and a circulatory system that allow for gas exchange.
■ Fish live in water and have specially adapted gills for gas exchange.
■ Amphibians are able to exchange gas through their mouth, lungs and skin.
■ Insects do not have blood, but they do have a system of tubes which allows oxygen to pass to all parts of their body.
■ Earthworms exchange gases through their skin, and they also have a blood system for transporting gases throughout their body.

Simpler organisms, such as *Amoeba* or *Hydra*, exchange gases directly with their surroundings by diffusion. Oxygen diffuses into the organisms from the surrounding air and carbon dioxide diffuses out. Plants exchange gases in a similar way.

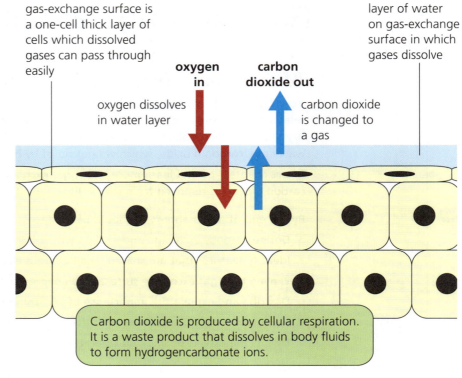

gas-exchange surface is a one-cell thick layer of cells which dissolved gases can pass through easily

oxygen dissolves in water layer

oxygen in

carbon dioxide out

carbon dioxide is changed to a gas

layer of water on gas-exchange surface in which gases dissolve

Carbon dioxide is produced by cellular respiration. It is a waste product that dissolves in body fluids to form hydrogencarbonate ions.

Figure 9.1 How gases are exchanged across a gas-exchange surface.

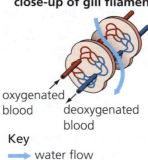

position of gills
gill arch

gill arch

rakers

filaments or lamellae

close-up of gill filament

oxygenated blood

deoxygenated blood

Key
→ water flow

Figure 9.2 How gas-exchange surfaces are adapted to increase their surface area.

Properties of gas-exchange surfaces

Most gas–exchange surfaces have the following characteristics:

- very thin (ideally only one-cell thick) and porous so that gases can diffuse across them quickly and efficiently
- large surface area to increase the rate of diffusion by allowing many molecules of gas to diffuse across the surface at the same time
- moist to prevent surface cells dehydrating and dying and to dissolve oxygen
- well ventilated (have a good supply of oxygen to maintain a concentration gradient for diffusion). More oxygen outside the cell allows oxygen to diffuse in and carbon dioxide to diffuse out
- close to a transport system (for example the circulatory system in large animals) so that gases can be transported to and from the cells in all parts of the body.

Maximizing the surface area for gas exchange

Gas–exchange surfaces are adapted so they have the largest possible surface area. Figure 9.2 shows some of these adaptations in a fish.

Activity 9.1 Drawing a bony fish gill

Aim: To draw and label a fish gill.

Equipment and apparatus
- preserved or fresh specimens of fish gill
- hand lens

Procedure

1 Using a hand lens, carefully observe the fish gill.
2 Following the guidelines for drawing on page 25, make a large annotated diagram of the fish gill on plain white paper. (Make sure you draw the one given to you and not the one in your textbook!)

Remember to note the magnification of your drawing. You can calculate the magnification using the equation:

$$\text{magnification} = \frac{\text{size of drawing}}{\text{size of specimen}}$$

Gas exchange in plants

Gas exchange in plants takes place during photosynthesis and during respiration.

We have learnt that when a plant is exposed to light, photosynthesis takes place. Carbon dioxide diffuses into the plant. The waste oxygen produced in photosynthesis diffuses out. In the daytime, the rate of photosynthesis is higher than the rate of respiration so no oxygen needs to enter the plant.

When there is no light, photosynthesis stops but respiration continues. During this stage, the plant uses oxygen to release energy by respiration and it produces carbon dioxide as a waste product. Gas exchange therefore takes place in the direction of oxygen diffusing into the plant and carbon dioxide diffusing out.

Check your knowledge and skills

1 Why is a continual supply of oxygen and the removal of carbon dioxide necessary in most organisms?
2 Gas-exchange surfaces in larger animals need to be moist with a large surface area and close to a blood supply. Explain why these three features are necessary.
3 Explain the mechanism used to increase the surface area of gas-exchange surfaces in a bony fish.
4 Humans have two lungs, each about the size of an inflated balloon, which fit into the chest cavity. How is it possible for balloon-sized lungs to have a gas-exchange surface of approximately 60 m^2?

Breathing in humans

We breathe all the time, even when we are asleep. Breathing involves inhaling (taking in) air rich in oxygen and exhaling (letting out) air rich in carbon dioxide. Breathing allows gases to be exchanged in the lungs and it allows oxygen to reach all the cells in the body.

Humans need an efficient system of gas exchange to provide enough oxygen for cellular respiration. The gas-exchange system is also called the respiratory system.

The respiratory system

In humans, the respiratory system consists of a system of linked air passages (the **trachea** and the **bronchi**) and the **lungs**. The main function of the respiratory system is to provide oxygen for respiration and to remove carbon dioxide from the body. Figure 9.3 shows you the main organs in the human respiratory system.

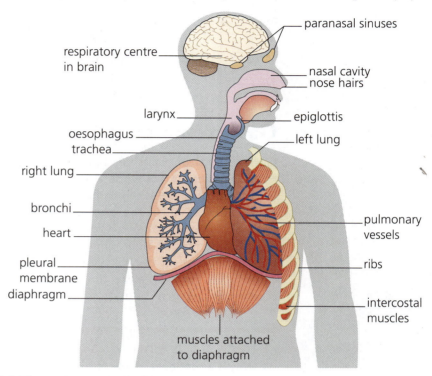

Figure 9.3 The main organs in the human respiratory system.

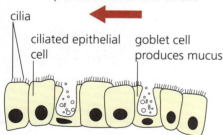

Figure 9.4 Respiratory surfaces are lined with a mucous membrane.

Inhaling air

When you breathe in, or inhale, air enters the respiratory system through the nose. The space behind the nose is called the nasal cavity. The nasal cavity connects the nose with the air passages that lead to the lungs.

The nasal cavity and air passages are lined with a mucous membrane like the one shown in Figure 9.4. The cells in this membrane produce slimy mucus which warms and moistens the air and traps dirt and germs carried in the air. This is important because dry air can damage or dry out moist gas-exchange surfaces in the lungs. The cells of the mucous membrane are covered by tiny hair-like structures called **cilia**. These move in a wave-like fashion all the time and they act to trap germs and dust and to prevent mucus from moving down into the lungs where it could block them. Instead, the cilia move the mucus up towards the nose and throat so that it can be swallowed or blown out of the nose.

From the nasal cavity, air moves down through the **larynx** towards the **trachea** (windpipe). The top of the trachea is covered by a small flap called the **epiglottis** which closes when you swallow to prevent food from entering the trachea.

The trachea is a tube which is kept open by C-shaped rings of cartilage along its length. The inner wall of the trachea is also lined with a mucous membrane. In the chest cavity, the trachea branches into two smaller tubes called the bronchi. One bronchus goes to the left lung; the other goes to the right lung.

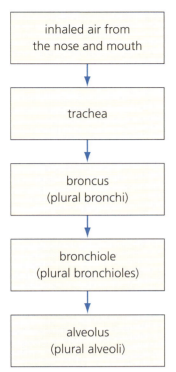

inhaled air from the nose and mouth

↓

trachea

↓

broncus
(plural bronchi)

↓

bronchiole
(plural bronchioles)

↓

alveolus
(plural alveoli)

Figure 9.5 Flow diagram showing the pathway of air into the lungs.

SBA skills
Observation/recording/ reporting (ORR)
Manipulation/ measurement (MM)
Analysis and interpretation (AI)

The lungs

The lungs are spongy sacs situated inside the chest cavity where they are protected by the ribs. Inside the lungs, the bronchi branch out many times into a number of smaller tubes called **bronchioles**. Each bronchiole ends in a bunch of tiny air sacs called **alveoli**. The lungs contain millions of alveoli and here gas exchange takes place.

The passage of air into the lungs and the relationship between the parts of the respiratory system are shown in Figure 9.5.

Gas exchange in the lungs

Each alveolus (see Figure 9.6) has a thin moist wall and is surrounded by a dense network of **capillaries**. When you inhale, oxygen-rich air reaches the alveoli. The oxygen dissolves in the water inside the alveoli. Because there is a greater concentration of oxygen inside the alveolus, this oxygen diffuses through the thin wall into the capillary network where it enters the blood.

In the same way, carbon dioxide diffuses from the blood into the alveoli where the concentration of carbon dioxide is lower. The oxygen is passed out of the lungs when you exhale (breathe out).

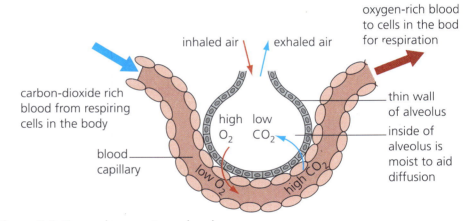

Figure 9.6 Gas exchange at an alveolus.

Activity 9.2 Investigating breathing and carbon dioxide levels

Aim: To detect carbon dioxide in exhaled breath; to investigate the effects of exercise on exhaled carbon dioxide levels.

Equipment and apparatus
- 4 conical flasks
- dropper
- 0.004% bromothymol blue indicator solution
- limewater
- 4 plastic drinking straws
- paper towel
- stopclock

Procedure
1 Working in pairs, place 30 ml of bromothymol blue indicator solution into each flask.
2 Place a straw into each flask; drape the towel over one of the flasks to prevent splashing.

3 Carefully exhale your breath through the drinking straw of one flask. (Do not suck or inhale through the straw.)

4 Using the stopclock, your partner should time how long it takes for the solution to change colour. (*Hint:* Use the solution in the other flask to compare the colour.)

5 Do some jumping jacks for three minutes; then repeat the experiment using the unused straw and indicator in the second flask.

6 Repeat the experiment using limewater.

7 Record your observations.

8 Compare your data with those of your classmates.

Questions

1 What was the longest time it took for a change in colour?

2 What was the shortest time?

3 How do you account for the difference?

4 Do you think that the length of time it takes to change colour relates to a person's physical characteristics? For example, their gender or athletic build? Explain your answer.

5 What is the scientific name for limewater?

How do we breathe?

As you breathe, your chest changes shape and size. As you inhale, the chest cavity expands and when you exhale it contracts. The lungs inside the chest cavity act like balloons, stretching and filling up as you inhale and sagging and emptying as you exhale.

The ribs and diaphragm

Lungs cannot inhale and exhale on their own – breathing requires the action of two sets of muscles: the intercostal muscles between the ribs and the **diaphragm**. The diaphragm is a large muscular sheet found below the lungs and heart. You can see these muscles and how they interact in Figure 9.7.

On inhaling:

- external intercostal muscles contract, pulling the ribs forwards (outwards) and upwards
- diaphragm contracts at the same time and it moves downwards
- contraction of the intercostal muscles and diaphragm increases the volume of the chest cavity and reduces the air pressure inside it
- lower air pressure causes air to be sucked into the lungs.

On exhaling:

- internal intercostal muscles relax and the ribs move downwards and together
- diaphragm relaxes at the same time and it moves upwards
- relaxation of the intercostal muscles and the diaphragm decreases the volume of the chest cavity and increases the air pressure inside it
- the higher air pressure forces air out of the lungs.

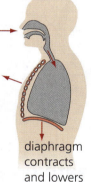

inhalation

air sucked in

external intercostal muscles contract pulling ribs forwards and up

diaphragm contracts and lowers

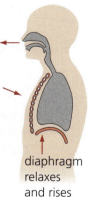

exhalation

air pushed out

internal intercostal muscles relax and ribs move together and down

diaphragm relaxes and rises

Figure 9.7 The intercostal muscles and the diaphragm control inflation and deflation of the lungs.

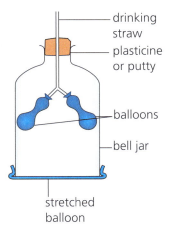

drinking straw

plasticine or putty

balloons

bell jar

stretched balloon

Figure 9.8 Apparatus used to build a model of breathing.

Activity 9.3 Building a model of breathing

Equipment and apparatus
- 3 balloons
- 2 L soda bottle
- drinking straws
- tape
- glue
- rubber bands
- Plasticine or putty
- nail

Procedure
1 Work in groups and assemble a model similar to the one shown in Figure 9.8.
2 Cut out the bottom half of the soda bottle; keep the top half.
3 Carefully punch a hole in the centre of the bottle cap.
4 Using tape and glue assemble the drinking straw into a 'Y' shape that will allow air to pass through. (Cut it to an appropriate length to fit into your chest model.)
5 Insert the long 'arm' of the 'Y'-shaped straw (trachea) through the hole of the bottle cap.
6 Secure it using Plasticine or putty to make it airtight.
7 Attach the balloons to the (bronchi) using rubber bands to secure them.
8 The third balloon should be cut and stretched across the cut underside of the bottle with an elastic band or sticky tape.

Questions
1 Which parts of your model represent the lungs, diaphragm, ribs and trachea?
2 What features of your model are *not* representative of how the actual chest cavity works during breathing?
3 Describe what happens when you pull and when you push on the stretched piece of balloon at the bottom of the model.

Activity 9.4 Using a model of the chest

Equipment and apparatus
- A model of the chest similar to the one shown in Figure 9.9.

Procedure
1 Pull down on the rubber sheet then push it back up again.
2 Repeat this a few times.
3 Observe what happens to the balloons.

Questions
1 Draw the model and label it to show what part of the body is represented by each part of the model.
2 Do you think this is an accurate model to show what happens as you inhale and exhale? Give reasons for your answer.
3 How could the model be improved?

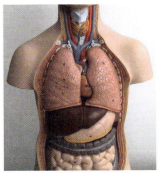

Figure 9.9 Model of a human chest.

Unit 3

Smoking and health

An adult applying for life insurance will be asked by the insurance company whether or not he or she smokes. The company asks this because it knows that smokers, on average, die younger than non-smokers. It also knows that smoking can impact on a person's health in many different ways. The graph in Figure 9.10 shows you the eight most common causes of death in the world today. The paler part of each column shows the proportion of deaths in each group that were related to tobacco use. These add up to almost 5.5 million deaths per year.

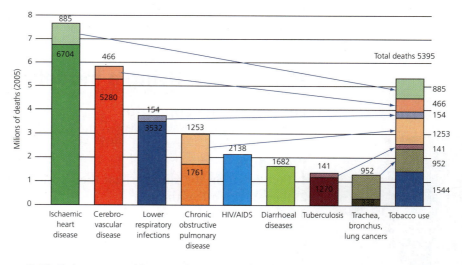

Figure 9.10 Tobacco smoking carries serious health risks.

Why is smoking harmful to health?

When a person smokes they are inhaling the smoke from burning tobacco and paper directly into their lungs. The smoke is hot and dry, so it can irritate the respiratory surfaces and dry them out. However, more seriously, cigarette smoke contains many toxic chemicals (see Table 9.1).

Chemical	Effect on the body
Nicotine	A strong toxin which causes **addiction**. Nicotine is a **stimulant** which increases heart rate and narrows blood vessels causing high blood pressure and long-term damage to the circulatory system An increase in heart rate increases oxygen demand, but smokers have less oxygen available so the heart muscle may become damaged
Carbon monoxide	A toxic gas which binds permanently with haemoglobin in the blood and reduces oxygen supply to the cells
Tar	Collects in the lungs as a sticky mass when smoke cools Causes 'smoker's cough' and eventually physical damage to the lungs Tar contains many chemicals which can cause cancer

Table 9.1 The three main toxins in cigarette smoke and their effects on the body.

The main toxins in cigarette smoke are:

- carbon monoxide
- nicotine
- tar
- in the smoke: cyanide, acetone, butane, methanol, ammonia, acetic acid, naphthalene, arsenic, phenols and radioactive polonium-210.

Diseases associated with smoking

Besides being a risk factor in the diseases mentioned above, cigarette smoking has been clearly linked to the following diseases.

Cancer

Cancer is a cellular disorder. Cancerous cells are abnormal cells which grow uncontrollably and invade and damage healthy cells around them.

Almost 90% of lung cancers occur in people who smoke, and smokers are ten times more likely to get lung cancer than non-smokers. Cancer of the mouth and larynx is five times more common in smokers than in non-smokers. Smokers are also more prone to cancer of the stomach, pancreas and bladder.

Emphysema

This is a serious illness in which chemicals destroy the walls of the alveoli. The walls break down and the surface area available for gas exchange is reduced. Breathing becomes difficult and the person may need to be supplied with oxygen. The incidence of emphysema in non-smokers is almost zero.

Bronchitis

Smoking irritates the lining of the air passages and destroys the cilia. Mucus, dirt and bacteria build up in the lungs as a result. The person coughs to get rid of this and this in turn irritates the lining of the passages even more.

Figure 9.11 A typical health warning on a cigarette packet would read 'The Minister of Health Advises that Smoking is Dangerous to your Health'.

Chapter summary

Do you know?

If you are unsure of any of the facts in the list, refer to the page number given in brackets.

■ Gas exchange is important in all living organisms because it provides the oxygen needed for respiration and removes the carbon dioxide produced by respiration. (page 95)

■ Gas exchange takes place at gas-exchange surfaces such as the lungs, moist skin and gills in animals and the spongy mesophyll layer in plant leaves. (page 95)

■ Gas-exchange surfaces are usually thin, moist, large in surface area and close to a transport system. (page 96)

■ In humans, gas exchange takes place in the respiratory system consisting of the air passages (trachea, bronchi, bronchioles) and lungs. (page 97)

■ Air passes from the nose and mouth through the trachea, bronchi and bronchioles to reach the alveoli where gas exchange takes place. (page 98)

■ The alveoli are well adapted for gas exchange and are surrounded by a dense capillary network. (page 99)

■ Breathing in is called inhalation and breathing out is called exhalation. (page 100)

■ Breathing depends on the intercostal muscles and the diaphragm which contract and expand to draw air into the lungs and to force air out of the lungs. (page 100)

■ Cigarette smoke is harmful because it is hot and dry and it contains harmful chemicals. (page 102)

■ The main toxins in cigarette smoke are nicotine which is addictive, carbon monoxide which reduces the oxygen-carrying capacity of the blood, and tar which collects in the lungs and can cause cancer. (page 103)

Are you able to?

If you have trouble doing these things, refer to the page numbers in brackets.

■ Locate and name the parts of the respiratory system. (page 98)

■ Use a model of a human thorax to demonstrate breathing. (page 101)

■ Read and interpret a bar graph. (page 102)

10 Transport in animals

Blood and circulation

You have learnt that the blood plays an important role by transporting substances around the body. But blood doesn't fill random spaces or move around randomly. It is contained in blood vessels and relies on the heart to move or circulate round the body along a pathway provided by the **circulatory system**.

In this chapter you will learn more about transport systems.

By the end of this chapter you should be able to:

- Explain the need for a transport system in multicellular organisms
- List the substances that are transported
- Identify and describe the different types of blood vessels
- Identify the parts of the heart and describe their functions
- Relate the structures involved in transport to their functions
- Describe the parts and functions of blood

Unit 1

The role of transport in animals

All cells in a living organism need a constant supply of digested food substances and oxygen so they can carry out their metabolic functions. As cells perform their functions, they produce metabolic wastes which have to be removed from the cell and transported to parts of the body where they can be excreted.

A system used to move substances into and out of cells and all around the body is called a **transport system**. In humans the transport system consists of:

- blood vessels – these form a network of transport tubes
- blood – the medium of transport
- the heart – this is the pump that keeps the substances moving.

The transport system in humans is also called the circulatory system because it circulates blood and dissolved substances around the body.

■ Transport in small organisms

Small unicellular organisms, for example *Amoeba* and *Paramecium*, do not need blood and a network of transport tubes. These organisms can get all the substances they need and get rid of waste products by diffusion. Although diffusion is a slow process, it is adequate for unicellular organisms because they have a large surface area compared to their volume. The small volume means that substances do not have to be transported very far (so no blood is needed) and the large surface area means that it is easy for substances to move into and out of the organism.

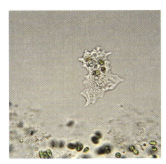

Figure 10.1 *Amoeba* (left) and *Paramecium* (right) have no need for a transport system.

Transport in multicellular organisms

Large multicellular organisms have a small surface area compared to their volume, so they cannot rely on diffusion to transport the substances they need to move. Diffusion is too slow and their smaller surface area is not large enough for enough of each substance to enter and leave.

Why diffusion isn't enough

Imagine what would happen in your body if the cells in your arms had to wait two weeks to receive digested food from the small intestine! If cells had to rely only on diffusion, they would not be able to operate at all. No respiration would be able to take place because it would take far too long for oxygen and glucose to reach the cells, and the cells would die.

In addition, humans and other mammals are so complex that some cells require certain substances in higher concentrations than others. Diffusion and osmosis are completely random processes so they would not able to provide for the different amounts of substances needed by cells.

What substances do organisms transport?

All the cells of the body need a continual supply of digested food and oxygen. They also need to get rid of waste chemicals and move these to the kidneys for excretion. In addition, chemicals such as hormones, antibodies and blood proteins have to be moved around the body.

Substance	Origin of substances	Destination of substances	Reason for transporting substance
Oxygen	Lungs	Mitochondria of cells	Reactant in respiration
Glucose	Small intestines	Cytoplasm of body cells	Reactant in respiration
Carbon dioxide	Body cells	Lungs	Excretion of waste products
Antibodies	White blood cells	Infected cells	To fight infection
Urea	Liver	Kidney	Excretion of waste products
Amino acids	Small intestines	Body cells	Used to build cell proteins
Hormones	Endocrine glands	Various parts of the body	Chemical control of processes
Heat	Liver and muscles	Rest of the body	To maintain a constant temperature

Table 10.1 Substances transported by the human circulatory system.

Unit 2

Blood vessels

Blood circulates round the human body in a network of blood vessels. There are three types of blood vessels in the human body:

- arteries
- veins
- capillaries.

The structure of each type of blood vessel is related to its function in the circulatory system. Table 10.2 overleaf shows the structure, features and functions of the three main types of blood vessels in humans.

Arteries and veins are linked by capillaries. The arteries branch into smaller and smaller vessels and veins branch in the same way. The smallest branches of both form a dense network of capillaries called a capillary bed.

Key fact

The main blood vessels are called arteries, veins and capillaries. However, there are parts of the body where you find smaller arteries and veins. These smaller vessels have the same features and functions of arteries and veins, but because of their smaller size they are called **arterioles** and **venules**.

Key fact

'Lumen' is the correct biological name for an empty space in a tube.

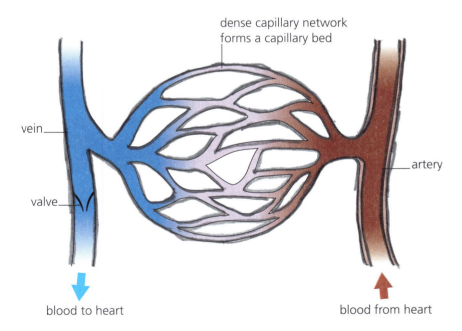

Figure 10.2 Arteries and veins are linked by thin-walled capillaries. Note that the vein has a valve to prevent blood flowing 'backwards'.

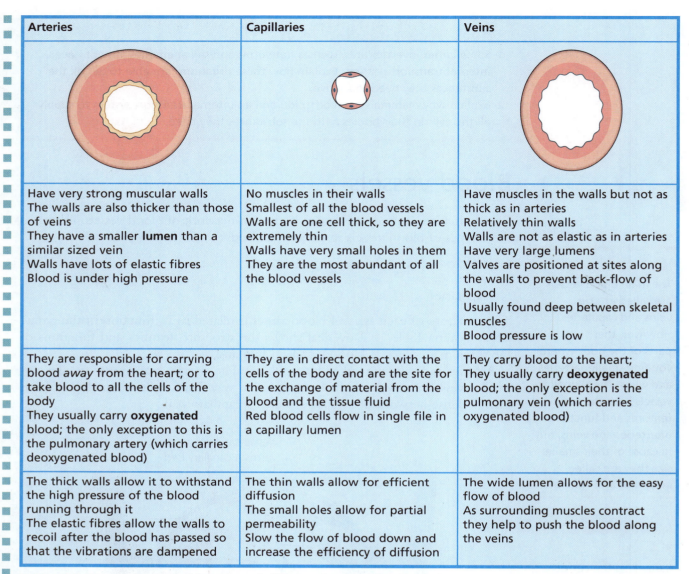

Arteries	Capillaries	Veins
Have very strong muscular walls The walls are also thicker than those of veins They have a smaller **lumen** than a similar sized vein Walls have lots of elastic fibres Blood is under high pressure	No muscles in their walls Smallest of all the blood vessels Walls are one cell thick, so they are extremely thin Walls have very small holes in them They are the most abundant of all the blood vessels	Have muscles in the walls but not as thick as in arteries Relatively thin walls Walls are not as elastic as in arteries Have very large lumens Valves are positioned at sites along the walls to prevent back-flow of blood Usually found deep between skeletal muscles Blood pressure is low
They are responsible for carrying blood *away* from the heart; or to take blood to all the cells of the body They usually carry **oxygenated** blood; the only exception to this is the pulmonary artery (which carries deoxygenated blood)	They are in direct contact with the cells of the body and are the site for the exchange of material from the blood and the tissue fluid Red blood cells flow in single file in a capillary lumen	They carry blood *to* the heart; They usually carry **deoxygenated** blood; the only exception is the pulmonary vein (which carries oxygenated blood)
The thick walls allow it to withstand the high pressure of the blood running through it The elastic fibres allow the walls to recoil after the blood has passed so that the vibrations are dampened	The thin walls allow for efficient diffusion The small holes allow for partial permeability Slow the flow of blood down and increase the efficiency of diffusion	The wide lumen allows for the easy flow of blood As surrounding muscles contract they help to push the blood along the veins

Table 10.2 The structure and function of different human blood vessels.

Plasma rich in nutrients and oxygen leaks out of the capillaries into the area surrounding the tissue cells. The fluid is now called tissue fluid.

blood capillary

The nutrients and oxygen will be absorbed from the tissue fluid by the cells they surround, while the waste materials such as carbon dioxide diffuse out of the cells into the surrounding fluid.

venule (small vein)

arteriole (small artery)

The tissue fluid will now be rich in carbon dioxide and wastes, and will start to flow into the lymphatic vessels and form lymph; the carbon dioxide and waste-rich tissue fluid that remains flow back into capillaries as the blood makes its way to the heart. The fluid once in the capillaries is now called plasma.

Figure 10.3 Plasma, tissue fluid and lymph.

Tissue fluid and the lymphatic system

You can see from Figure 10.2 on page 107 that capillary beds form the sites of exchange of materials between the blood and the cells. When blood reaches a capillary bed some of the liquid from the blood leaks out through the capillary walls. This liquid flows into the spaces between cells, where it forms **tissue fluid** (see Figure 10.3). Because tissue fluid is a liquid, it allows substances such as nutrients, gases, wastes and water to be dissolved and exchanged between the blood in the capillaries and the internal contents of cells. Some tissue fluid re-diffuses into the blood via the capillaries but most of it is returned to the bloodstream via the lymphatic system.

The lymphatic system is a network of vessels similar in structure to blood vessels. Lymph vessels carry a clear fluid called **lymph**. Tissue fluid that drains into the lymph vessels is called lymph. The lymphatic system has three main functions:

- it carries tissue fluid back into the circulatory system
- it transports fatty acids and fats to the circulatory system
- it plays a role in assisting the immune system.

The contents of all the lymph vessels eventually drain back into the blood vessels.

Blood vessels and major organs

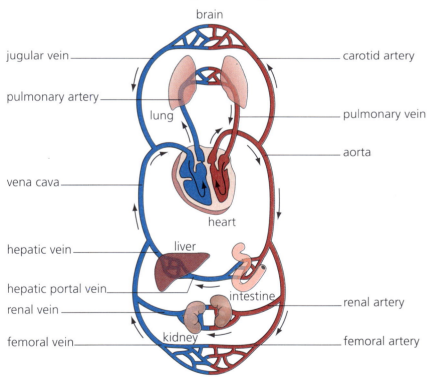

Figure 10.4 The main blood vessels in humans and the major organs they supply.

Check your knowledge and skills

1 Explain why the walls of arteries and veins are different thicknesses.
2 Why are the walls of capillaries only one cell thick?
3 Draw simple labelled cross-sections of the three main types of blood vessels to show the differences between them.
4 Name the blood vessel responsible in each case for carrying oxygenated blood to:
  a the lungs   **b** the liver   **c** the brain

The human heart

The human heart is a muscular organ made mainly of **cardiac muscle**. Cardiac muscle is unique in that it constantly contracts and relaxes without becoming tired. It is said to be **myogenic**.

The heart is situated in the thorax, it lies between the lungs and is slightly tilted to the left. The heart consists of four chambers; two upper and two lower ones. The upper chambers are called **atria** (singular atrium) and the lower chambers are called **ventricles** (singular ventricle). Figure 10.5 shows the structure of the heart and the directions of the blood flow through it, and Table 10.3 lists the functions of each side of the heart.

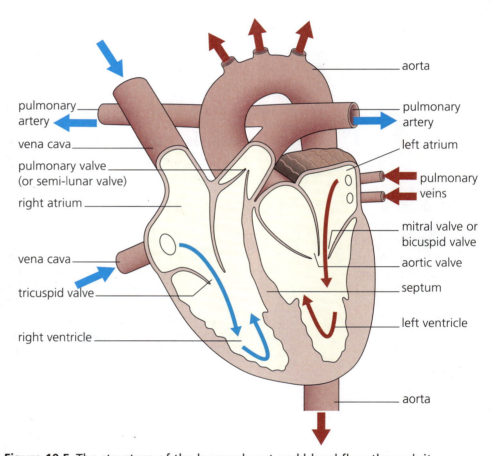

Figure 10.5 The structure of the human heart and blood flow through it.

Check your knowledge and skills

1 Why is the wall on one side of the heart thicker than the other wall?
2 How do the valves help the heart to function better?
3 Draw a labelled sketch to show the internal structure of a mammalian heart.

Right side	Left side
Pumps only deoxygenated blood	Only oxygenated blood is found here
Walls of ventricles are thinner	The ventricle has the thickest wall of all the chambers
Vena cava bring the blood from the rest of the body	The pulmonary vein brings blood to this side of the heart from the lungs
Blood leaves through the pulmonary artery	Blood leaves through the aorta
Blood goes to the lungs from the right side to become oxygenated	Oxygenated blood from the lungs goes to the rest of the body from this side of the heart

Table 10.3 Functions of the heart.

Heart facts

- The **vena cava** is the largest vein in the body. It even has its own blood supply to keep it alive.
- The **aorta** is the largest artery in the body. It also has its own blood supply to keep it alive.
- Oxygenated blood refers to blood containing a high concentration of oxygen. It is bright red in colour.
- Deoxygenated blood refers to blood containing a low concentration of oxygen. It usually has a purplish or darker red colour.
- The **septum** is a dividing wall that separates the left side of the heart from the right side, preventing the mixing of oxygenated and deoxygenated blood.

Circulation in the heart

The heart is a double pump. Blood is pumped from the right side of the heart to the lungs where carbon dioxide is removed from the blood and oxygen is added to it. The oxygenated blood travels back to the left side of the heart to be pumped to the rest of the body.

Oxygen is removed from the blood as it reaches the various parts of the body. The deoxygenated blood then travels back to the right side of the heart and then to the lungs to pick up more oxygen.

Understanding blood flow

Study Figure 10.5 carefully and read the notes in the list to understand how blood circulates through the heart.

- Beginning at the left side of the heart, oxygen-rich blood coming from the lungs enters the heart through the **pulmonary vein**.
- This empties into the left atrium.
- When it is filled, the **bicuspid valve** that separates the atrium from the ventricle opens and the blood flows into the left ventricle.
- The valve then closes to prevent the blood from going back into the atrium.
- The extremely powerful muscle of the left ventricle contracts.
- Blood is then pumped into the massive artery called the aorta.
- The force causes the aorta to expand and recoil as the blood passes by.
- The blood continues through the arteries to all the cells in the body.
- Cells throughout the body remove oxygen from the blood and use it in cellular respiration.
- Metabolic processes in the cells cause them to create carbon dioxide and other wastes.
- These must be removed to keep the cells healthy.
- The blood is taken to the kidneys where all the wastes are removed with the exception of carbon dioxide.
- The blood loaded with carbon dioxide returns to the right side of the heart.
- Blood enters the heart in the vena cava.
- The blood empties into the right atrium.
- When it is filled, the tricuspid valve that separates the atrium from the ventricle opens and the blood flows into the right ventricle.
- The valve then closes to prevent the blood from going back into the atrium.
- The muscle of the right ventricle contracts.
- The deoxygenated blood is then pumped into the **pulmonary artery**.
- The blood goes to the lungs where it loses carbon dioxide and picks-up oxygen.
- The oxygen-rich blood goes back to the left side of the heart and the cycle continues.

Did you know?

The entire process described happens in less than half a minute. Every day your heart pumps blood through the blood vessels so often that the blood covers a distance of about 100 000 km.

How the heart works

The heart works by making the chambers contract and relax in a rhythmic way. When the heart muscles contract it is called **systole**. When they relax it is called **diastole**. During systole blood is squeezed out of the chambers. And when they are in diastole the chambers are being filled. Both sides of the heart contract and relax at the same time. Therefore, when the left ventricle is squeezing oxygen-rich blood out of the heart into the aorta, at the same time the right ventricle is squeezing oxygen-poor blood out of the heart into the pulmonary artery.

Both of the blood vessels which transport blood out of the heart, namely the aorta and the pulmonary artery, have valves which close after the blood exits the heart. This prevents blood from flowing back into the heart. These valves are called **semi-lunar valves**.

Although the cells that make up the heart can beat without any outside stimulation, there is a need for a controlled rhythmic beat in the heart so that the cells contract and relax in synchrony (in time) with each other. A small group of specialized cells called the **sino-atrial node** is responsible for initiating the heart beat. These specialized cells are sometimes called the pacemaker.

Coronary heart attack

Sometimes sections of the coronary artery can become blocked by fatty deposits. This prevents blood from flowing to a part of the heart. The absence of blood to heart tissues means that they will be deprived of oxygen, and if the blockage is not removed quickly enough then the affected section of the heart muscle will be permanently damaged.

Activity 10.1 Examining a mammalian heart

Aim: To observe and draw a typical mammalian heart.

Equipment and apparatus
- your teacher will supply a dissected heart from a sheep or goat
- protective gloves
- protective clothing
- safety glasses

Procedure
1 Examine the outside of the heart carefully.
2 Note down the difference in thickness of the walls of the heart.
3 Try to identify the main blood vessels leaving and entering the heart.
4 Now examine the inside of the heart carefully.
5 Locate the valves between the atria and the ventricles.
6 Locate the semi-lunar valves on the aorta and pulmonary artery.
7 Can you find the septum?
8 Dispose of your specimen as instructed by your teacher. Wash your hands with soap and water when you have completed this practical.

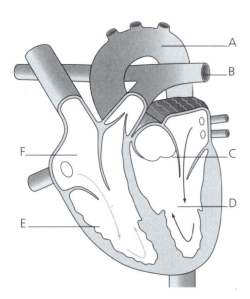

Unit 4 Blood and its functions

You have about 5 litres of liquid blood circulating through your body. About half of the blood is made up of blood cells. The rest is a liquid called plasma which consists of water and dissolved substances. Remember that every substance that is transported around the body has to be dissolved in the blood before it can be circulated. We say that blood is the medium of transport.

What is blood?

Biologically, blood is considered to be a tissue. Because it connects the parts of the body, it is classified as a connective tissue.

Figure 10.6 shows how blood can be broken down, level by level, into individual components and shows some of the components.

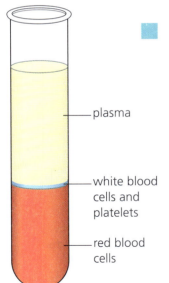

Figure 10.6 The components of blood.

- **Plasma** – the fluid part of blood. It is mostly water (about 90%) but it also contains dissolved substances such as nutrients, hormones and metabolic wastes.
- **Red blood cells** – a drop of blood contains millions of red blood cells. These are produced in the marrow of large bones. They are red disc-shaped cells which look like a flat dumpling with a sunken centre. They lack a nucleus so they have a very short life span (two to three months). When they die, red blood cells are broken down by the body and replaced with new cells.
- Red blood cells contain the chemical **haemoglobin** and it is this compound that gives them their colour. Haemoglobin combines with oxygen to transport it in the blood.
- **White blood cells** – these are irregularly shaped cells which are larger than red blood cells. They live much longer than red blood cells because they contain a nucleus, but they are not found in nearly such large numbers in the blood. They are produced in the bone marrow and in lymph nodes. The two main types of white blood cells are **lymphocytes** and **phagocytes**.
- **Platelets** – these are fragments of cells. They have no nucleus but they can release enzymes which help blood to clot at a wound site.

The different cells in blood have different functions. Each type of cell is adapted to suit its function (see Table 10.4, overleaf).

Blood component	Function	How component is adapted
Red blood cells (erythrocytes)	Transport oxygen to body cells Transport small amounts of carbon dioxide to the lungs	Narrow and flexible so can move through capillaries easily Contain haemoglobin which picks up oxygen in lungs and releases it to cells Lack of nucleus allows more space for haemoglobin
White blood cells	Phagocytes surround invading microbes and prevent them from causing infection	Thin, irregular shape allows them to squeeze out through the walls of capillaries to the site of infection Surface of cell is sensitive to presence of microbes Contain enzymes that digest harmful microbes
	Lymphocytes produce antibodies to defend against disease	Large nucleus produces many antibodies
Platelets	Assists with the clotting of blood at the site of a wound	Able to release enzymes that make blood clot

Table 10.4 The components of blood – their structures and functions.

Biology at work

Chapter 10 – Donating blood

The role of blood in transport

Blood has four main transport functions.

- Carries dissolved food substances (glucose, amino acids, fatty acids, vitamins, minerals) from the small intestine to the liver and then to the rest of the body.
- Picks up metabolic wastes (urea and others) from the cells where they are produced and transports them to the liver and kidneys for removal.
- Carries oxygen and carbon dioxide between the body cells and the alveoli in the lungs.
- Transports hormones from the endocrine glands to the organs where they are needed.

How to answer an essay question

When you need to write a long-answer essay question in an exam you need to remember the following helpful hints.

- *The number of marks required indicates the number of points you need to make.* So, for a question with three available marks you should not be writing over five lines of answer, and alternatively you should not write three or less lines for a question that is worth 10 marks.
- *Focus your answer to the parameters of the question being asked using the points that you made during your reading time.* Only answer the question being asked; irrelevant information (even if correct) will not get you additional marks. Writing irrelevant information also wastes your valuable time.
- *Learn how to spell scientific terms correctly.* You don't need to be a spelling-bee champion but you need to pay keen attention when spelling commonly misspelled homonyms, for example their, there; hear, here.
- *This type of examination not only evaluates your understanding of the subject but also your level of expression.* Avoid using colloquial terms in your writing. You need to write your answers in Standard English using correct grammar and you should never write in patwa (patois) on an exam paper.

Check your knowledge and skills

1 The movement of blood through the circulatory system of a mammal is described as a double circulation.
   a Explain how the structure of the heart maintains the double circulation.
   b Describe what is meant by a coronary heart attack.
   c Describe how a molecule of glucose passes from the liver to the heart and then to the muscle of your leg. Ensure that you mention the names of the blood vessels that carry the molecule.

2 In Biology we often say that the structure of a cell, tissue or organ is closely related to its function. Write an essay (worth 10 marks) in which you explain why this statement is true for blood cells.

Chapter summary

Do you know?

If you are unsure of any of the facts in the list, refer to the page number given in brackets.

- Multicellular organisms cannot rely on diffusion and osmosis to transport materials to and from their cells. (page 106)
- Arteries have the thickest muscular walls of all the blood vessels. (page 108)
- Veins have very large lumens and they contain valves to make sure blood flows in one direction only. (page 108)
- Capillaries are leaky. Their walls are one cell thick. (page 108)
- The heart is a muscular pump responsible for pumping blood to all parts of the body. (page 110)
- Deoxygenated blood goes through the right side of the heart only. (page 110)
- Oxygenated blood goes through the left side of the heart only. (page 110)
- The left ventricle has the thickest muscular wall in the heart. It is responsible for pumping the blood to all parts of the body. (page 111)
- Blood is a tissue composed of plasma, red blood cells, white blood cells and platelets. (page 113)

- Red blood cells transport oxygen to the cells. They contain haemoglobin which oxygen binds to. (page 113)
- White blood cells play a role in fighting bacteria and disease, and platelets help the blood to clot at wounds. (page 113)
- Blood is the medium for transporting dissolved food, gases, hormones and water products around the body. (page 114)

Are you able to?

If you have trouble doing this, refer to the page numbers in brackets.

- Draw labelled diagrams to show the difference in structure between veins, arteries and capillaries. (page 108)
- Relate the components of the circulatory system to their functions and state how each is adapted to suit its function. (page 108)
- Name the blood vessels which supply the major organs of the body. (page 109)
- Identify the external features of a mammalian heart. (page 110)
- Identify the internal features of a mammalian heart. (page 110)

11 Transport in plants

Moving substances around plants

You know that plants take in water and minerals from the soil through their roots. Plants also make food substances in their leaves during photosynthesis. This food is needed by cells all over the plant and extra food is stored in storage organs. So, how does water from the roots reach the cells at the top of the plant? And how do food substances move from the leaves to all parts of the plant?

In this chapter you will learn about moving water and food substances in plants.

By the end of this chapter you should be able to:

■ Describe the structure and function of xylem vessels, sieve tubes, companion cells

■ Draw and label the vessels involved in transport in plants

■ Outline the processes involved in transpiration

■ Discuss the factors that affect the rate of transpiration

■ State the function of phloem as the food-carrying vessels of plants

Unit 1 The role of transport in plants

You have read that plants, like humans and other larger animals, need to transport water and food substances to all its parts. All cells of the plant need water, but water can only enter the plant through its roots. Plants therefore need to move or transport the water from their roots to all the cells. Also, plants synthesize food mainly in their leaves but all cells need the food substances. Again, the plant needs a mechanism to move the manufactured food to all parts of the plant.

The transport system in plants consists of specialized cells and vessels made up of **vascular tissue**. Vascular tissue consists mainly of a collection of narrow vessels or tubes arranged together in vascular bundles (see Figure 11.1). A vascular bundle contains two types of vessels: **xylem** and **phloem**. The xylem and phloem tubes are found throughout all parts of the plant and they are arranged in specific ways. The arrangement of vascular tissues allows for efficient movement of substances (from where they are taken up or made) to where they are needed. It also strengthens and supports the plant.

> **Key fact**
> 'Vascular' means veins or vessels.

Check your knowledge and skills

1 Why do plants, such as sugar canes, need a transport system whilst small plants, such as mosses, can deliver water to all their cells by diffusion?
2 Name the main components of the vascular system in dicotyledonous plants.
3 How do plant cells which do not photosynthesize get their energy?
4 Describe where vascular bundles are found in plant stems.

In effect, plants have two separate transport systems.

■ The xylem tubes transport water and dissolved minerals upwards from the roots to all parts of the plant.

■ The phloem tubes transport food substances from the leaves in all directions around the plant to where they are needed.

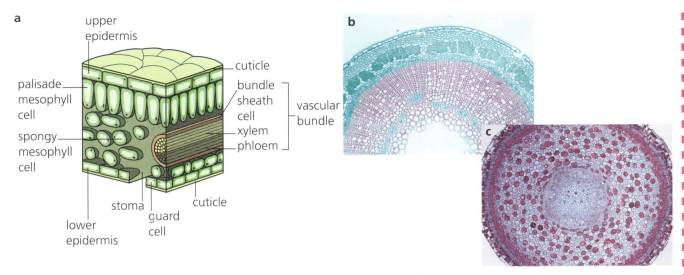

Figure 11.1 a A diagram of the cross-section of a dicotyledonous leaf, **b** a cross-section of a stem, and **c** a root as seen through a light microscope.

Unit 2 — Transport of water

Plants cannot live without water. In order to grow and survive they need to take in water and minerals from the soil and then transport them throughout their cells.

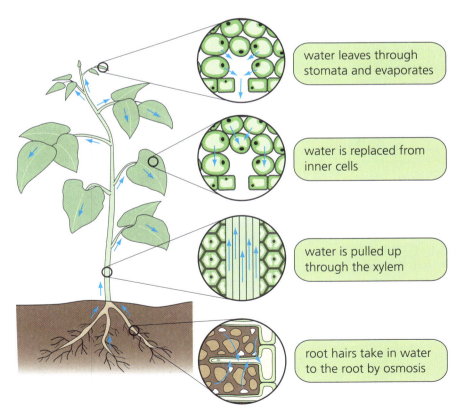

water leaves through stomata and evaporates

water is replaced from inner cells

water is pulled up through the xylem

root hairs take in water to the root by osmosis

Figure 11.2 Uptake and transport of water in a plant.

The transport of water in plants is important for three main reasons.

- For photosynthesis – water is one of the raw materials for photosynthesis which is a very important process for plants and all living things.
- To support the plant – when the cells of a plant have a good supply of water, they swell up (become **turgid**) and press tightly against each other to make the plant firm but flexible and to help keep the plant upright. When plant cells do not receive enough water they shrink and the plant wilts.
- For transporting dissolved materials in solution – mineral ions and the products of photosynthesis are dissolved in water and transported.

Approximately 99% of the water that a typical plant takes in is lost by evaporation from the leaves during the day. The other 1% is used for photosynthesis and other metabolic processes. The evaporation of water from the leaves and other surfaces of plants is called **transpiration**. You will learn more about this process on page 120.

Water uptake in plants

Plants take up water through their root hairs. Each root hair is a specialized cell with a section of the cell membrane extending from the root into the soil. The purpose of root hair cells is to provide a large surface area over which water can be absorbed from the soil. The root hairs take up water when the concentration of water in the soil around the roots is higher than the concentration inside the roots. This creates a **concentration gradient** and water leaves the soil and enters the root hairs. Figure 11.3 shows this process.

Did you know?

As water enters the root hairs it continually moves towards the xylem that carries it away to other parts of the plant. This maintains a steep concentration gradient between the root hairs and the soil ensuring the constant uptake of water into the root.

Key fact

As water enters the root hairs, mineral ions from the soil also enter the roots. However, their entry is by the process of active transport.

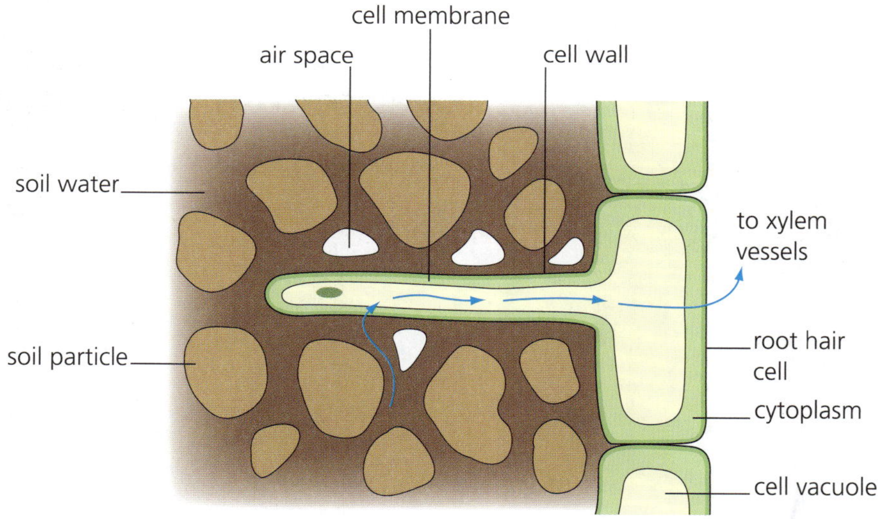

Figure 11.3 Water is taken up by the root hairs by osmosis. It is taken through the root cortex and then into the xylem.

As the water enters the root hairs it moves across the cortex cells by osmosis. This continues until the water reaches the xylem vessels which are specially designed for the transport of water.

Movement of water up the stem

Xylem cells are stacked end to end from the roots of the plants to the topmost leaf or branch of the plant. Water moves upwards through the plant without being pushed or pumped. There are three factors that contribute to the movement of water up the stem without the use of a pump.

- **Root pressure** – this forces water which is absorbed from the soil to move through the roots and up through the stem of the plant. Root pressure is caused by the osmosis of water from the soil into the root cells, and the active transport of salts into the xylem vessels which maintains a concentration gradient for the water to move along.
- **Capillary action** – this is the tendency of a liquid to rise up narrow tubes or small openings. It is sometimes called capillarity. A simple example of capillary action can be seen if you place the end of a strip of paper towel in water. You will notice that the part of the paper strip that was not immersed in the water eventually becomes wet as water rises up the paper. The smaller the space or tube the higher the water will rise. Xylem vessels are very thin and this helps water rise up the xylem vessel.
- **Transpiration pull** – most of the water which reaches the leaves of plants evaporates from the cells' surfaces and diffuses out of the leaf through the stomata (transpiration). This loss of water creates a steep gradient in the leaf which is filled by water rising up the xylem. So, as more water evaporates from the leaf it is replaced by the constant stream of water rising up from the roots through the xylem. The resulting transpiration pull is similar to the action of sucking on one end of a drinking straw; this removes liquid from the straw, creating a vacuum which is then filled by more liquid from the container.

Key fact
There is a constant stream of water from the root of the plant to the top of it. This is called a transpiration stream and it is maintained by the transpiration pull and strength of the xylem vessels.

Activity 11.1 Observing the movement of coloured water through a stem

Aim: To observe the movement of water through a plant stem.

Equipment and apparatus
- stalk of celery or callaloo or spinach
- beaker
- eosin or food colouring
- hand lens
- microscope
- scalpel blade
- cutting mat

SBA skills
Observation/recording/reporting (ORR)

Procedure
1 Cut a thin layer from the bottom section of the plant material. Be careful when using the scalpel blade.
2 Using the hand lens, observe the cut section. Make a sketch of what you see.
3 Stand the plant material in the beaker of coloured water for several hours.
4 Using the scalpel blade, carefully cut a thin layer from the plant material.
5 Using the hand lens, observe the cut section. Make another sketch of what you see, especially note where you see colour from the dye.
6 Your teacher may cut a very thin section from the stalk and mount it on a slide to observe it under a microscope.

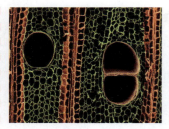

Figure 11.4 An electron-micrograph of a cross-section through xylem vessels. Can you identify the structures listed in Table 11.1 in the photograph?

Structure of xylem vessels

Xylem vessels (see Table 11.1) are well suited to their function in plants.

Feature	How the structure relates to function
Cells are dead and hollow	There are no cytoplasm and cellular organelles to impede the flow of water
Cell walls are extra thick with a strengthening substance called **lignin** and extra cellulose	Keeps the vessels rigid so that they do not collapse under the extreme pressure created by the transpiration pull
Cells are elongated and the ends of the cells are eroded so that a continuous column is created	There is no barrier to the flow of water between neighbouring cells
There are small holes along the sides of the cells	This allows for the sideways movement of water to non-xylem cells
The lumen of the cells is very narrow	This increases the distance that water can rise up the tube by capillary action

Table 11.1 The structure and function of xylem vessels.

More about transpiration

In leaves, water diffuses out of the cells into the air spaces in the spongy mesophyll layer as water vapour. This increases the concentration of water in the air spaces. This increased concentration creates a concentration gradient with the layer of air just outside the leaf. So water diffuses out of the leaves through small pores on the underside of the leaf called **stomata**. Each stomata is surrounded by a pair of **guard cells** which control the opening and closing of the stomata.

Transpiration can also occur through loosely packed cells on the surface of woody stems called **lenticels**. Gases are lost through these openings too.

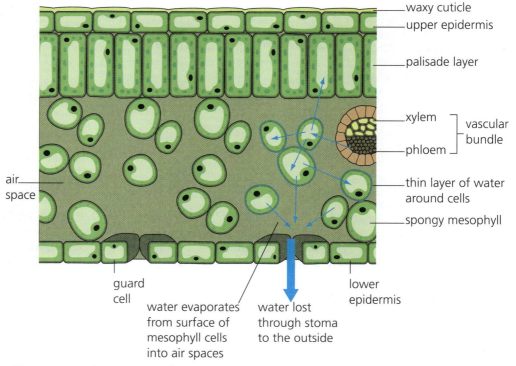

Figure 11.5 The passage of water through a leaf.

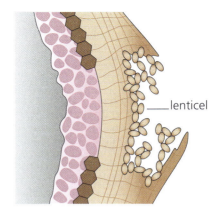

Figure 11.6 A lenticel in a woody stem and its structure.

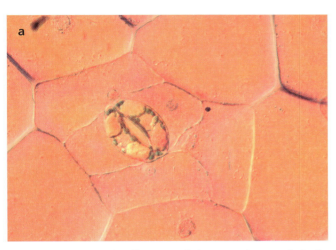

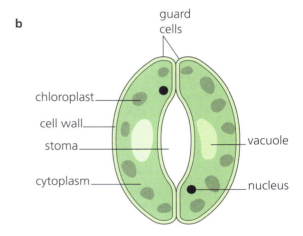

Figure 11.7 Guard cells control the opening and closing of stomata on the underside of the leaf; **a** shows a stomatal pore as seen through a light microscope; **b** shows a diagram of a stomatal pore.

The role of transpiration

Less than 1% of the water that passes through a plant in the transpiration stream is actually used by the plant. So, is the process of any importance to the plant?

The answer is 'yes'. Transpiration has several functions.

- It provides water to the very highest parts of a plant.
- It helps to keep a plant cool on hot days.
- It allows for the movement and absorption of minerals in plant cells.

Factors that affect the rate of transpiration

Both internal and external factors can have an effect on the rate of transpiration. Table 11.2 overleaf shows some of these factors.

Influencing factor	Explanation	Effect
Light	During the day stomata are open to allow for photosynthesis. In the night stomata are closed	Increasing the light intensity increases the rate of transpiration
Humidity	A humid environment means that there is an increase in the water vapour in the atmosphere. Therefore the concentration gradient between the inside of the leaf and the outside will not be as steep	Increasing the humidity decreases the rate of transpiration
Temperature	At high temperatures the rate of evaporation increases	Increasing the temperature increases the rate of transpiration
Wind speed	In windy conditions the air around the leaf is continuously moving. This means that the gases that diffuse out of the leaf will be quickly taken away. This will keep the concentration gradient steep, thus increasing diffusion out of the leaf	Increasing the wind speed increases the rate of transpiration
Water availability	An increase in water in the soil will increase the movement of water into the roots of the plant. This is because the concentration gradient between the soil and root hairs is steep	A well watered soil increases the rate of transpiration
Leaf size	The larger the leaf the more surface area there is for evaporation to take place	Larger leaf size can increase the rate of transpiration
Cuticle thickness	The waxy cuticle on the leaf surface prevents excessive water loss. Therefore the thicker the cuticle the less water will be able to evaporate from the leaf	Thick cuticle lessens the rate of transpiration
Number of stomata	Most of the water is lost through the stomata	Increasing the number of stomata increases the rate of transpiration

Table 11.2 Factors that affect the rate of transpiration.

How are plants adapted to prevent water loss?

Plants which live in hot dry conditions, such as deserts, experience high transpiration rates so they are usually adapted to reduce excessive water loss. Such plants are called **xerophytes** and their adaptive features are shown in Table 11.3.

Adaptation	How they work
Leaves are small, sometimes reduced to spines	Reduces surface area, reducing transpiration rate
Extremely thick cuticles	Prevents water from evaporating from leaves
Leaves are arranged in such a way to increase humidity around them, or the stomata are 'sunken' to increase humidity, or leaves are hairy	Increases humidity and decreases transpiration rates

Table 11.3 How xerophytes are adapted to reduce water loss.

Plants that live in very wet conditions, such as in water (for example water lilies), have their own adaptations for increasing transpiration rates. These features include: stomata on upper epidermis and no cuticle.

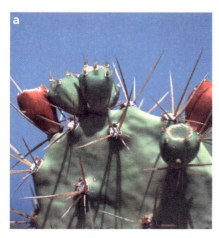  

Figure 11.8 Plants are adapted for the areas they live in: **a** cactus growing in the desert; **b** marram grass growing on sand dunes; **c** water lilies. Can you list the ways in which these plants are adapted to their environments?

Activity 11.2 Investigating the rate of uptake of water by a plant

Apparatus and materials
- 2 measuring cylinders (100 ml)
- water
- soft plant (for example calaloo, spinach or any other soft stemmed plant)
- scalpel blade
- cutting mat

Procedure
1 Fill both cylinders with 90 ml water.
2 Make a slanting cut at the end of the stem and then place it in one of the cylinders.
3 Wait 24 hours, and record the volume of water in each cylinder. (Make sure to carefully remove the stem from the cylinder before taking your readings.)
4 Repeat step 3 each day for two more days.
5 Calculate the rate of water uptake by the plant.

Questions
1 What was the purpose of the cylinder that had only water in?
2 What caused the water level to go down in the cylinder that contained only water?
3 What caused the water to go down in the cylinder with the plant stem?
4 Explain how you calculated the rate of uptake of water by the plant.
5 Name two factors that could have caused different uptake rates for each day.
6 How do these factors affect the transpiration rate?

Activity 11.3 Measuring the uptake of water by a plant using a simple potometer

Apparatus and materials
- capillary tube
- plastic or rubber tubing
- ruler
- small plant shoot (for example calaloo or any other soft stemmed plant)
- scalpel blade
- clamp stand
- stopclock
- marker pen

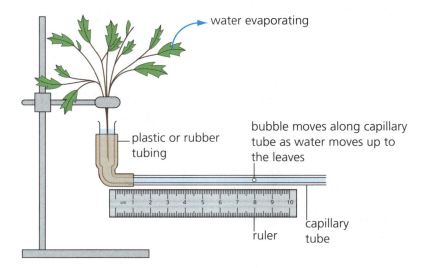

Figure 11.9 A simple potometer.

Procedure
1 Attach the tubing to the capillary tube as shown in Figure 11.9.
2 Fill the apparatus with water.
3 Cut the end of the shoot under water at an angle.
4 Insert it into one end of the rubber tubing. Fasten it securely.
5 Use a clamp to keep the plant upright.
6 Place the ruler along the capillary tube.
7 Every five minutes, mark and record how far the water level in the capillary tube has moved.

Questions
1 Calculate the rate at which water was taken up by the plant.
2 If a working fan was placed near to the plant shown in Figure 11.9, what do you think would happen to the rate of water uptake?
3 Predict what would happen, and why, if you were to place a plastic bag over the shoot. (This would increase the humidity around the plant.)

Activity 11.4 Investigating transpiration

Equipment and apparatus
- 2 conical flasks
- soft plant shoot (for example calaloo, spinach or any other soft stemmed plant)
- balance
- cotton wool
- oil
- marker pen

Procedure
1 Place the same amount of water in both conical flasks.
2 Add a small amount of oil to cover the surface of the water.
3 Place the shoot in one of the flasks, securing it with the cotton wool.
4 Stopper the mouth of the other flask using the cotton wool.
5 Weigh each flask and record the measurement.
6 Using a marker pen, mark the level of water in each flask.
7 Wait 24 hours.
8 After this time period re-weigh each flask.
9 Record and explain your observations.

Developing and testing a hypothesis

Most investigations start with an observation, in other words, something you notice. But observations on their own are not enough for an investigation. When you make an observation you have to be able to turn it into a hypothesis.

A hypothesis is a statement that can be tested scientifically. For example: 'Leaves will lose more water through transpiration in moving air than in still air'. This is a reasonable hypothesis because you can test it by placing leaves in still air and in moving air, and by measuring transpiration levels to compare them.

Here are some guidelines for writing a good hypothesis.

- Your hypothesis should be about something you can observe. If you cannot see it and measure it, then you cannot really test it.
- Your hypothesis should be clearly stated and the data you collect in your investigation must be able to support or disprove your hypothesis.
- Your hypothesis must be about something you can measure. You can measure the amount of cement used in a pillar and you can measure the strength of a pillar.
- You should be able to test your hypothesis. In other words, you must be able to find results that can help you to support your hypothesis or disprove it. If you cannot disprove a hypothesis then it is normally not a good hypothesis.

Scientists never say they have proven their hypothesis because they are never 100% sure they are correct. There is always a chance that some other information might become available in the future, or that someone else might do an experiment that proves them wrong. It is better to say things like: 'our results suggest ...' or 'our findings show ...' when you are talking about the results of a test.

Once you have stated your hypothesis and planned your investigation, you need to test the hypothesis and collect data to support or disprove it.

Activity 11.5 Investigating the rate of water loss from plants

Aim: To plan and carry out an investigation using similar apparatus to that used in Activity 11.4. Using two shoots and a fan, you will investigate how wind speed affects the rate of transpiration.

Check with your teacher first before carrying out your investigation.

Activity 11.6 Investigating water loss from leaves

Equipment and apparatus
- 4 healthy leaves of the same size from the same plant, labelled A to D
- Vaseline
- Scotch tape
- string
- 2 clamp stands

Procedure
1 On leaf A place a thick coat of Vaseline on both sides of the leaf.
2 On leaf B place a thick coat of Vaseline on the upper surface of the leaf.
3 On leaf C place a thick coating of Vaseline on the under surface of the leaf.
4 Leaf D does not have any Vaseline put on it.
5 Tape the leaves by their petiole onto a length of string at equal distances from each other.
6 Tie the string on two stands to suspend the leaves.
7 Observe and record the leaves over the next two to four days.

Questions
1 Write up your experiment and explain your observations of the four leaves. Relate your explanations to whether water exits leaves from the stomata, which are mostly found on the underside of the leaf.
2 Discuss how the results could be made more reliable or valid.

Check your knowledge and skills
1 Give three reasons why plants need water.
2 Draw a labelled sketch of a xylem vessel.
3 Describe how the structure of xylem makes it suited for its functions in plants.
4 Explain the following terms:
   root pressure   capillarity   transpiration pull
5 How can leaf structure reduce transpiration? Explain with reference to an example.
6 Name three atmospheric conditions that can affect transpiration.

Unit 3 — Transport of manufactured foods

Food is manufactured in the leaves of a plant during photosynthesis. After the food is made it needs to be taken to other parts of the plant. Food substances are carried in solution though vascular tissue known as phloem (see page 116). The transportation of manufactured organic materials is called **translocation**.

Phloem tissue

Phloem tissue consists of two main types of cells – **sieve-tube elements** and **companion cells**.

Sieve-tube elements

Sieve-tube elements are tubular cells similar to xylem vessels; they are also arranged end to end. However, the end walls of sieve tubes are not fully removed as in the case of xylem. Instead, the end walls are partially removed so they look perforated, like a kitchen strainer. Sieve tubes are not dead like xylem – but they lack a nucleus when they are mature. The walls of sieve tubes are thicker than typical plant cells but they are not as thick as those found in xylem tissue. The walls of sieve tubes do not contain lignin. They are composed of cellulose.

Companion cells

Companion cells are found pressed up against sieve-tube elements. They look like a typical plant cell containing all the usual organelles. It is thought that companion cells provide substances that the sieve tube cannot manufacture since sieve tubes do not have a nucleus when mature.

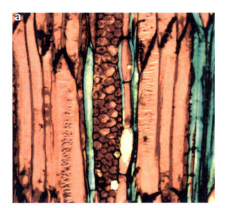
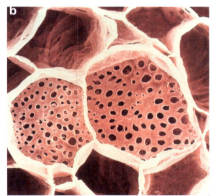
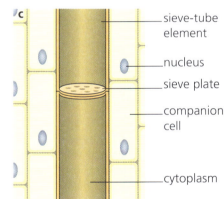

Figure 11.10 The structure of phloem: **a** shows a longitudinal section of a plant stem with sieve-tube elements; **b** shows a longitudinal section of a plant stem with sieve plates; **c** shows sieve-tube elements in phloem.

Translocation

The substance most usually transported in the phloem is sucrose in solution with water. Sucrose is loaded into the phloem by a series of steps that involve active transport. The dissolved substances flow through the tubes to the other branches and roots and stem. **Mass–flow hypothesis** is the term used to describe how the substances flow through the sieve tubes.

Dissolved food substances are transported both upwards and downwards in the phloem. Excess food substances are transported to storage organs until they are needed. When the plant needs these substances, for example in spring when it is growing strongly and producing flowers, food substances are transported from the storage organs to the buds where they supply the energy needed for growth.

Biology at work

Chapter 11 – Eating themselves to death

Figure 11.11 Aphids' method of feeding makes them effective vectors (carriers) of viral diseases. The virus is carried by the aphid and injected into the phloem of the plant via its mouthparts.

Chapter summary

Do you know?

If you are unsure of any of the facts in the list, refer to the page number in brackets.

- Water and dissolved substances are transported round plants in a specialized system of vascular tissues consisting of xylem and phloem vessels. (page 116)
- Xylem is responsible for the transport of water and mineral ions in solution. Phloem is responsible for the transport of dissolved food substances. (page 116)
- Water relations are important in plants. Plants need water for photosynthesis, support and as a medium for transporting solutes. (page 117)
- Water enters and leaves root cells by osmosis, mineral ions enter root cells by active transport. (page 118)
- Plants lose water from their stems and leaves as a result of transpiration. (page 118)
- Transpiration plays an important role in water movement by creating a transpiration pull to draw water upwards. It also helps to regulate the temperature in plants. (page 119)

- Light, humidity levels and air movement are external factors which affect transpiration rates. Structural features of the plant itself may also affect transpiration rates. (page 122)
- Both xylem and phloem are well suited to their functions as transport tubes. (page 127)

Are you able to?

If you have trouble in doing these things, refer to the page number in brackets.

- State the function of phloem in the transport system of plants. (page 116)
- Explain how water enters a plant. (page 118)
- Demonstrate water movement in plants using coloured water. (page 119)
- Explain how the structure of xylem makes it suited to its purpose. (page 120)
- Define transpiration and describe factors which affect transpiration rate. (pages 118 and 122)
- Carry out controlled investigations to show how external factors affect transpiration. (pages 123 to 125)
- Draw and label xylem vessels, sieve-tube elements and companion cells. (page 127)

Food storage

Chapter

Unit

Dealing with excess food

Animals and plants get energy from food. They need energy all the time, even when they are dormant or asleep. However, humans and other large animals don't eat all the time. We take breaks between meals and we don't eat while we are asleep. Plants only make food by photosynthesis when the sun is shining. Yet organisms still need energy even when they are not eating or making food.

In this chapter you will learn why organisms store excess food, and find out how and where food is stored in the organs of animals and plants for later use.

By the end of this chapter you should be able to:

■ Discuss the importance of food storage in living organisms
■ Identify the products stored and the sites of storage

The importance of storing food

Animals do not use all the energy from the food that they eat immediately. Similarly, plants do not use all the energy they produce during photosynthesis. These organisms store their excess energy in various forms and in various places. They keep some energy stores for a short time before using it; other energy stores are longer term.

Living organisms store excess food so that they do not need to eat or make food continuously. The energy from stored food also allows organisms to survive during periods when food is scarce. Food stores are particularly important for animals that hibernate. These animals rely on energy stored in their bodies during the long winter periods when they sleep and do not eat.

Plants store food for many reasons. Some plants, particularly those in dry areas, store food so that they have enough energy to use for rapid growth when they need it. For example, they may use their energy to produce flowers during spring.

Plants and animals also store energy for reproductive purposes. Growing embryos need a constant and safe supply of food. In animals, this supply comes from the mother's body or from food stores, for example the yolks in eggs. In plants, the seeds contain a store of food for the embryo. During germination the young embryo can get energy from the store of food in the seed until it is able to photosynthesize and produce its own food.

The way in which plants store food is very important for humans and other animals. Plants are the producers in all food chains, so the stored energy in plants is essential for the health and survival of all animal species.

Figure 12.1 Emperor penguins do not eat for several months while they protect their eggs in freezing conditions. Without a good store of body fat (energy) they would not survive.

Check your knowledge and skills

1 State two reasons why it is important for organisms to store food.
2 Why do the seeds of plants contain a store of food?
3 In the northern hemisphere, animals such as the ground squirrel and hedgehog sleep in burrows for the whole winter. How do they survive without eating?

Figure 12.2 This euphorbia stores energy so that it can produce a massive flower stalk in the springtime.

Unit 2 Food storage in plants

Plants store excess food in their roots, stems, leaves, fruits and seeds. Plants that use roots, stems or leaves as storage organs show adaptations of these organs to suit the storage function. Usually, the storage organ swells up to store the food.

Plant adaptation

What do we mean when we say that a part of a plant has 'adapted'? Remember, a plant cannot simply decide or choose to change its structure or appearance! However, if a specific feature helps a plant to survive and reproduce in a particular environment, then individual plants that have the feature survive better than those that don't. In time, they begin to outnumber the individuals that do not have the feature, until it eventually becomes a usual feature of the plant. We call this kind of feature an adaptation. You will learn more about the way adaptations develop in Chapter 20.

The photographs in Figures 12.3 to 12.5 show you some of the ways in which roots, stems and leaves are adapted for food storage.

a A **root tuber** is the swollen end of a root as in this sweet potato.

b Food can also be stored in a swollen tap root as in this carrot.

Figure 12.3 Roots can be modified to store food.

a A **rhizome** is a swollen part of the main stem, as in this ginger plant.

b A **stem tuber** is a swollen rhizome, as in this Irish potato.

c A **corm** is a swollen underground stem, shown by this taro plant.

d Sugar cane and cacti store food in their stems above ground.

Figure 12.4 Modified stems are used as food storage organs in some plants.

a A **bulb**, such as this onion, is an underground shoot with dry outer leaves and fleshy inner leaves.

b Leafy vegetables, such as spinach, store food in their leaves.

Figure 12.5 Plants can store food in leaves beneath or above the ground.

Seeds and fruits

Plants can store food in their seeds. This stored food is used to sustain the growing embryo until its root system and leaves are developed enough for it to photosynthesize and make its own food. Food can be stored in the **endosperm** of seeds or in the **cotyledons**.

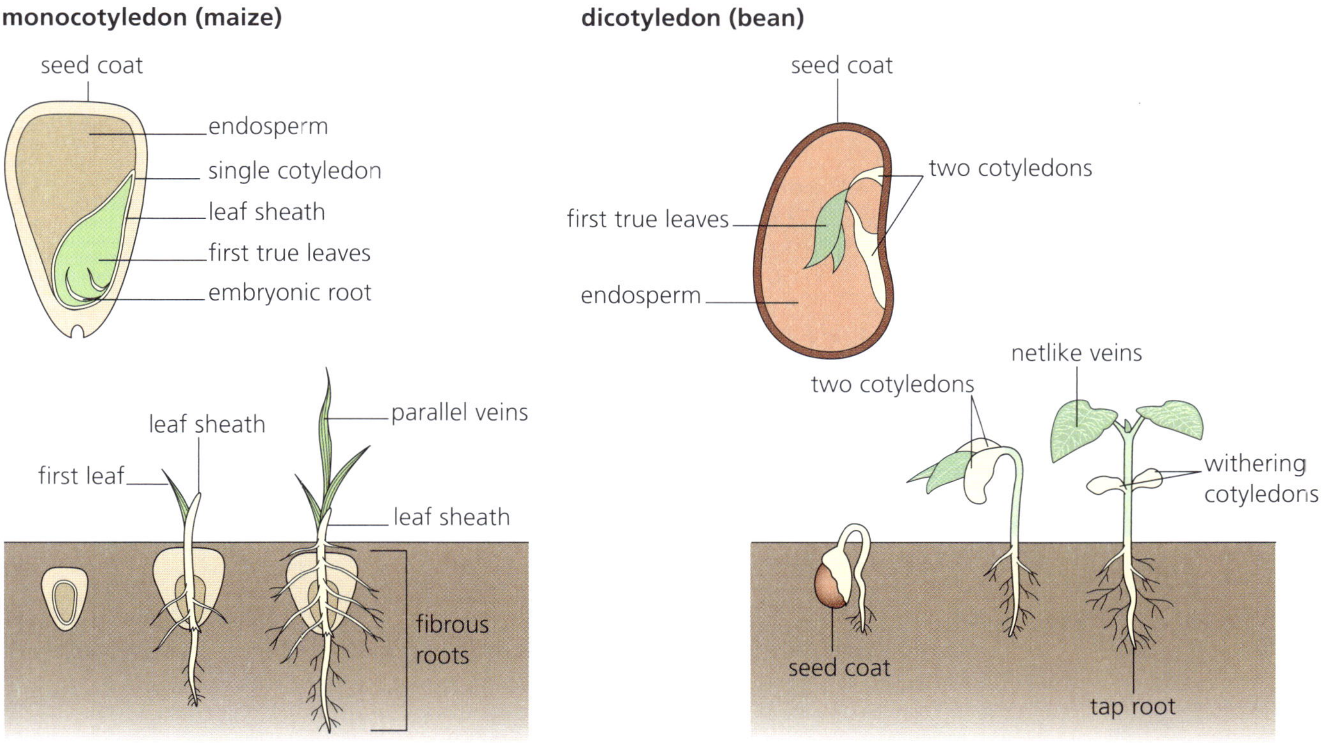

monocotyledon (maize)

dicotyledon (bean)

Figure 12.6 In maize seeds the food store is found in the endosperm. In bean seeds the food store is found in the cotyledons (seed leaves).

Some plants store food in their fruits. Food can be stored in the form of starch, sugars or lipids.

■ Plantains, breadfruits and jackfruits are all good sources of starchy carbohydrate.
■ Mangoes, peaches and berries are high in sugars.
■ Avocado and ackee fruits are rich in oil.

Activity 12.1 Drawing storage organs

SBA skills
Drawing (D)

Equipment and apparatus
■ uprooted ginger lily
■ pigeon peas or beans

Procedure
1 Make a clear labelled sketch to show the parts of a rhizome.
2 Make a clear labelled sketch to show where food is stored in the seeds of plants using pigeon peas or beans as an example.

Products stored in plants

Plants can store food in the form of sugars and lipids, but the main storage form is starch. When the plant needs to use the stored energy in the starch, the starch is converted back to glucose molecules by the process of **hydrolysis**.

Check your knowledge and skills

1 Identify the sites of food storage in the following plants:

   a tomato   **b** bean   **c** coconut   **d** cassava   **e** pigeon pea.

2 Name two plants which store food in the form of:

   a sugars   **b** lipids.

3 What is the difference between:

   a a corm and a bulb   **b** a swollen tap root and a rhizome?

Unit 3 — Food storage in animals

Animals store food in almost all parts of their bodies, particularly in the form of fat. However, the main storage organ for excess food is the liver.

The liver

Excess glucose is converted into **glycogen** (see page 45). Glycogen is stored mainly in the liver. When the body needs glucose, the glycogen is converted back to glucose and transported via the bloodstream to where it is needed.

Body fat

Animals also store lipids in the form of body fat. When an animal does not have enough glucose or glycogen to provide energy, body fat is broken down to produce energy. You can see this in hibernating animals. These animals build up a large store of body fat before they go into hibernation. During the **hibernation** period, the animal does not eat. It uses the energy stored in its body fat to sustain its metabolic processes. The animal usually emerges from hibernation having lost up to half its body mass. It then needs to eat to build up its energy stores.

Humans may eat more food than their bodies can use. When this happens, the excess food is stored in the adipose tissue in the body in the form of fat. Unless this energy is used up, it builds up on the body as fat and the person may become **obese** (see also page 84).

Eggs

Egg yolk is rich in fat and protein. This is an important food store which is used to feed the developing animal as it grows to hatching age. Birds, turtles, reptiles and some mammals lay eggs which contain a rich store of food.

Biology at work

Chapter 12 – Calculating BMI

Figure 12.7 Mammals that live in cold climates, such as the polar bear, have a thick layer of stored food in the form of fat beneath their skin. This keeps them warm and stores energy.

Check your knowledge and skills

1 Vegetarians usually rely on plant sources for their protein intake. They usually use beans, peas and nuts for this purpose.

a Describe how you could compare semi-quantitatively the protein content of red beans and peanuts.

b Explain how peanut and pea plants store food.

c Suggest two more types of food that may be found in the peanut and describe how you would test for the presence of each.

Answering 'use of knowledge' questions in the exam

'Use of knowledge' (UK) questions take you beyond the simple recall of facts and basic information to situations where you have to use and apply what you know. You need four skills to answer these types of questions:

1 Application of knowledge – use what you have learnt in new and unknown situations and use the correct formulae and procedures to solve a problem.

2 Analysis and interpretation of facts and relationships – work out cause and effect based on the relationship between factors in a given situation. You also need to be able to make inferences and predictions and to draw conclusions based on what you have learnt.

3 Synthesize information – be able to draw together what you have learnt in different sections and combine it to answer questions. You also need to be able to combine knowledge and skills to make predictions and solve problems.

4 Evaluation of information – be able to read critically and make judgements and recommendations based on what you have learnt. In these questions you may have to make value judgements based on the implications for the environment or human well-being.

Ask your teacher to give you some sample UK questions from past exam papers.

Question

1 Look carefully at the questions in an exam paper and make a list of the different 'instruction words' that appear – for example, explain, suggest, discuss.

Chapter summary

Do you know?

If you are unsure of any of the facts in the list, refer to the page number in brackets.

- Plants and animals store food so that they do not need to eat or manufacture food on a continuous basis and so they have reserves for times when food is scarce. (page 129)
- Food is also stored to provide reserves for special functions, such as rapid growth, reproduction and embryo development. (page 129)
- Plants store food in their roots, stems, leaves, seed and fruits. (page 130)
- Food is stored in swollen tap roots and tubers in roots, in tubers, rhizomes and corms in stems, and in bulbs and fleshy leaves. (page 130)
- Plants store most food in the form of insoluble starch. However, food is also stored as sugar in some fruits and as lipids in seeds and some fruits. (page 131)
- The liver is the most important site of food storage in animals. Glucose is converted to glycogen for storage in the liver. The glycogen is hydrolysed to form glucose when it is needed. (page 132)
- Excess food is also stored in the form of adipose tissue (fat) under the skin of many larger animals, for example in hibernating animals and those who live in very cold climates. (page 132)

Are you able to?

If you have trouble in doing these things, refer to the page number in brackets.

- Discuss the importance of food storage in living organisms. (page 129)
- Identify the sites of storage in plants. (page 130)
- Give examples of the parts of plants that are adapted for food storage. (page 130)
- Identify the substances stored in plants. (page 132)
- Identify sites of storage in animals. (page 132)
- Identify the substances stored in animals. (page 132)

13 Excretion

Unit 1 Plants and animals must excrete wastes

Within the cells of every plant and animal, chemical processes go on all the time to keep the organism alive. The word **metabolism** is used to describe these chemical processes. Whilst the metabolic processes (such as photosynthesis and respiration) provide useful new compounds for an organism, they also produce unwanted substances. The organism has to get rid of these waste products. If they accumulate they become toxic and interfere with on-going metabolic processes. The process of removing waste products from the body is called **excretion**, and the materials that are excreted are called **excretory products**.

■ Excretion in plants

Excretion in plants is different to the process in animals. This is mainly because animals produce more waste products than plants do. Plants are in general more efficient with materials and tend to recycle them. However, plants do produce some waste products. These include:

■ oxygen – waste product of photosynthesis
■ carbon dioxide – waste product of respiration
■ water – waste product of photosynthesis and respiration
■ chemical compounds, such as gums, resins and tannins.

Plants do not have specialized organs of excretion. They excrete gaseous wastes by diffusion from their stomata and other surfaces. They excrete waste or excess water by transpiration (see page 120) or they store it in their cells to maintain turgidity.

    Plant cells have large vacuoles in which they store toxic wastes. When these wastes are not excreted they build up in the form of harmless crystals in the vacuoles.

Figure 13.1 Deciduous trees lose their leaves in autumn. This allows them to excrete all the waste products stored in the leaves. Other waste products get moved to the bark.

Larger plants, such as trees, translocate (move substances using energy) wastes into the dead cells of the outer xylem vessels. For example, these dead cells form the bark in trees. Once the wastes have moved to the outer xylem vessels, they may perform a useful function for the plant. The tough bark makes it difficult for bacteria, fungi and herbivores to reach the softer tissues inside the plant. Thus the waste products can have a protective function, preventing other organisms from eating or destroying the plant.

Other plant wastes are excreted when parts of the plant fall off or die. For example, old or yellowed leaves fall from the plant as they age and die. Plants also shed their leaves during certain seasons. Fallen leaves which contain chemical wastes may also help to repel insects in the soil around the tree.

Excretion in animals

The most important metabolic wastes produced by animals are carbon dioxide (produced in respiration) and **urea** (produced in the liver as a waste product of **deamination** which is the break down of old proteins). These metabolic wastes cannot be allowed to build up to dangerous levels in the body – they must be continually excreted.

Excretory product	Source	How is it harmful?
Carbon dioxide	Aerobic respiration	Can change the pH of the blood and other body fluids Enzymes no longer able to work effectively
Water	Respiration	Changes the osmotic gradient inside and out of cells Cells, such as the red blood cells, can swell and could block small blood vessels
Salts/ions	Many metabolic reactions	Changes the osmotic gradient of cells Cells shrivel and become irregularly shaped Red blood cells stick to each other, causing a clump of cells that could block small blood vessels and increase blood pressure, increasing the risk of a stroke
Nitrogenous compounds	Deamination of proteins	Nitrogenous compounds tend to be very reactive and could get involved in unwanted reactions Could also change the pH of the blood, which would have disastrous effects on the action of enzymes

Table 13.1 Excretory products of animals, their source and what happens if they are allowed to build up to high levels in the body.

Animals get rid of unwanted substances in several ways. Some can be so dangerous that the animal may need to modify the substance before eliminating it – for example, metabolic reactions producing ammonia, a nitrogenous compound, as a waste product. Ammonia is so toxic to the body that it cannot be tolerated – even at low levels. The liver converts ammonia to a less toxic substance called urea, which is produced by a series of reactions in the liver called the urea cycle. Urea can remain in the body for a while without endangering the animal.

Simple unicellular organisms, such as *Amoeba*, are able to excrete wastes by diffusion into the surrounding environment. They have a large surface area to volume ratio, so the waste product ammonia diffuses easily into the surrounding water. This form of excretion is efficient for very small organisms, but larger animals have to use other ways of getting rid of their excretory products.

Key facts

■ **Secretion** is different from excretion. Secretion is usually the release of useful substances, for example hormones from cells.
■ The liver detoxifies harmful substances. Examples include drugs, alcohol and poisons, but they also include the harmful or indigestible parts of the food we eat. Our bodies store the useful components of these substances, and excrete their less toxic remnants with urea.

How unwanted substances are removed from the body

Carbon dioxide is removed from mammals during exhalation. During gaseous exchange, carbon dioxide accumulated in the deoxygenated blood diffuses from the capillaries into the lungs. As the person exhales, this carbon dioxide along with other gases already in the lungs gets expelled into the atmosphere (see page 99).

We do not usually think of water as a product of excretion, because it is vital to all living organisms. However, too much water can harm an organism. Excess water is removed from the body in several ways. A small amount leaves the body during exhalation as water vapour. As we perspire we also lose water. Urine also contains a large amount of water, which is expelled during urination.

Salt and ion wastes are also released from the body during perspiration and urination.

Animals release nitrogenous wastes via their kidneys when they urinate. Humans also release small amounts of these wastes in our sweat.

Activity 13.1 Making your own deodorant

Equipment and apparatus

For a basic powdered deodorant:
■ ½ cup baking soda
■ ½ cup cornstarch
■ few drops antibacterial essential oil, such as lavender or cinnamon
■ glass jar
■ clean washcloth

For a basic liquid deodorant:
■ ¼ cup witch hazel
■ ¼ cup aloe vera gel
■ ¼ cup mineral water
■ 1 teaspoon vegetable glycerin
■ few drops antibacterial essential oil, such as lavender or cinnamon (optional)
■ spray bottle

Procedure

1 Choose to make either the powdered deodorant or the liquid deodorant.
2 To make the powdered deodorant, place the substances in a glass jar.
3 Shake the glass jar to blend the substances.
4 Sprinkle the powder on a damp washcloth. Pat on. Don't rinse.
5 To make the liquid deodorant, place the substances in a spray bottle.
6 Shake the spray bottle to blend the substances.
7 Spray a little on to the skin and allow it to dry.

Unit 2 Excretion in humans

The human body has a **urinary system** that carries out the process of excretion. This system consists of the kidneys, a pair of ureters, the bladder and urethra.

■ The human kidney

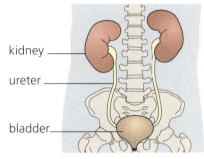

Figure 13.2 The location of the kidneys

The kidney is a specialized organ that carries out excretion. In humans, the kidneys are bean-shaped organs located near the vertebral column at the small of the back. Humans and other mammals have two kidneys, one on either side of the vertebral column. The kidney to the left of the vertebrae is usually higher than the one to the right. They are usually dark red in colour.

A cross-sectional view of the kidney shows three distinct areas (see Figure 13.3). The outer layer is called the cortex. The layer inside this is called the medulla. There is a small section towards the tube leading out of the kidney called the pelvis. Leading from each kidney is a tube called the **ureter**. This tube empties its contents into a bag-like structure called the bladder.

■ Cortex – the outer part of the kidney. In the cortex, blood is filtered by a process known as **ultra-filtration** (or high pressure filtration) because it only works if the blood entering the kidney in the renal artery is at high pressure.
■ Medulla – where water and salt levels in the blood are controlled.
■ Pelvis – the region of the kidney where urine collects.

A closer view of the kidney shows that it is a collection of thousands of tube-like structures called **nephrons**. A nephron is the functional unit of the kidney.

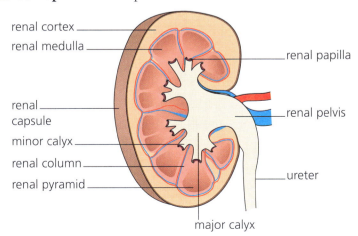

Figure 13.3 A cross-section of a human kidney.

The kidney as the major excretory organ

The kidney's main functions are to:

- remove metabolic/toxic waste products from the body
- regulate the water content of body fluids
- regulate the pH of body fluids
- regulate the chemical composition of body fluids by removing excess substances not needed by the body.

The kidney has a rich blood supply. It keeps the composition of the blood at a steady level. In this way, it contributes to **homeostasis** (see Chapter 14).

The structure of a nephron

The kidney is mainly made up of nephrons. Each nephron (see Figure 13.4) is a tiny structure that filters the blood and produces urine.

Each human kidney has an estimated 1 000 000 nephrons, each approximately 3 cm in length. Together, the nephrons provide a very large surface area and are able to filter about 125 cm³ of fluid per minute. About 99% of the filtered water is returned to the blood, so only about 1 cm³ of urine is made per minute. This varies according to other factors, such as how much a person sweats or drinks.

Each nephron consists of five main regions. Each region has a different function. Refer to Figure 13.4 when reading this list:

- renal cortex, made up of the Bowman's capsule and glomerulus
- proximal convoluted tubule
- loop of Henle
- distal convoluted tubule
- collecting duct.

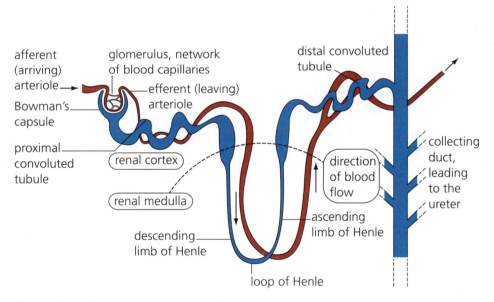

Figure 13.4 The structure of a nephron.

Blood enters the kidney from the renal artery which branches into finer and finer arterioles before entering the glomerulus. The blood vessels that enter the glomerulus are called the afferent arterioles. The blood vessel that leaves the glomerulus is called the efferent arteriole. Blood is filtered in the glomerulus. The filtered blood leaves the glomerulus through the efferent arteriole.

The path through a nephron and the processes that take place

Ultra-filtration

Blood is filtered under high pressure in the glomerulus. This is called ultra-filtration. You have learnt that the glomerulus is a dense knot of blood capillaries in the Bowman's capsule. The blood enters the capillaries in the glomerulus at high pressure because it comes direct from the heart.

The artery that brings blood to the glomerulus (the afferent arteriole) is much bigger in diameter than the capillaries in the glomerulus. As a result, the pressure rises as the blood enters the **glomerulus**. This squeezes the blood so that small molecules, such as water, glucose, salts and urea, get pressed through the tiny holes in the capillary walls into the Bowman's capsule. Larger molecules, such as proteins, as well as blood cells and platelets, stay in the blood.

The filtered fluid in the Bowman's capsule is now called glomerular filtrate. It has a similar composition to blood plasma. It contains small molecules, such as glucose, amino acids, vitamins, ions, nitrogenous waste, some hormones and water. Because of this, the blood leaving the glomerulus has a higher concentration than the blood that entered.

Selective reabsorption

After filtration, useful substances, such as water and glucose, may be reabsorbed into the blood as it passes from the Bowman's capsule towards the collecting duct. This is called selective reabsorption. Ultrafiltration produces about 180 L of filtrate per day, yet only about 1.5 L of urine are produced each day. This means that a great deal of reabsorption must occur.

So the main functions of the nephron are to:

- filter the blood to remove unwanted excretory products
- selectively reabsorb useful materials back into the blood
- actively secrete waste products from capillaries surrounding the tubules.

The cells that make up the proximal convoluted tubule are adapted for reabsorption in the following ways:

- large surface area provided by microvilli
- high numbers of mitochondria (to provide energy for the active process)
- blood capillaries close by.

Over 80% of the glomerular filtrate is reabsorbed here – all the glucose, amino acid, vitamins, hormones and about 80% of the sodium chloride and water.

With the exception of water, the reabsorption occurs by the process of diffusion and active transport. The water is removed by osmosis.

- Glucose, amino acid and ions diffuse into the cells of the first convoluted tube from the filtrate.
- The active uptake of sodium and other ions makes the tubular filtrate have a higher water potential so the water diffuses out until the water potential inside and outside is the same.
- About 40 to 50% of the urea from the filtrate is reabsorbed, by diffusion, into the blood capillaries and is passed back into circulation.

Blood under pressure

Think about what happens when a big crowd of people enters a small room through a doorway. Once all the people are inside the room, everyone gets very squashed. It is this effect that makes the blood under pressure in the capillaries in the glomerulus.

A little maths!

If ultrafiltration produces 180 L of filtrate per day yet only 1.5 L of urine is produced, calculate the percentage of filtrate that becomes urine.

The loop of Henle

The function of the loop of Henle is to conserve water. The longer the loop of Henle, the more concentrated the urine that can be produced. Birds and mammals are the only vertebrates with loops of Henle. They are also the only vertebrates whose urine is more concentrated than their blood. The drier the animal's habitat, the longer their loop of Henle. Animals that live in wetter habitats tend to produce more urine, so they need a shorter loop of Henle.

The descending limb of the loop of Henle is highly permeable to water and to most solutes. Its function is to allow substances to diffuse easily through its walls.

Both parts of the ascending limb are almost totally impermeable to water. The cells in the thick part are able to selectively reabsorb sodium, potassium, chloride and other ions from the tubule.

Normally, water would follow by osmosis, but this cannot happen because the loop is impermeable to water. The fluid in the loop of Henle is very dilute when it reaches the distal convoluted tubule.

The distal convoluted tubule and the collecting duct

The distal convoluted tubule and the collecting duct are responsible for the fine-tuning of the body fluid composition. They also control pH levels in the body.

The cells of the distal convoluted tubule are similar to those in the proximal convoluted tubule. Microvilli line their inner surface, increasing the surface area for reabsorption. There are also numerous mitochondria for active transport. The collecting duct carries the fluid towards the ureter.

As the fluid moves down the collecting duct the tissue fluid surrounding the duct gets more concentrated; water therefore leaves the collecting duct by osmosis.

The flow diagram in Figure 13.5 summarizes the process from when blood enters the kidneys until it is excreted as urine.

Ureter and urethra

The **ureter** is the tube that leads from the kidney to the bladder and the **urethra** is the tube that leads from the bladder to outside the body. Don't get them confused!

Biology at work

Chapter 13 – Living with kidney disease

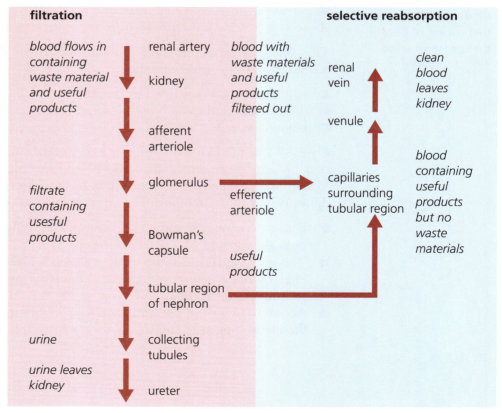

Figure 13.5 The movement of blood through the kidneys.

How to draw a bar graph

In a bar graph, the data are represented as a series of bars or columns. Normally the columns are equally wide and equally spaced. Figure 13.6 shows you an example of a bar graph.

The following guidelines will help you to construct a bar graph.

- ■ Organize your data in a table if one is not provided.
- ■ Decide whether you are going to use horizontal or vertical bars depending on the space available for the graph.
- ■ Draw axes of a convenient length.
- ■ Label the axes (remember any units) and give the graph a title.
- ■ Choose an appropriate scale for the axis with the numerical values. Divide the axis equally using the scale you have chosen and label the divisions.
- ■ Decide how wide the bars will be and how much space you will leave between them.
- ■ Construct the bars using the numerical scale to decide how tall or long each one will be.

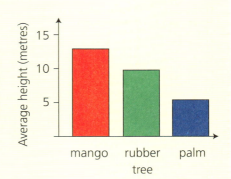

Figure 13.6 A bar graph comparing the average heights of different tree species.

Check your knowledge and skills

1 Choose from the following words to write the correct answers below.

    kidney   renal
    artery   ureter
    urethra   bladder

  a Stores urine. ____
  b Filters urea and other waste chemicals out of the blood.

    ____

  c Carries blood with a high concentration of urea. ____
  d Carries urine down to the bladder. ____

2 Figure 13.7 shows the human urinary system.

  a Name the parts labelled A, B and C.
  b Name the functional unit of the kidney.

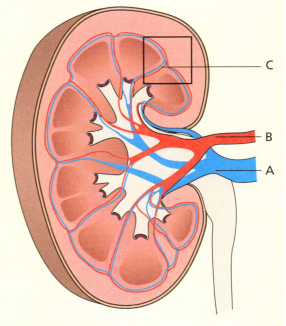

Figure 13.7

3 The table below shows the concentrations of five substances present in the blood entering a kidney, in a nephron, and in a sample of urine. (All values are in mg/dm³.)

Substance	Blood entering kidney	Nephron	Urine
Urea	0.4	20	20
Glucose	1.5	1.5	0
Amino acids	0.8	0.8	0
Salts	8.0	8.0	16.6
Protein	82	0	0

a Draw a bar graph for the information in the table above.
b Which substances pass from the blood into the nephron?
c By what process do they pass into the nephron?
d Which substances are reabsorbed into the blood from the nephron? Explain why this happens.
e Explain the results for protein.

4 Eating which of the following foods would make the kidneys work the hardest to excrete nitrogenous wastes? Explain your answer.
    a jerked chicken   b cake   c plantain chips   d vegetable soup.

Chapter summary

Do you know?

If you are unsure of any of the facts in the list, refer to the page number in brackets.

- Excretory products are produced from various metabolic reactions in the body. (page 134)
- Plants have simpler and fewer products to excrete than animals. (page 134)
- Excretion is not the same as egestion. (page 135)
- The nephrons are the functional unit of the kidney. (page 137)
- Nephrons are responsible for the filtering of blood. (page 138)

- Urine is the result of the filtration of the blood and the reabsorption of important substances from the filtrate, such as all glucose, some water, amino acids and salts. (pages 139–140)
- The afferent arteriole has a larger diameter than the efferent arteriole. This builds up pressure in the glomerulus thus increasing the efficiency of the ultrafiltration process. (page 139)

Are you able to?

If you have trouble in doing these things, refer to the page number in brackets.

- Draw a bar graph from given data. (page 141)
- Interpret and analyse information from a bar chart. (page 142)

14 Regulation and control

Maintaining an internal balance

Our environment changes all the time. Some days are bright and hot, and other days are cold. All living organisms need to keep their internal environment constant. This involves controlling metabolic processes so that they run constantly.

In this chapter you will learn about some of the principles of regulation and control that animals and plants use to maintain a constant internal environment.

By the end of this chapter you should be able to:

■ Understand the need for homeostasis

■ Relate negative feedback mechanisms to various mechanisms of control

■ Explain osmoregulation

■ Understand what is meant by a positive feedback system

■ Explain the role of the hormone ADH in osmoregulation

■ Discuss adaptations in plants to conserve water

1 Homeostasis

Plants and animals stay alive because of metabolic processes at work all the time within their cells. All these processes take place as a result of enzymes working together. Enzymes are easily affected by each other and by other substances in the tissue fluid. Tissue fluid is drained plasma that flows between and around body cells. This fluid constantly bathes the cells of the body, forming the watery environment in which enzyme activity can take place. The composition of this fluid affects functioning of the cells. The following variables must remain within specific limits to ensure that each enzyme can continue to function:

■ pH
■ water concentration
■ salt concentration
■ glucose concentration
■ temperature.

The world around an organism forms its external environment. The cells and tissues of the organism itself form its internal environment. Cells function best in a stable environment. That means the body must keep the internal environment stable, even when the outside conditions are changing. This process of maintaining a stable internal environment in the living body is called **homeostasis**. Homeostasis allows an organism to regulate its internal environment so that it can survive changes in its external environment.

■ Changes in the internal environment

Homeostasis is one of the most important functions of any multicellular organism. If homeostasis is disrupted, cells can suffer or even die. Table 14.1 overleaf shows what happens if there are changes to the internal environment.

Biology at work
Chapter 14 – Why do you feel hot when you have a cold?

Factors that are normally kept at a steady state	Effect on the body if too high	Effect on the body if too low
Water concentration	If there is excess water in the body, it flows into cells by osmosis. This causes cells to become swollen Blood cells can become too big to travel through small blood vessels High water concentration may also prevent cells from losing excretory solutes	Many metabolic reactions require an aqueous environment. If there is too little water available, the cells become dehydrated, preventing these processes
Temperature	If the temperature gets too high, enzymes may get destroyed	As the temperature drops, enzyme activity slows down. This can make it more difficult for the body's immune system to fight infections
Glucose levels	If glucose levels rise sharply, water leaves cells by osmosis. This can cause dehydration and tissue damage Excess glucose in the blood can affect the immune system's ability to fight infections. Most pathogens thrive in a condition that has lots of sugar. They will multiply at a faster rate, putting strain on the immune system Long-term effects could include diabetes	Cells use glucose to provide energy for respiration. If the body's glucose levels drop sharply, there is no energy available for cell respiration; A person with a condition of abnormally low blood sugar is described as **hypoglycaemic**. Signs include hunger, nervousness, shakiness and sleepiness. Hypoglycaemia may lead to unconsciousness

Table 14.1 How changes in the internal environment affect the body.

Negative feedback and homeostasis

Think of a refrigerator set to maintain a temperature between 5 °C and 8 °C. The refrigerator contains a thermostat. When you switch on the refrigerator, the motor in the refrigerator works to cool the air inside. A sensor in the thermostat registers as soon as the temperature drops below 5 °C. This triggers a mechanism that turns the motor off. Once the motor is off, the air gradually gets warmer. As soon as the temperature goes above 8 °C, the thermostat triggers a mechanism that turns the motor on again. This is a typical example of a **negative feedback** system. The flow diagram in Figure 14.1 shows this type of system.

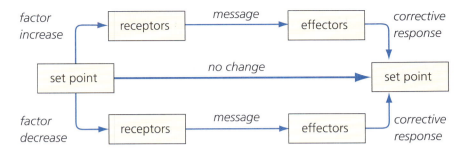

Figure 14.1 A negative feedback mechanism.

In order to maintain homeostasis, the body needs to keep various factors within a tolerable range. The body attempts to maintain a normal internal environment, called the **set point**. This is the ideal or normal range for the specific condition or state. For example, the set point for blood glucose levels in the human body is 90 mg per 100 ml of blood. Figure 14.2 illustrates how the body maintains a steady concentration of glucose in the blood. As the glucose levels in the blood increase, the pancreas detects the increased sugar concentration and releases a hormone called **insulin**. The insulin acts on the liver, which starts taking up glucose and storing it. When the glucose levels drop, the pancreas responds by releasing a different hormone called **glucagon**, which acts on the liver to cause it to convert glycogen to glucose and release it into the blood again.

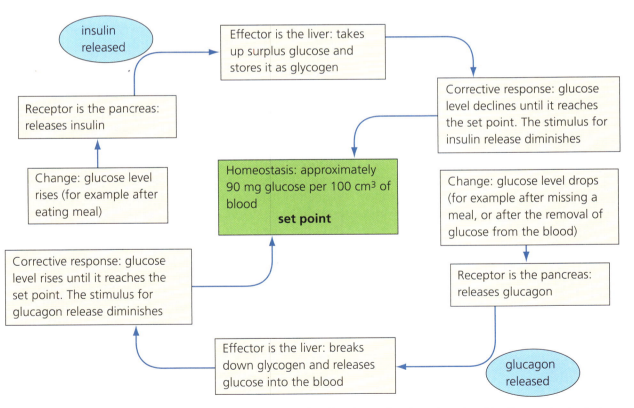

Figure 14.2 This flow diagram illustrates the negative feedback mechanisms involved in the control of glucose levels in the body.

Receptors detect a change and send a signal to a control centre. The control centre will make a decision on what must be done to rectify the situation. **Effectors** are responsible for bringing the condition back to normal upon instructions from the control centre. In our example of glucose regulation (see Figure 14.2), the pancreas is the receptor and the liver is the effector.

Negative feedback mechanisms are responsible for maintaining:

- a balance between the levels of oxygen and carbon dioxide in the blood
- a steady heart rate
- tolerable glucose levels in the blood
- osmoregulation (balance of water in the living system)
- tolerable pH in the cells, tissues and tissue fluid
- a normal body temperature
- normal blood pressure.

Extreme conditions can disable a negative feedback mechanism. In sudden extreme changes, death can result, unless medical treatment is administered in time to bring about the natural occurrence of these feedback mechanisms. For example, your body regulates its temperature (thermoregulation) so if your temperature drops one or two degrees below normal, your body will respond with strategies to reduce heat loss, e.g. shivering. These strategies are the first stage of hypothermia. If you lose one or two more degrees, the symptoms become more serious: violent shivering, muscle contractions, slowed-down movements. If your temperature drops below 32°C, cellular metabolic processes begin to shut down; at a body temperature below 30°C, organs fail and death occurs.

Check your knowledge and skills

1 Write your own definitions of the following terms:
  a negative feedback mechanism   **b** homeostasis.

Did you know?

Negative feedback mechanisms promote stability. However, sometimes a system needs to change. In general, change is driven by **positive feedback** mechanisms. In a positive feedback system, the change to the internal environment leads to even more change in the *same direction*. Whereas a negative feedback mechanism helps a system to return to its normal conditions, positive feedback systems help the system to change. Here are some examples.

- During labour, a woman's contractions get stronger and stronger. The contractions are stimulated by the release of a hormone called oxytocin and each contraction stimulates more oxytocin to be released. Eventually the birth of the baby inhibits the release of oxytocin and the contractions stop.
- When cells are ready to divide, they synthesize a molecule that triggers the increase of proteins called G_1 cyclins. The more of these within a cell, the more get produced, which helps the cell to divide quickly and to keep dividing.

Unit 2 Osmoregulation

Have you noticed that after you drink a lot of water that you tend to go to the bathroom more frequently? Or, if the weather gets cold and you do not sweat much, that your trips to the bathroom are also increased? Or, when it's hot and you sweat a lot, that you do not need to go as often? Your brain (the control centre) keeps track of the concentration of water in your blood and your tissues and sends messages to your kidneys (effectors). Your kidneys control the balance of water in your body in a process called **osmoregulation**. Osmoregulation takes place in all living organisms.

■ Osmoregulation in animals

As the blood circulates through the body, it passes an area in the brain called the hypothalamus. When there is a lot of water in the body, the blood has a low concentration, as it is diluted with water. When there is less water in the body, the blood is more concentrated. The hypothalamus contains special cells called osmoreceptors that detect the concentration levels of the blood and send messages to the pituitary gland also in the brain, which controls the release of a hormone called **antidiuretic hormone (ADH)**.

■ On a hot day – you may lose water through sweating. Your blood becomes more concentrated. The osmoreceptors detect this, and send a message to the pituitary gland to secrete ADH. The hormone travels in the blood to the kidneys, where it makes the distal convoluted tubules and collecting ducts become more permeable to water. This allows these tubules and ducts to reabsorb more water from the filtrate in the nephron. The body collects less urine in the bladder.

- On a cold day or if you have been drinking a lot of water – the blood will be more dilute (lower in concentration). The hypothalamus sends a message to the pituitary gland to stop secreting ADH. Since there is a decreased concentration of ADH in the blood, the distal convoluted tubules and collecting ducts become less permeable to water. Less water gets reabsorbed from the filtrate. The body produces much dilute urine. This helps to remove excess water in the blood.

Table 14.2 shows the changes in urine when it is dilute and concentrated, and Figure 14.3 summarizes the stages in osmoregulation in the body.

	In the presence of ADH	In the absence of ADH
Characteristics of the urine	Dark in colour Low volume Very pungent odour	Lighter in colour or clear High volume Milder odour

Table 14.2 Changes in the composition of urine.

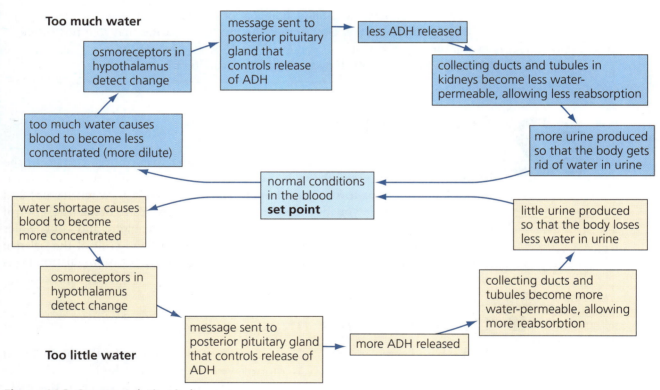

Figure 14.3 Osmoregulation in humans.

■ Osmoregulation in plants

You have learnt that plants lose water through transpiration (see page 120). If the plant lives in an area where water is freely available, it can take up more water through its roots. However, many plants live in environments where there is a limited supply of water. For example, plants near beaches have to survive in sandy soil which does not retain water. These plants have adaptations that assist with osmoregulation. Plants that have these adaptations are known as xerophytes. You looked at some of the adaptations of xerophytes in Chapter 11 on page 122.

Activity 14.1 Adaptations in plants

Aim: To describe plant adaptations that help them survive in their environment.

Copy the table below. In the table, write a fuller description of each adaptation and illustrate it to complete the table. If possible, find an area that has xerophytes (for example, a coastal area where there are plants growing in a very sandy habitat). Observe the plants growing there and draw them.

Adaptation	Description	Illustration
Increased root length		
Closure of stomata		
Increased cuticle thickness		
Water storage in swollen leaves or stems		
Reduced leaf area		
Sunken stomata		
Rolled leaves		

Check your knowledge and skills

1 Explain the relationship between the amount of ADH in the blood and the quantity of urine produced.
2 Answer the following with true or false. Give reasons for your answer.
  a Osmoregulation only takes place during hot weather.
  b The kidneys release ADH into the blood.
  c On cooler days, when a person drinks a lot of water and does not sweat much, they will probably urinate more frequently.

Unit 3 Temperature regulation

The temperature outside your body changes all the time. The outside temperature may drop at night, cooling your body down. Then it gets warm again during the day, heating your body up. Perhaps you may go into a cold, air-conditioned room, or open a cold refrigerator. Then you may walk outside into hot sunlight.

There are also temperature changes inside your body. Some metabolic reactions give off heat as a by-product, leading to an increase in body temperature. This heat travels mainly in the blood.

No matter how much the temperature changes outside your body, a healthy body maintains a constant internal body temperature within a small range. The enzymes that control your metabolic processes work best at a steady temperature of approximately 37 °C. So, whenever your temperature rises or falls outside of its optimal range (also known as its set point), your body has special processes that help to bring it back to normal.

What happens when body temperature changes?

As the internal temperature begins to increase, the thermal centres in the brain detect this change and use a negative feedback mechanism to bring the temperature back to normal. Corrective mechanisms are shown in Table 14.3.

When the body gets too hot	When the body gets too cold
Blood vessels at the surface of the skin dilate (get wider). This lets blood flow more easily near the surface of the skin so that more heat is lost by heat radiation	Blood vessels near the skin surface constrict. This reduces the flow of blood near the skin surface, so less heat is lost by heat radiation
The sweat glands in the skin are stimulated, causing the person to perspire more. When the sweat evaporates from the skin surface, it has a cooling effect on the body	The sweat glands stop making sweat
The hair on the skin lies flat, allowing heat to escape easily from the body. This mechanism is particularly important in animals with thick fur such as dogs and cats. See Figure 14.6 for a cross-section of the skin.	The hair on the surface of the skin stands on end. This traps the air inbetween and prevents too much heat from escaping from the body
	Shivering begins. This is when the muscles start to contract and relax rapidly. This increases the heat generated by the body through metabolic reactions

Table 14.3 What happens when the body gets too hot or too cold.

Figure 14.4 When an athlete performs their metabolic rate increases. The increased metabolic activity releases heat, which is why they get hot and perspire.

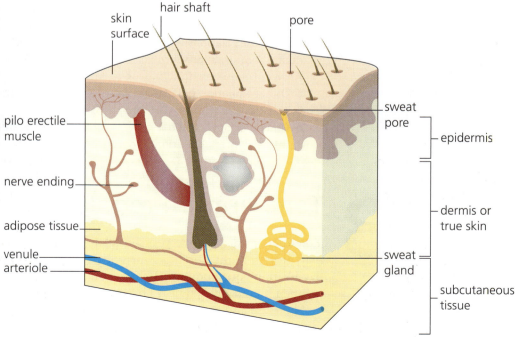

Figure 14.5 Cross-section of human skin.

Activity 14.2 Drawing a flow diagram to illustrate negative feedback

Aim: To draw a simple diagram to represent temperature regulation in humans.

You have seen flow diagrams that illustrate the negative feedback systems controlling osmoregulation and glucose levels. Draw your own flow diagram to illustrate the negative feedback system that controls temperature in the body. You do not need to include all the detailed examples of ways by which the body raises or lowers its temperature.

Check your knowledge and skills

1 Sweating is one way the body tries to lose heat. Explain why, during exercise, the body tries to lose more heat.
2 Explain the role of negative feedback in homeostasis.
3 How is sugar stored when your body is at rest?
4 The coach of the national football team may ask the players not to drink any caffeinated drinks such as coffee and some sodas before a game. Knowing that caffeine is a diuretic (causes more water to be removed from the body in urine), explain fully why he may make this request.
5 Using a flow diagram, outline the negative feedback mechanism that takes place when the concentration of water in the body is low.

Chapter summary

Do you know?

If you are unsure of any of the facts in the list, refer to the page number in brackets.

- Cells need a stable environment in which to function. (page 143)
- The process of regulating and controlling the internal environment of a living organism is called homeostasis. (page 143)
- Homeostasis involves the control of factors such as glucose levels, pH, water concentration, salt concentration and temperature. (page 143)
- A negative feedback mechanism is a process that keeps returning a set of variables to a set point or optimum range. (page 144)
- Positive feedback is feedback that reinforces or amplifies the response that causes it. (page 146)
- Osmoregulation is the process of controlling the concentration of water in the blood. (page 147)
- The hormone ADH controls osmoregulation by stimulating or inhibiting water reabsorption. (page 147)
- Osmoregulation in plants is assisted by adapta- tions that allow them to survive in environments with a limited water supply. (page 149)
- In mammals, the skin has various functions, including functions that assist in regulating body temperature. (page 150)
- The skin allows us to lose heat by sweating and vasodilation. We can conserve heat by vasoconstriction, shivering and the erection of hairs on the skin surface. The fat layer under the skin also conserves body heat. (page 150)

Are you able to?

If you have trouble in doing these things, refer to the page number in brackets.

- Define homeostasis and osmoregulation. (pages 143 and 147)
- Describe, giving examples, a negative feedback mechanism that assists in the process of homeostasis. (page 145)
- Explain the role of the kidneys and ADH in controlling osmoregulation. (page 147)
- List the ways that plants perform osmoregulation. (page 149)
- Describe the control and regulation of temperature in the human body. (page 150)

151

15 Sensitivity and coordination

> ## *Responding to the environment*
>
> Every living organism is sensitive to its environment. This means it must constantly detect and respond to changes in its environment.
>
> In this chapter you will learn about the ways in which the body receives messages and how it responds to them in an organized way.
>
> By the end of this chapter you should be able to:
>
> ■ Define sensitivity and coordination, stimulus and response, receptors and effectors
> ■ Describe the response of green plants to unilateral stimuli of light and gravity
> ■ Explain why the response to stimuli is important for the survival of organisms
> ■ Describe the response of invertebrates to light, temperature and moisture
> ■ Compare the endocrine system with the nervous system and list endocrine glands
> ■ Describe the functions of the parts of a typical neurone
> ■ Describe a reflex arc
> ■ Describe the functions of the parts of the brain
> ■ Describe how the eye works

Unit 1

What are sensitivity and coordination?

In order to survive, all living things need to respond to changes that happen around them and within them. For example, in the morning, a lizard may move into a sunny place in order to get warm. During the day, when the weather gets too hot, the lizard may move into the shade to cool down and at night it may move into a sheltered place in order to hide from predators.

Sensitivity is the name given to our ability to respond to stimuli. A stimulus is any detectable change from the internal or external environment. Animals and plants have sense organs which contain receptors. These detect stimuli and send information to control centres in the brain. The control centre activates an effector, which may be a muscle or a gland and this carries out the response.

An organism's ability to respond to stimuli is important because it:

■ protects the organism or parts of it from dangerous situations
■ helps to ensure that reproduction occurs
■ allows for growth and development of the organism, for example ability to move.

■ Sensitivity in plants

Plants do move, but at a much slower pace than animals. Although plants do not have sense organs like humans do, they are able to detect changes and respond to stimuli in their environment. Plants respond to the presence, absence or direction of:

■ light
■ water
■ pressure (touch)
■ surfaces

■ heat
■ gravity
■ nutrients.

The response may be very quick: for example, a Venus fly trap snaps shut as soon as it detects an insect in its leaves. More commonly, however, the response is slow. The plant's leaves and stems grow gradually upwards towards the light, and the roots grow downwards in the direction of gravity's pull. These movements are called **tropisms**. You will learn more about plant tropisms in Chapter 16. Plants also respond to stimuli by using growth movements.

Some flowers produce flowers that respond to light intensity by opening during the day and closing during the night.

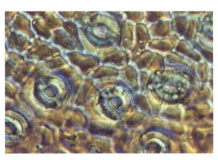

The stomata on leaves close (at night) and open (in the day) in response to light intensity.

Mature seed pods respond to water by exploding and dispersing their contents.

The leaves of the 'Shame O' Lady' *Mimosa pudica* plant respond to touch by closing their leaves quickly.

This ivy plant responds to a hard rough surface by 'sticking' to it and growing up it.

Figure 15.1 Examples of plant tropisms.

■ Sensitivity in invertebrates

Invertebrates are animals that do not have an internal skeleton, and include sponges, molluscs and worms. Like all organisms, these animals respond to stimuli, such as light intensity, heat and moisture.

Activity 15.1 Observing responses to stimuli in invertebrates

Aim: To observe responses to stimuli in invertebrates.

Equipment and apparatus

- approximately 80 to 100 mealworms
- 4 shoeboxes
- 4 lamps
- shredded paper
- water
- marker pen

Questions

1 **a** Where do mealworms usually live?
   b What conditions do you think mealworms prefer in terms of light, temperature, smell, gravity, sound, moisture and so on?
   c Formulate three predictions about what mealworms would do if put into different environments. Use the structure of 'If … then … ' statements.

Procedure

1 Take four shoeboxes and place some shredded paper in the bottom of each box.
2 Using the marker pen, label the boxes A to D (see Figure 15.2).
3 In boxes A and B, wet the paper on one side of the box, and leave the paper on the other side dry. In box C, wet all the paper; in box D, leave all the paper dry.
4 Set up the lamps so that there is a light source on one side of each box. For box A, the light source should be next to the dry side of the box. For box B, the light source should be next to the wet side of the box. For boxes C and D, the light source can be on either side.
5 Place 20 to 25 mealworms in the centre of each box.
6 Leave the mealworms for a few hours.

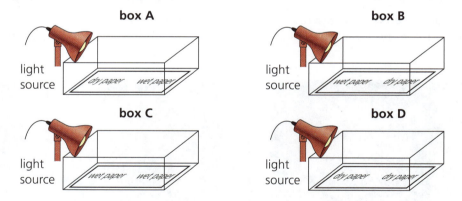

Figure 15.2 Experimental set-up for investigating what conditions are favoured by mealworms.

Questions

2 Write your observations about where the mealworms moved to in each box.
3 Choose from the list below to complete the following statement correctly. Give reasons for your answer. 'Mealworms respond to …'.
   light but not moisture   moisture but not light   both moisture and light
   neither moisture nor light
4 **a** Write your own conclusions about what the investigation demonstrates about mealworm sensitivity.
   b Suggest any improvements that could be made to the investigation.

Stimulus and response in humans

Humans have two systems that respond to stimuli: the nervous system and the hormonal system. The nervous system sends its messages using electrical impulses. The hormonal system, also known as the **endocrine system**, sends its messages using chemicals called **hormones**. Table 15.1 compares the two systems.

Endocrine system	Nervous system
Uses chemical messengers to transmit information	Uses electrical impulses to transmit messages
Produces and releases hormones	Sends impulses along nerves
Has a slower response	Has a very fast response
Tends to have long-term effects (for example growth)	Tends to have short-term effects (for example blinking the eye)
Uses specialized secretory cells found in endocrine glands	Uses specialized cells called neurones which can transmit impulses
May affect target cells far away from the glands	Target cells tend to be in close proximity to the nerve end
Only acts on target cells	Electrical impulses affect any cells that they contact
Hormones travel through the blood	Impulses travel through the neurones

Table 15.1 Endocrine and nervous systems.

Endocrine glands secrete hormones. These are released directly into the blood–stream. Examples include the pituitary gland, which secretes ADH into the blood. There are also other glands in the body that do not belong to the endocrine system – these are called the **exocrine glands** and they secrete juices that do not go directly into the bloodstream. Examples include the salivary gland, which releases saliva into the mouth when you eat.

Check your knowledge and skills

1 a Define sensitivity and draw a simple flow diagram to illustrate your definition.
  b Why is sensitivity important for animals and plants?
2 Name three examples of stimuli that cause a mealworm to move to a new area.
3 Compare and contrast the endocrine and nervous systems of the human body.

Unit 2

Nervous coordination

The nervous system performs three main functions:

- it receives sensory input (stimuli) from internal and external environments
- it integrates the input
- it responds.

The nervous system is made up of two main parts:

- **central nervous system (CNS)** – made up of the brain and spinal cord
- **peripheral nervous system (PNS)** – made up of the sense organs and the nerve cells (neurones) that are not part of the brain or spinal cord.

■ The peripheral nervous system

There are five sense organs (see Table 15.2).

Sense organ	Sensory cells	Function
Eyes	Cones and rods in the retina and the optic nerve	Sight, respond to colour and light
Ears	Hair cells in the inner ear and the auditory nerve	Hearing, respond to vibrations
Nose	Hair cells and olfactory nerve	Smell, respond to odours and fragrances
Tongue	Taste buds	Taste, respond to flavours
Skin	Various nerve endings for the different types of 'feels'	Touch/feelings, respond to texture, temperature, pressure, pain

Table 15.2 The five sense organs and their functions.

Neurones and their structure

The nervous system is made up of nerve cells or neurones. These are highly specialized cells that carry messages throughout the nervous system. A neurone can send impulses in one direction only.

- ■ **Sensory neurones** – these peripheral neurones collect information from the body and transmit it *towards* the brain and spinal cord (the central nervous system).
- ■ **Motor neurones** – these peripheral neurones transmit information *away* from the CNS.

Sensory and motor neurones are made up of a cell body, an axon, dendrites and a nerve ending (see Figure 15.3). The **cell body** forms the largest part of the neurone, containing the nucleus and much of the cytoplasm. Most of the metabolic activity of the cell is carried out here, including making ATP. The **axon** is a long fibre that carries impulses away from the cell body. Each neurone has only one axon. The axon ends in a series of small swellings called axon terminals. The **dendrites** are short branch extensions spreading out from the cell body. Dendrites receive stimuli and carry impulses from the environment or from other neurones towards the cell body.

Neurones may have dozens or even hundreds of dendrites but usually only one axon. Each axon is covered with a lipid layer known as the myelin sheath, which both insulates and speeds up transmission of action potentials through the axon. Gaps (nodes) in the myelin sheath along the length of the axon are known as the nodes of Ranvier. Impulses travel through those neurones by 'jumping' from one node of Ranvier to the next for quick transmission. The nodes of Ranvier are especially important to longer neurones, as they allow impulses to move along the axon faster.

Relay neurones

There are neurones found in the CNS that form a link between sensory and motor neurones. These are called **relay neurones** (or intermediate neurones) and they are part of the CNS. They are short, unlike the very long neurones in the peripheral nervous system.

Did you know?

An axon may be up to one metre long!

sensory neurone

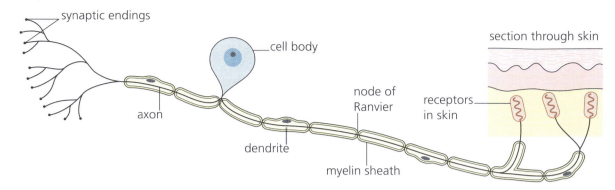

motor neurone

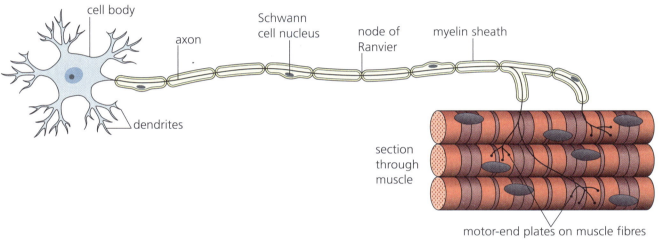

relay neurone

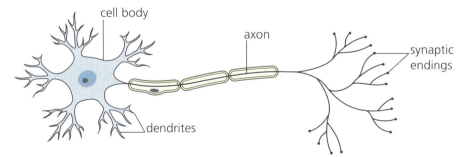

Figure 15.3 Types of neurones.

■ **Reflex actions**

The simplest type of nervous response is a **reflex action**. This is an automatic response, over which you have no conscious control. It happens very quickly because it involves only a few neurones. The simplest of all reflex actions consists of only one sensory neurone and one motor neurone – this is known as a monosynaptic reflex. Reflexes with more components are known as polysynaptic reflexes.

bright light

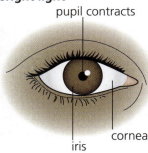

pupil contracts

iris

cornea

dim light

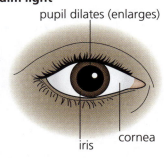

pupil dilates (enlarges)

iris

cornea

Figure 15.4 The cranial reflex of pupils dilating in dim light is an example of a reflex action.

There are two types of reflex actions (see Figures 15.4 and 15.5):

■ cranial reflexes – these are reflexes which transmit information to and from the brain, for example dilation of the pupils in dim light
■ spinal reflexes – these are reflexes which transmit information to and from the spine, for example the knee-jerk reflex.

The term reflex arc describes the nerve pathway involved in a reflex action.

Reflex responses often have a protective function. For example, if you accidentally touch a hot object, such as a hot iron, you automatically pull your hand away very fast to prevent the object from burning you (see Figure 15.5).

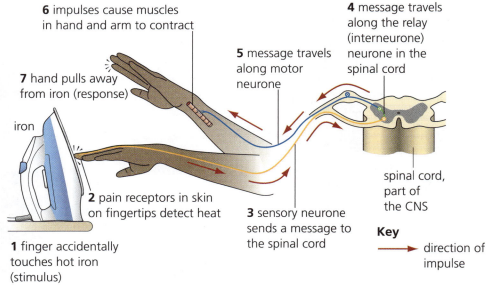

6 impulses cause muscles in hand and arm to contract

4 message travels along the relay (interneurone) neurone in the spinal cord

5 message travels along motor neurone

7 hand pulls away from iron (response)

iron

2 pain receptors in skin on fingertips detect heat

3 sensory neurone sends a message to the spinal cord

spinal cord, part of the CNS

1 finger accidentally touches hot iron (stimulus)

Key

→ direction of impulse

Figure 15.5 A spinal reflex can have a protective role in the body and is an example of a reflex action. The reflex arc is illustrated here.

Activity 15.2 Test some of your reflexes

Procedure

Work with a partner. Read the instructions that follow before you carry out each test. Observe carefully how your partner responds to each stimulus. Make notes of your observations.

Test 1 One person sits on a bench or table with their legs hanging and relaxed. Their partner taps firmly on the person's leg, just below the knee.

Test 2 One person looks straight ahead and their partner suddenly waves their hands in front of the person's eyes.

Test 3 One person kneels on a chair with their feet hanging loose. Their partner taps the back of the person's foot just above the heel.

Questions

1 What happened when you tapped your partner's knee?
2 What happened when you waved your hands suddenly in front of your partner's eyes?
3 What happened when you tapped your partner's foot?

Synapses

A **synapse** is the place where the nerve ending of one neurone meets the dendrite of the next neurone. The two neurones do not physically touch each other. There is a tiny gap between the two, called the synaptic cleft. Impulses pass from one neurone to the next across the synaptic cleft via chemical messengers called neurotransmitters.

Synapses allow impulses to travel faster between neurones and also to go in one direction only. Sometimes the ending of a neurone meets with a muscle fibre. This is called a neuromuscular junction.

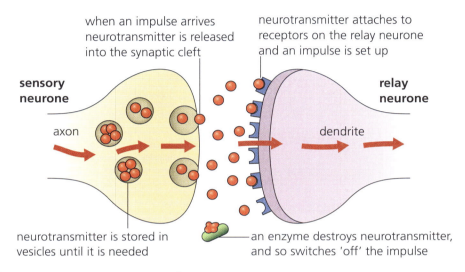

when an impulse arrives neurotransmitter is released into the synaptic cleft

neurotransmitter attaches to receptors on the relay neurone and an impulse is set up

sensory neurone

relay neurone

axon

dendrite

neurotransmitter is stored in vesicles until it is needed

an enzyme destroys neurotransmitter, and so switches 'off' the impulse

Figure 15.6 How a synapse works.

Check your knowledge and skills

1 The following is a list of actions. From your knowledge indicate whether each action is a voluntary action or a reflex action.

- Scratching your head.
- Pulling your hand away from a hot object.
- Producing saliva when you smell food.
- Blinking when dust gets in your eye.
- Crossing your legs.
- Writing a letter.
- Coughing if food goes down your windpipe.
- Clapping your hands.
- Speaking.
- Sweating in hot weather.
- Sneezing.
- Reading a book.
- Shivering in the cold.

2 Describe what happens at a synapse.

Unit 3

The brain

The central nervous system consists of the brain and spinal cord. The spinal cord runs directly into the base of the brain. The brain controls all voluntary action and some involuntary action. Figure 15.7 shows the main parts of the brain and Table 15.3 lists the functions and actions of the brain.

Area of the brain	What does it do?	Type of action
Cerebrum (or cerebral hemispheres)	Coordinates incoming or sensory messages and sends out motor messages that produce appropriate responses Also functions in speech, reasoning, emotions and personality	Voluntary
Cerebellum	Muscular coordination, balance and posture, muscle tone	Involuntary
Medulla oblongata	Contains centres which control the involuntary activities of internal organs such as heartbeat, breathing, peristalsis, temperature regulation	
Mid-brain	Functions mainly as a message centre Also involved with some sight, hearing and some responses	

Table 15.3 Parts of the brain – their functions and actions.

Did you know?

Scientists record brain activity by attaching electrodes to a person's scalp. They then connect these electrodes to a machine called an electroencephalograph (EEG). The EEG records the electromagnetic signals of brain activity as a series of wavy lines. EEGs have helped scientists study what happens in our brains while we are sleeping, thinking or performing different activities.

Discussion

Before the invention of EEG machines, how do you think scientists worked out which part of the brain was responsible for which actions?

frontal lobe – reasoning, planning, speech, movement, problem solving

parietal lobe – movement, balance, recognition, sense perception

occipital lobe – sight

temporal lobe – hearing, memory, speech

cerebellum

spinal cord

brain stem

front   back

Figure 15.7a Areas of the brain.

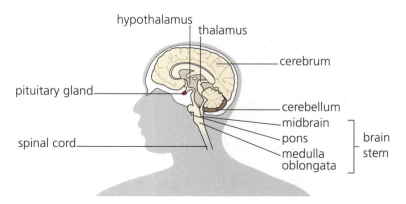

hypothalamus
thalamus
cerebrum
pituitary gland
cerebellum
midbrain
pons
spinal cord
medulla oblongata
brain stem

Figure 15.7b Parts of the brain

Figure 15.8 An EEG machine.

Biology at work

Chapter 15 – Juggle your way to a new brain in one week!

Check your knowledge and skills

1 Copy the diagram of the central nervous system in Figure 15.9. Write in the correct names of the parts A to E.
2 George was in a car accident in which he sustained injuries to his head. After the accident, he kept losing his balance and falling over. After some months of working with a physiotherapist, he gradually became able to keep his balance.
   a Which part of his brain do you think was injured?
   b What do you think had to happen in order for George to relearn how to keep his balance again?

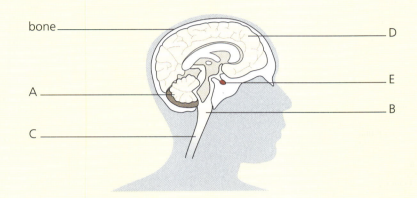

Figure 15.9 The central nervous system.

Concept mapping

Concept mapping is a useful tool in which we draw a sort of map or plan of the connections our brains make between ideas.

You will need:
■ pens   ■ small index cards or a small pad of sticky notes   ■ paper.

Choose a key word or topic for your concept map. You might want to choose a topic you have already studied in Biology: for example, sensitivity, nutrition, food storage. Remember that you can also use concept mapping for your other subjects. Now follow the steps below to create your concept map.

1 **Brainstorm and make notes.** Write your chosen key word in big letters near to the top of the page and draw a circle or oval round it. Think of as many words as possible that relate to your key word. Write all the words on a piece of paper. For this first step, don't get stuck on a single word or idea, though you might want to discuss each one briefly as you note it down.
2 **Read over and add.** Read over the words you have written down, and add any more that come to mind.
3 **Make some cards.** Write each word, including your key word, on individual index cards or small sticky notes.
4 **Rank the cards.** Arrange your cards or notes in order – from the broadest to the most specific.
5 **Cluster related information.** Group the cards together with others that relate to the same specific area.
6 **Sort into an array.** Arrange the concepts into a logically ordered array.

Remember, there are many different ways of mapping the same information, so there is no 'right' or 'wrong' map for any topic or idea. Aim to arrange your information in a way that makes logical sense to you.

Unit 4

The eye

The eye is a complex sense organ that controls our sense of sight. Light rays bounce off objects in our external environment. Inside our eyes, these rays are converted into impulses that are sent to the brain via the optic nerve. The brain then translates this information into images that we can recognize.

Did you know?

Some students get confused and think that we see because the eyes send out rays of light. This is incorrect – we see because rays of light are reflected off objects into our eyes. Hence in a darkened room we can't see objects.

The structure of the eye

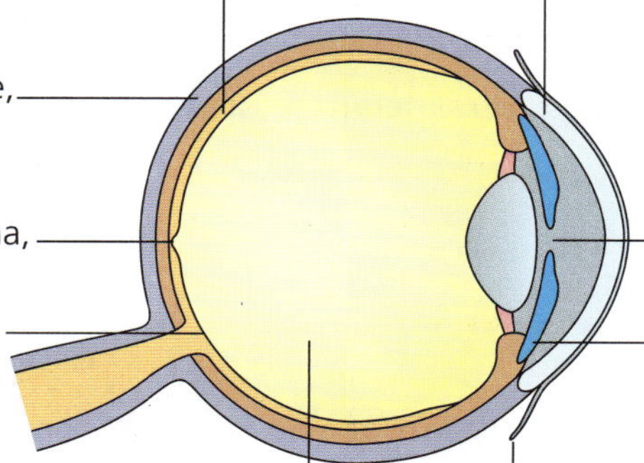

retina, the light-sensitive inner lining at the back of the eye. Rays of light enter the eye and are focused on the retina by the cornea and lens. The retina produces an image which is sent to the brain for interpretation along the optic nerve.

sclera, the tough white of the eye, which forms the outer coating of the eyeball.

fovea, most sensitive area of retina, containing only cone cells.

blind spot, part of the retina that contains no light-sensitive cells. This is where the optic nerve leaves the eye.

vitreous humour, clear, jelly-like substance that fills the inside of the eye, helping the eye to maintain its shape.

cornea, a transparent membrane which forms the outer coating at the front of the eyeball and covers the iris and pupil. It also helps to focus light on the retina.

pupil, the dark circular hole in the centre of the iris.

iris, the ring of coloured tissue surrounding the pupil. It changes the size of the pupil and allows different amounts of light to enter the eye.

conjunctiva, the mucous membrane that covers the exposed front portion of the sclera and lines the inside of the eyelids.

Figure 15.10 A cross-section of the eye.

How we see

When we see, we are responding to light rays bouncing off objects we are observing. As these light rays enter the eye, they pass through the **cornea**, the lens and the vitreous humour. The rays refract (bend) as they pass through these parts of the eye, causing the rays to converge (cross each other) at the focal point. The focal point lies exactly on the surface of the **retina** in order for our brain to process a focused image of the object (see Figure 15.11).

Did you know?

The image that falls on the retina is actually upside down, but the brain inverts the image so that we see the world the right way up!

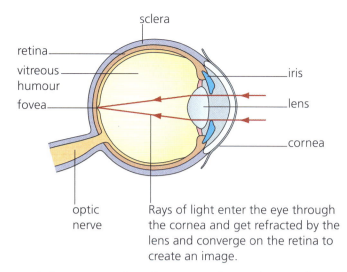

Rays of light enter the eye through the cornea and get refracted by the lens and converge on the retina to create an image.

Figure 15.11 How light rays entering the eye are focused on to the retina.

Accommodation

A healthy eye can focus on objects that are close and objects that are far away. This ability is known as **accommodation**. When the eye views objects at close distances, the muscle holding the lens relaxes and the lens become fatter. This causes the light to bend (refract) more, focusing the image on the retina. When the eye views a faraway object, the muscle holding the lens contracts and the lens is stretched thin and thus causes light not to bend as much. This will ensure that the image again falls on the retina, creating a focused image.

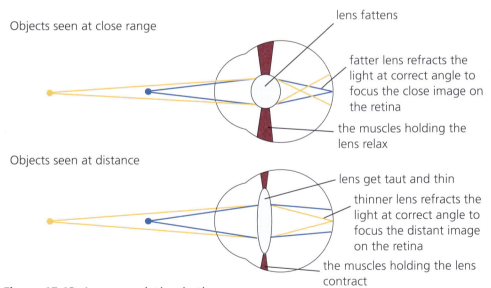

Figure 15.12 Accommodation in the eye.

The eye responds to changes in light intensity

Have you ever seen a cat's eye in daylight? Its pupil is so tiny that it is almost invisible. But at night its pupil appears so huge that the iris seems to disappear. This is because the iris is sensitive to light. In bright light, circular muscles in the iris contract to make the pupil smaller. This prevents too much light from getting to the retina and damaging it. In darker conditions, circular muscles in the iris relax to make the pupil larger (it 'dilates') allowing maximum light to get to the retina so the cat can see in dim light (see Figure 15.13, overleaf).

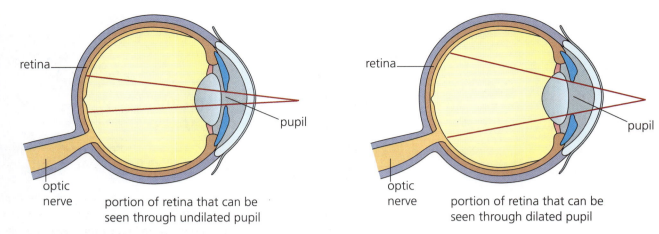

Figure 15.13 How the pupils change size in different light conditions.

Some common eye defects

Eye defects may cause a person not to be able to see properly without the aid of glasses, contact lenses or – in some cases – surgery. Problems include short sightedness (myopia), far sightedness (hypermetropia), astigmatism, cataracts and glaucoma.

Short sightedness

A person who is short sighted is able to focus clearly on nearby objects but not on objects that are further away. There may be two reasons for this:

- the eyeball may be too long from front to back, or
- the lens may be too thick, even when the ciliary muscles are relaxed.

Both these defects cause the light rays to bend too much before they reach the retina. Because of this, the light rays coming from distant objects come to a focal point in front of the retina (Figure 15.14). This is the reason the image seen is blurred.

This problem can be corrected by wearing glasses that have a concave lens. Concave lenses cause light to diverge when it passes through the lens. This means that the light rays spread out before they enter the eyes. They take further to come to a focal point and thus will do so on the retina.

Conjunctivitis

Also known as red eye or pink eye, conjunctivitis is inflammation of the conjunctiva. It is highly contagious.

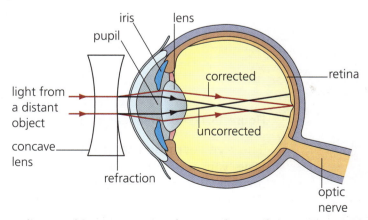

Light from a distant object comes to a focus in front of the retina. With a concave corrective lens light from a distant object is focused on the retina.

Figure 15.14 Short sightedness and how it is corrected.

Far sightedness

A person who is far sighted is able to see distant objects clearly but is unable to focus on near objects.

There may be two reasons for this:

- the eyeball may be too short from front to back, or
- the lens may be too thin, even when the ciliary muscles are contracted.

Both these defects cause the light rays to not bend enough so light rays coming from near objects come to a focal point beyond the retina (see Figure 15.15). Thus the image seen is blurred.

This problem can be corrected by wearing glasses with a convex lens. Convex lenses cause light rays to converge (bend the light rays inwards) before they enter the eyes. This means that they will come to a focal point quicker and thus do so on the retina (see Figure 15.15).

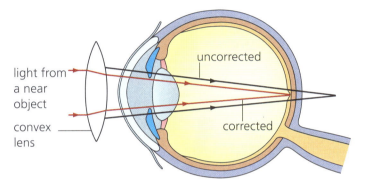

Light from a near object comes to a focus behind the retina. With a convex corrective lens light from a near object is focused on the retina.

Figure 15.15 Far sightedness and how it is corrected.

Astigmatism

An irregularly curved cornea causes visual images to be blurred. This is because the light rays are bending at different angles as they pass through the cornea, so it is difficult for all of them to come to a focal point. To correct the problem, an optometrist can prescribe specially designed glasses which have lenses with a particular curve that corrects the angle of the ray as it passes through the lens. It is also possible to have laser surgery to correct astigmatism. This is when an ophthalmic surgeon uses a laser beam to re-shape the cornea.

Cataract

As it gets older, the lens can become cloudy or opaque, making light unable to pass through it. This clouding is called a cataract, and it can cause partial or full blindness. An ophthalmologist can perform surgery to remove the lens and replace it with a synthetic one. In some cases, the surgeon removes the lens completely and the patient uses special glasses to help them to see.

Glaucoma

Glaucoma is a disease of the eye that occurs when too much fluid builds up in front of the lens. This build-up puts pressure on the aqueous humour. The pressure may damage the optic nerve, causing the affected person to have a reduced field of vision. This condition is associated with age and heredity, and other medical problems such as stroke. It can be corrected with the use of medication. Glaucoma is the most common cause of blindness.

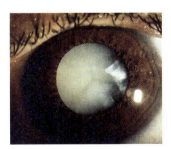

Figure 15.16 In a person suffering from cataracts the cloudiness of the eye indicates that light will not be able to pass through.

Check your knowledge and skills

1 Look at the diagrams below that show the front of the eye.
   a Copy the diagrams and label the pupil and iris on your drawing.
   b Which diagram shows an eye in bright light? Explain how you worked out your answer.
   c Why is this reflex (called the pupil reflex) important for protecting the eye?

undilated pupil   dilated pupil

Figure 15.17

Unit 5 Chemical coordination

The body has two main ways of coordinating or responding to stimuli: through electrical impulses and through chemicals known as hormones. Chemical coordination involves the endocrine system. The endocrine system is made up of different endocrine glands that secrete hormones (see Figure 15.18).

Endocrine gland	Location	Hormones released and their functions
Thyroid	In the neck	Calcitonin, regulates the calcium levels in the body Thyroxine, increases metabolic rates
Pancreas*	Behind the stomach, below the liver	Insulin, reduces glucose levels in the blood Glucagon, increases glucose levels in the blood
Gonads – the testes (males)	In the genital area, hanging outside of the body	Testosterone, develops secondary sexual characteristics in boys, stimulates the production of sperm
Gonads – the ovaries (females)	Internal, just below the abdominal area	Oestrogen, develops secondary sexual characteristics in girls, such as breasts, etc. and prepares the body for possible pregnancy every month Progesterone, inhibits the development of new follicle, preventing further ovulation
Pituitary**	In the brain, beneath the hypothalamus	Thyroid-stimulating hormone (TSH), stimulates the making of thyroxine Growth hormone, stimulates growth Antidiuretic hormone (ADH), regulates water levels in the body Follicle-stimulating hormone (FSH), luteinising hormone, prolactin and oxytocin, work in association with the gonads to perform various roles in the menstrual cycle, pregnancy, birth and lactation
Adrenals	On the upper portion of each kidney	Adrenaline, prepares the body for immediate action in stressful situations (also known as the fight-or-flight response)

Table 15.4 Some of the endocrine glands, their hormones and the functions.
* The **pancreas** is both an endocrine and an exocrine gland. It produces pancreatic juice which is released into the small intestine as part of digestion; and it also produces insulin, a hormone which helps to regulate the body's glucose level.
** The **pituitary** is often referred to as the 'master gland' as it produces so many hormones. The hypothalamus controls much of the functioning of the pituitary gland, either directly through nerve stimulation or indirectly through hormonal activation. The hypothalamus itself produces many more hormones.

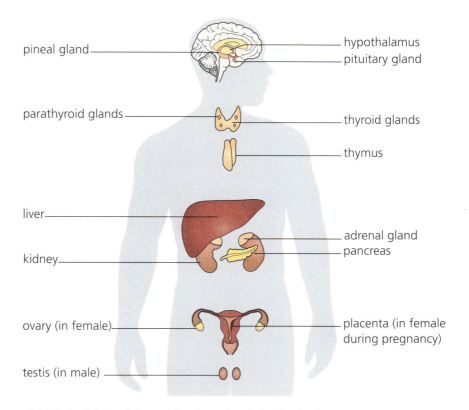

Figure 15.18 Position of the endocrine glands in the body.

The endocrine system broadcasts its messages to all cells by secreting hormones into the blood and extracellular fluid. Like a radio broadcast, a receiver is needed to pick up the message – in the case of endocrine messages, cells must have a receptor for the hormone being 'broadcast' in order to respond.

Most hormones circulate in the blood and come into contact with all cells. However, a given hormone usually affects a limited number of cells, which are called **target cells** – only these cells have receptors to the specific hormone.

Check your knowledge and skills

1 Name four endocrine glands, their secretions and importance in the body.
2 Copy the outline of a person on the right and draw in the positions of the endocrine glands.
3 Copy and complete the following paragraph.

The central nervous system (CNS) is made up of the brain and the _____. These are protected by the _____ and ____ ____ respectively. The ____ neurones take electrical impulses to the central nervous system. The impulses are then passed on to the _____ neurones. When a decision is made the impulse leaves the CNS via the _____ neurone. The junction between two neurones is called a _____.

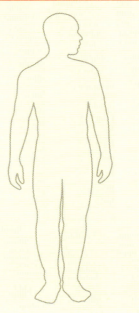

Figure 15.19

4 a Name the parts labelled **A** to **G** in the diagram.
  b Explain what is happening at points **A** to **G**.
5 Draw and label a typical motor neurone. Include the following labels on your diagram: cell body, axon, myelin sheath, node of Ranvier, dendrites.
6 Name the parts of the nervous system described below.
  a carries impulses from the spinal cord to an effector
  b controls learning and memory
  c gap between two neurones
  d area responsible for controlling balance
  e area responsible for controlling breathing, heartbeat and blood pressure.
7 Describe four ways in which the nervous system differs from the endocrine system.
8 Name the hormone responsible for each response below, and name the gland that secretes the hormone.
  a development of male secondary sexual characteristics
  b flight or fight response
  c lowering of glucose level in the body
  d lowering the water content of the body
  e growth.

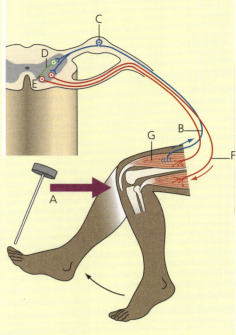

Figure 15.20

Chapter summary

Do you know?

If you are unsure of any of the facts in the list, refer to the page number in brackets.

■ The nervous system controls and coordinates your actions. (page 155)
■ Sense organs are responsible for detecting stimuli and sending messages to the CNS via the sensory neurones. (page 156)
■ A reflex arc is one of the simplest nerve pathways. It is usually protective and gives a very fast response. (page 157)
■ The junction between two neurones is called a synapse. Synapses help to speed up nerve impulses and ensure that impulses travel in only one direction. (page 159)
■ Different parts of the brain are responsible for specific things. (page 160)
■ The endocrine system is also a control and coordinating system, involving the secretion of hormones. Hormones normally have a slower response. (page 166)
■ Hormones are produced by endocrine glands and secreted in response to a stimulus which is usually internal. (page 166)
■ Hormones travel through the blood and only affect their target cells. (page 166)

Are you able to?

If you have trouble in doing these things, refer to the page number in brackets.

■ Compare the endocrine system and the nervous system. (page 155)
■ Draw and label a typical neurone from memory. (page 157)
■ Describe examples of reflex actions, and explain what happens during a reflex action. (page 158)
■ Name the main areas of the brain and list the functions for which they are responsible. (page 160)
■ Label a diagram showing the structure of the eye. (page 162)
■ Draw diagrams of the eye showing how short and long sight can be corrected. (pages 164 and 165)
■ Name the main endocrine glands. (page 166)

Chapter

16 Support and movement

Keeping upright and moving around

When a builder constructs a house, there is always a frame or support structure that gives the building its shape, and then helps it to hold this shape. If parts of the building need to move (for example the doors, windows), the shape and materials of the structure need to allow for this movement. Similarly, every living organism has a specific structure that holds it, or supports its shape and movement.

In this chapter you will learn about the systems that provide support and movement in plants and animals.

By the end of this chapter you should be able to:

- ■ Describe support systems in animals and plants
- ■ Relate the structure of the skeleton to its functions in humans
- ■ Describe movement in a human limb (including joints and muscles)
- ■ Distinguish between cervical, thoracic and lumbar vertebrae
- ■ Describe phototropism and geotropism

Unit

1 Support systems in plants and animals

Usually, we use the word **skeleton** to refer to animal skeletons. However, all living organisms, from the smallest unicellular organisms to the largest animals and plants, need a physical structure that can support them in two ways:

- ■ to keep their shape
- ■ to enable movement.

Did you know?

The smaller the organism, the less support it needs. Unicellular organisms, such as *Amoeba*, have a tiny body mass which is easily supported by the plasma membrane that encloses the cell fluid and also by the water in which the organism lives.

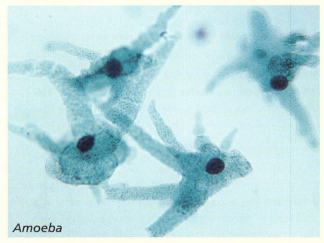
Amoeba

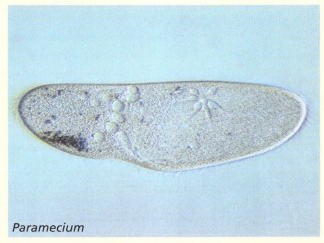
Paramecium

Figure 16.1 These unicellular organisms are small enough to be supported by their plasma membranes and by the water that surrounds them.

Did you know?

Small aquatic organisms rely on the water in which they live for support. However, most organisms require a physical support structure or frame for support and movement. When this frame of supporting tissues is formed external to the animal's soft organs, it is called an **exoskeleton** (external skeleton). When the frame is inside the animal's body, it is called an **endoskeleton** (internal skeleton) (see Figure 16.2).

Insects, crabs and worms are examples of animals that have exoskeletons. Vertebrates such as mammals, fish, birds, reptiles and amphibians all have endoskeletons.

Figure 16.2 Which of these animals have an exoskeleton and which have an endoskeleton?

Plant skeletons

Plant skeletons do not need to move as much as animal skeletons do. Their main function is supporting the organism – keeping its shape and position. The stems, roots and leaves of most plants contain a woody tissue called **xylem**. (In Chapter 11, you learnt about xylem and its role in transporting nutrients around the plant.) Xylem is made up of cells that have very thick cell walls, reinforced with cellulose and a substance called lignin. As plants grow bigger and taller, they need more support to hold up their weight. The wood that makes up the trunk, bark and branches of large trees is composed of xylem. The roots also grow outwards underground to help form a stable base that anchors the plant in the ground.

Did you know?

The world's tallest tree species is the California redwood (*Sequoia sempervirens*), found along the Pacific coast of North America. Some redwoods reach over 100 metres in height. These giant trees have huge root systems that anchor the tree.

Key fact

The parts of a plant that have more xylem tend to be hard and woody. Parts that have less xylem are softer and fleshier. These parts of the plant hold their shape in a different way, known as **cell turgor**. When a plant gets enough water, each cell has enough cytoplasm to fill its walls tightly. The liquid inside the cells presses outwards on the cell walls, keeping the shape of the cell firm or turgid. The cells, in turn, press against each other, keeping the plant firm and holding its shape.

Figure 16.3 A California redwood (also known as a Coast redwood).

Activity 16.1 Studying support in plants

Aim: To investigate the roll of cell turgidity in plant structure.

Questions

1 Describe the appearance of the leaves and stems of the plant in Figure 16.4.
2 Why do you think the main stem of the plant is still upright?
3 What do you think will happen if the plant in the picture:
   a is watered?
   b does not get water for several more days?
   c does not get water for several more months?
4 Besides the internal cell structure of the stems and leaves, what other important structure helps to keep this plant upright?

Figure 16.4 A drooping pot plant.

Did you know?

In most animals the axial skeleton is horizontal.

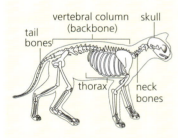

Figure 16.5 Skeleton of a cat, showing its horizontal axial skeleton.

The human skeleton

Our skeletons are made from bone and cartilage. Bone is a hard kind of tissue that makes up most of the skeleton. **Cartilage** is the softer bony tissue that makes up our ears and noses. It is also found at the ends of bones, where bones meet to form joints. Here it acts as a kind of shock absorber and also prevents friction between the bones as they move over or against each other. **Tendons** are tough, fibrous strands of tissue that join muscles to bones. **Ligaments** are elastic fibres that join one bone to another and also allow movement of the joints.

The skeleton can be divided into two main parts:

- **axial skeleton** – this is the bones of the skull, thorax and vertebral column
- **appendicular skeleton** – this is the bones of the arms and legs, the pectoral girdle and the pelvic girdle.

In humans, the axial skeleton is vertical; whereas in most other animals it is horizontal.

Functions of the human skeleton

The main functions of the human skeleton (see Figure 16.6 overleaf) are to:

- protect the soft internal organs
- support the body and provide its shape and form
- enable movement (locomotion)
- produce blood cells.

The vertebral column

The central support structure of the human body is a column of 33 small bones called **vertebrae** (singular 'vertebra'). The vertebrae are stacked on top of each other. Discs of fibrous cartilage lie between each of the cervical, thoracic and lumbar vertebrae. These discs cushion the vertebrae against shock and allow the bones to move against each other without friction.

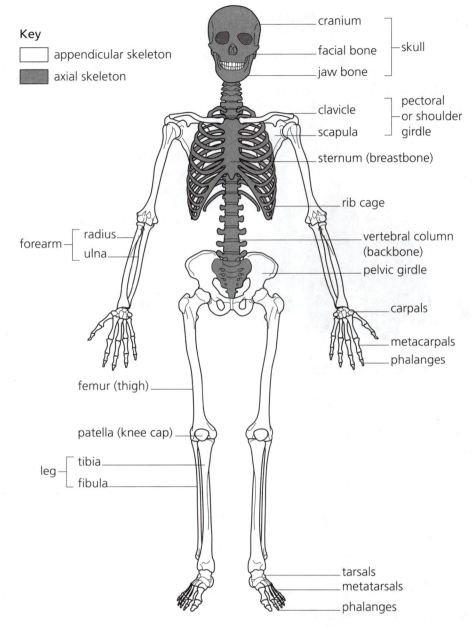

Key

☐ appendicular skeleton

■ axial skeleton

cranium — skull
facial bone —
jaw bone —

clavicle — pectoral or shoulder girdle
scapula —

sternum (breastbone)

rib cage

vertebral column (backbone)

pelvic girdle

carpals

metacarpals
phalanges

forearm — radius, ulna

femur (thigh)

patella (knee cap)

leg — tibia, fibula

tarsals
metatarsals
phalanges

Figure 16.6 In humans, the axial skeleton is vertical (upright).

From top to bottom, we can divide the vertebral column into:

- seven cervical vertebrae (C1 to C7)
- 12 thoracic vertebrae (T1 to T12)
- five lumbar vertebrae (L1 to L5)
- five fused vertebrae of the sacrum (S1 to S5)
- four fused vertebrae of the coccyx (CO1 to CO4).

These are shown in Figures 16.7 and 16.8.

atlas (C1)

The top two vertebrae are different to the others. The atlas (C1) is the topmost vertebra. It is a ring of bone that allows the head to nod up and down.

dens

axis (C2)

The second vertebra is the axis (C2). It has a blunt upward projection like a thumb or tooth. This is called the dens, and it allows the atlas and the skull to rotate.

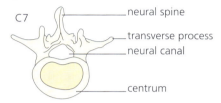

C7

neural spine

transverse process

neural canal

centrum

The rest of the cervical vertebrae increase in size from C3 to C7. They are small, with articulating surfaces that allow the head to move in different ways. They have two small holes for arteries.

a cervical vertebrae

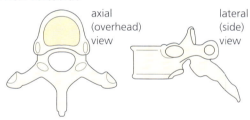

axial (overhead) view

lateral (side) view

The thoracic vertebrae increase in size from T1 to T12. Each thoracic vertebra has a long neural spine that attaches to the upper back muscles. Long transverse processes join the thoracic vertebrae to the ribs.

b thoracic vertebrae

Figure 16.7 Vertebrae of the human spine.

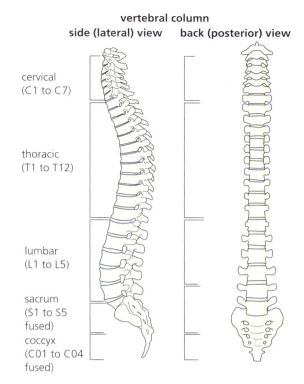

vertebral column

side (lateral) view   **back (posterior) view**

cervical (C1 to C7)

thoracic (T1 to T12)

lumbar (L1 to L5)

sacrum (S1 to S5 fused)

coccyx (C01 to C04 fused)

The lumbar vertebrae form the biggest and strongest vertebrae. They have a larger centrum in order to support the weight of the body.

axial (overhead) view

lateral (side) view

c lumbar vertebrae

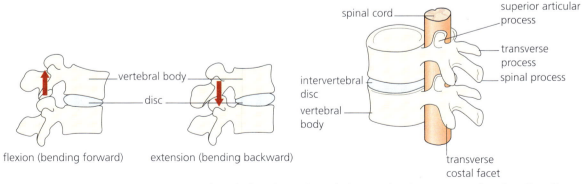

vertebral body

disc

flexion (bending forward)

extension (bending backward)

spinal cord

superior articular process

transverse process

spinal process

intervertebral disc

vertebral body

transverse costal facet

The spinal cord runs through the neural cavity. A spongy disc of cartilage lies between the vertebrae, cushioning them against shock and friction.

a vertebrae allow movement   **b** vertebrae link together to form the vertebral column

Figure 16.8 How vertebrae connect and move.

Activity 16.2 Studying the vertebral column

Aim: To study the vertebral column.

Questions

1 Copy and complete the table below.

Part of the skeleton	Soft organs it protects
Skull	
Ribs	
Vertebrae	
Pelvis	

2 Match each of the vertebrae listed below to one of the following functions and then describe how its shape is adapted to its function:

supporting the weight-bearing part of the body
supporting the head and neck
anchoring the ribs

a Cervical vertebra   **b** Thoracic vertebra   **c** Lumbar vertebra.

Bone tissue

Your bones are made from living tissue. They store minerals such as calcium and phosphorus. Deposits of calcium salts, mainly calcium phosphate, give bone its rigidity and strength. Protein fibres help prevent the bone from breaking easily. Your red blood cells and some white blood cells are produced in the soft, inner part of the bone, called the **bone marrow**.

Check your knowledge and skills

1 Explain how unicellular organisms hold their shape.
2 Give two examples of each of the following:
   a animals that have an endoskeleton
   b animals that have an exoskeleton.
3 Briefly describe the following and explain how they help plants to keep their shape:
   a xylem   **b** cell turgor.
4 How do the axial skeletons of animals differ from those of humans?
5 Make a sketch of Figure 16.9 and name the parts labelled **A** to **E**.

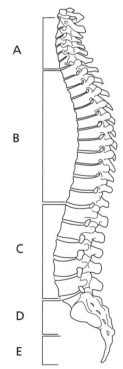

Figure 16.9

Plant movements

We can see plant movements most easily in a germinating seedling. If all the conditions for germination are satisfied, the shoot grows towards the light and the root grows downwards in the direction of gravity. Most plant movements are therefore usually **tropisms** such as phototropism or geotropism (see below). Growth towards a stimulus is called a positive response. Growth away from a stimulus is a negative response. Plant shoots respond positively to light and negatively to gravity; plant roots respond positively to gravity and negatively to light.

Phototropism and geotropism

Plants are sensitive to light, gravity and moisture. **Phototropism** is growth in the direction of a light source. Plant shoots respond positively to light. When light shines on a plant from one direction, the shoot will grow towards the light source. Special chemicals called **auxins** are responsible for this effect. Auxins found in the growing areas of the plant stimulate the elongation of cells. Auxins tend to move away from light, towards the shady areas of the shoot. These cells elongate faster than the other cells, causing the shoot to bend towards the light source.

Geotropism is plant growth in response to the pull of gravity. This is one of the reasons the roots of plants grows downwards. The auxins responding to the pull of gravity settle on the underside of the root. Auxins found in the root cells inhibit cell elongation. The cells in the upper part of the root will grow faster than those in the underside; so if the root is lying horizontally, the tip of the root will start to turn downwards.

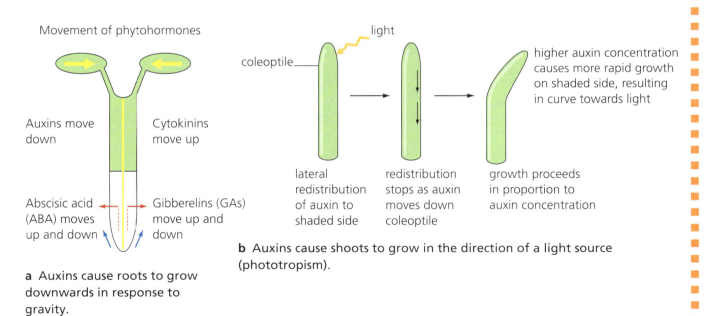

Movement of phytohormones

Auxins move down

Cytokinins move up

Abscisic acid (ABA) moves up and down

Gibberelins (GAs) move up and down

a Auxins cause roots to grow downwards in response to gravity.

light

coleoptile

higher auxin concentration causes more rapid growth on shaded side, resulting in curve towards light

lateral redistribution of auxin to shaded side

redistribution stops as auxin moves down coleoptile

growth proceeds in proportion to auxin concentration

b Auxins cause shoots to grow in the direction of a light source (phototropism).

Figure 16.10 The way auxins affect plant growth.

Check your knowledge and skills

1 Name three adaptations shown by plants that enable them to move.

Animal movements

Animals move far more than plants do, and in a wider range of ways. Walking, running, swimming, jumping, flying, crawling, and burrowing – these are just a few of the ways that animals move. Animals need to move for several reasons.

- Nutrition – animals move in order to find food. Herbivores move around to find fresh plant matter to feed on; carnivores have to hunt their prey.
- Reproduction – animals move around to find a mate and reproduce. They also move to protect their young from predators and competition, and as they grow, the young move away from their parents.
- Safety – animals move to get away from dangerous situations. They may need to escape predators or environmental dangers such as heat, cold, fires or floods.
- Environmental conditions – animals move away from extreme conditions such as extreme heat, cold, dry or wet in order to find conditions more suitable for survival.

Joints

A joint is any point where two bones (or parts of the skeleton) attach to each other. There are three types of joint.

- Immoveable joints (also known as sutures) – the bones in these joints are fused together and do not allow movement, for example cranium, pelvic girdle.
- Partially moveable joints – these joints allow limited movement. They include the gliding joints, for example ankles and wrists, and pivot joints, for example the joint between the axis and atlas at the top of the spine.
- Synovial joints or moveable joints – these joints allow the widest range of movement. They include hinge joints, for example elbow, knee and finger joints, and ball-and-socket joints, for example shoulder and hip joints.

Synovial joints connect in a way that permits movement. This is also known as an articulation. Joints perform three important functions.

- They hold the ends of the bones together, preventing the bones from separating or dislocating.
- They reduce friction between two moving bones.
- They absorb shock between two bones.

The structure of synovial joints

A joint that can move is known as a synovial joint (see Figure 16.11). It contains an oily liquid called synovial fluid which lubricates the ends of the bones. At the end of each bone, there is a layer of cartilage that acts as a rubbery cushioning layer between the bones. The synovial fluid fills the joint cavity (space) around the cartilage. Each joint is enclosed in a fibrous capsule, with strong ligaments that connect one bone to another across the joint. The ligaments are extremely strong so that they can withstand significant pulling and pushing forces without breaking. They are also able to stretch slightly to allow the joint to bend without dislocating the bones.

Ball-and-socket joint

A **ball-and-socket** joint gets its name from its shape. One bone has a ball shape at its end, which fits into a socket shape on the other bone. This kind of joint allows movements in three planes (up and down, side to side, round). It can easily be dislocated (pulled out of position) if it is pulled too forcefully (see Figure 16.12).

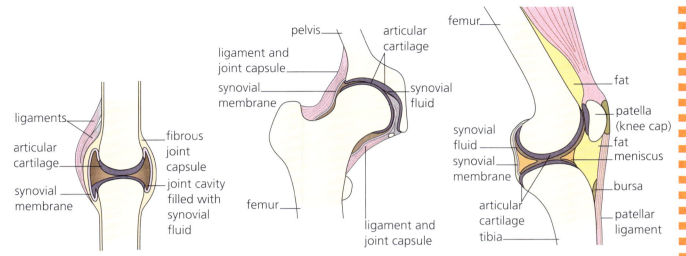

Figure 16.11 The structure of a synovial joint such as the wrist joint.

Figure 16.12 The structure of a ball-and-socket joint, the hip joint which connects the pelvis and the femur (thigh bone).

Figure 16.13 The structure of a hinge joint, the knee, which connects the femur and the tibia.

Hinge joint

A **hinge joint** allows movement in only one plane, for example up-down movement or side-to-side movement. It does not dislocate as easily as a ball-and-socket joint and is therefore more suitable for carrying heavy loads.

◼ Muscles

In order to move, we need to contract and relax pairs of **antagonistic muscles**. These are pairs of muscles that work in a coordinated way – one contracts and the other relaxes. A contraction involves shortening and tightening of a muscle. When a muscle relaxes, it lengthens and softens.

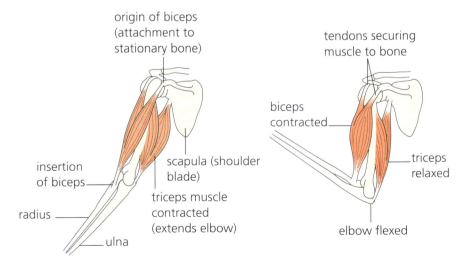

Figure 16.14 How antagonistic muscles facilitate movement.

Activity 16.3 Creating a model to demonstrate muscle action

Aim: To create a model to demonstrate muscle action.

Equipment and apparatus

- cardboard or ice-cream sticks
- split pin, paper fastener or short length of wire
- small elastic band

Procedure

1 Cut the cardboard to represent the bones of the upper and lower arm.
2 Carefully punch a hole in the cardboard where you will join the two bones.
3 Cut a small groove at the far back end of the upper part, and another small groove in the middle of the bottom part, as shown in Figure 16.15.
4 Use the split pin or length of wire to connect the two parts.
5 Arrange the elastic band around the connected 'arm' as shown in Figure 16.15.

Questions

1 a What happens when you pull the 'bones' in your model to an open, L-shaped position, and then close them to an acute angle?
  b Explain how this demonstrates the muscle action of the arm.
2 In what ways does your model differ from the structure of a real arm?

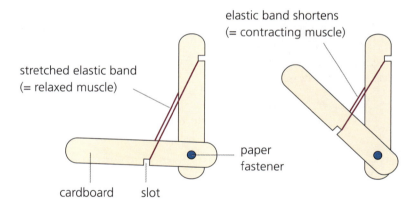

Figure 16.15 Preparing a model of the upper and lower arm.

Check your knowledge and skills

1 Explain the role of the following in helping plants to keep their shape:
   a xylem   b cell turgor.
2 List the three main reasons that animals need to move.
3 Explain briefly how auxins cause plant movements. Draw simple diagrams to illustrate your answer.
4 Draw a labelled sketch of each of the following:
   a a cervical vertebra   b a thoracic vertebra   c a lumbar vertebra
5 What is the main difference between cartilage and bone?
6 What is a synovial joint?
7 Explain what we mean by antagonistic muscles.

Chapter summary

Do you know?

If you are unsure of any of the facts in the list, refer to the page number in brackets.

- All living organisms rely on some sort of structure to support them and help them keep their shape. (page 169)
- An external skeleton is called an exoskeleton. (page 170)
- An internal skeleton is called an endoskeleton. (page 170)
- The stems, roots and leaves of most plants contain woody xylem that helps them to hold their shape. (page 170)
- The softer parts of a plant stay firm and in shape because of cell turgor. (page 170)
- Human skeletons are made from bone and cartilage. (page 171)
- The axial skeleton includes the central bones of the skull, thorax and vertebral column. (page 171)
- The appendicular skeleton includes the bones of the arms and legs, the shoulder girdle and pelvic girdle (the appendages). (page 171)
- The vertebral column protects the spine and helps us to stay upright. It is composed of 33 vertebrae – the cervical vertebrae, thoracic vertebrae, lumbar vertebrae, sacrum and coccyx. (page 173)
- Plants move in response to gravity, water and light. These movements are called tropisms. (page 175)
- Auxins are hormones that cause plant tropisms. (page 175)
- Animals have three kinds of joints – immoveable joints, partially moveable (or gliding) joints and synovial joints. (page 176)

Are you able to?

If you have trouble in doing these things, refer to the page number in brackets.

- List the importance of the human skeleton in terms of protection, support, locomotion and blood formation. (page 171)
- Identify the parts of the human skeleton, and label sketches of it. (page 172)
- Compare the differences between the different kinds of vertebrae, and relate their form (size, surface, projections) to their functions. (page 174)
- Observe how plants respond to different conditions and describe the response of their stems and roots. (page 175)
- Describe how the muscles and bones interact when we move. (page 177)

Chapter 17 Growth and development

Getting bigger and more complex

When organisms grow, they get bigger. But growth does not just mean increasing in size. Think about a frog puffing up to attract a mate. These are temporary changes in size. Growth involves a more permanent increase in size and mass.

Most living organisms begin life as a single cell. Animals and plants, too, begin as a single fertilized cell, and then develop and grow into bigger, more complex organisms, with trillions of cells and complex organs and tissues. This is all possible because of growth and development.

In this chapter you will learn about the processes that enable plants and animals to grow and develop.

By the end of this chapter you should be able to:

- Outline the process of mitosis
- Describe the role of mitosis in growth
- Perform investigations that demonstrate growth in living organisms
- Describe the structure of a dicotyledonous seed
- Describe the processes taking place within a seed during germination
- State the functions of hormones in controlling growth and development

Unit 1 Basic mechanisms of growth

Growth is a permanent increase in the number of cells that makes up an organism and thus an increase in its mass. During your life, you grow from a tiny foetus to a small baby to a child, and your body is still growing. Even when your body has reached its adult size, you will continue to grow new cells throughout your life. A plant may begin life as a tiny germinating seed. As its cells divide and grow, it develops shoots and roots, and gradually increases its size and mass. Later in their life, organisms may experience a decline in their growth rate as they age.

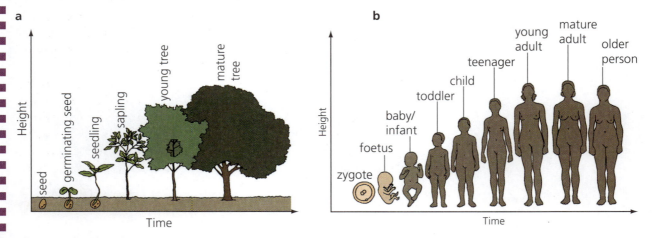

Figure 17.1 Living organisms grow throughout their lives; **a** shows stages in plant growth from germination to a mature tree; **b** shows stages in human growth from foetus to adult.

Cell division and cell enlargement ensure that an organism can grow (increase its number of cells and its mass). Cell differentiation ensures that an organism can develop (become more complex).

For growth to occur, both cell division and cell elongation are required. Cell division happens when a cell splits into two whole new cells. In order for cell division to take place, the cell must undergo a process called **mitosis**. In mitosis cell division creates two new cells, each half the size of the original.

By itself, cell division is unable to produce growth. For growth to occur, the two new cells produced by cell division must now increase in size. Cell elongation occurs following synthesis of new material and continues until each new cell has reached the size of the original cell. The organism is now twice the size of the original as illustrated in Figure 17.2.

Similarly, cell elongation cannot cause growth on its own, because the cell is likely to become too large to function efficiently. As cells become larger, the surface area to volume ratio decreases and diffusion occurs more slowly. The cell therefore divides after reaching the maximum size that allows for efficient diffusion. Efficient diffusion is necessary to supply materials such as oxygen to cells for respiration, and to remove toxic materials such as carbon dioxide.

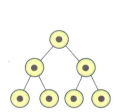

cell division increases the number of cells

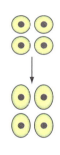

cell elongation increases the size of cells

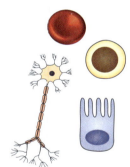

cell differentiation changes the shape and function of the cells

Figure 17.2 The general mechanisms of growth.

Unit 2

Mitosis

Mitosis is a type of nuclear division that takes place when a single cell copies its genetic material and splits into two identical cells with identical nuclei. The production of identical cells makes mitosis important for growth.

Did you know?

Another name for mitosis is karyokinesis. Karyokinesis refers specifically to the division of the cell's nuclear material (as opposed to its cytoplasm). The name comes from 'karyo', referring to eukaryotes and 'kinesis' meaning movement. Eukaryotes are organisms such as animals and plants whose cells contain protein complexes including a nucleus and organelles. Prokaryotes are simpler organisms, such as bacteria, which are unicellular and lack a nucleus and the other complex proteins found inside eukaryotic cells. Prokaryotes grow by a process called binary fission.

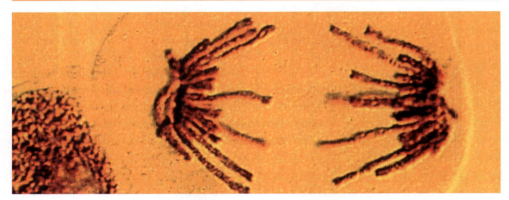

Figure 17.3 A photograph of a dividing cell during mitosis. The cell isn't really these bright colours – the specimen has been stained in order to show the structures.

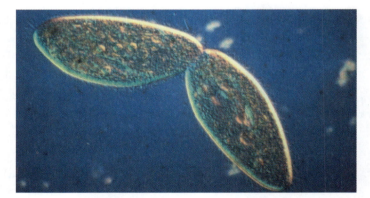

Figure 17.4 Binary fission is a simple form of cell division which takes place in unicellular organisms such as bacteria.

Mitosis has five main stages:

- **prophase**
- **prometaphase**
- **metaphase**
- **anaphase**
- **telophase.**

At the beginning and end of each cycle, the cell is in **interphase**.

Activity 17.1 Investigating mitosis in plant roots

Aim: To investigate mitosis in plant roots.

SBA skills

Observation/recording/
reporting (ORR)
Manipulation/
measurement (MM)
Analysis and
interpretation (AI)

Equipment and apparatus

- large garlic clove
- conical flask or boiling tube
- water
- small beakers
- scissors
- cutting tile
- acetic alcohol
- distilled water
- 1 M dilute hydrochloric acid
- microscope with magnifications of ×100 and ×400
- toluidine blue stain
- safety glasses

- hollow glass block
- watch glass
- filter paper
- toilet tissue
- pair of fine forceps
- dropping pipette
- clean microscope slide
- cover slip

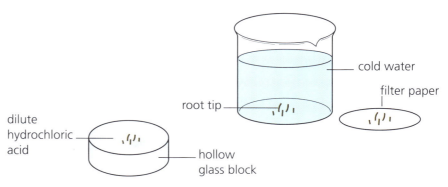

1 Support a garlic clove over water. Leave until roots develop (3–4 days).

2 Cut off the root tips 1–2 cm long. Put them in a small volume of acetic alcohol for 10 minutes.

3 Wash root tips in cold water for 4–5 minutes then dry them on filter paper.

Figure 17.5 Steps in the experiment to investigate mitosis in garlic root tips.

Procedure

1 Place a garlic clove in the neck of a conical flask or boiling tube over some water for a few days until it develops roots.
2 Using scissors, cut off the root tips, about 2 cm from the tips. Place them in a small volume of acetic alcohol for 10 minutes.

Notes

- Use 3 parts of absolute alcohol to 1 part acetic acid. You may also use 95% ethanol instead of absolute alcohol, but the chromosomes may not be as clearly visible in your results. Be careful when working with acid.
- The root tip contains meristematic tissue – the tissue where mitosis takes place. In a normal root this tissue is three-dimensional and you cannot easily see individual cells. To observe individual cells, we separate the cells out into a thin layer.
- Plant cells are connected by a layer of calcium pectate. Hydrochloric acid dissolves this calcium pectate but leaves the cellulose cell wall intact. The acid also kills and fixes the cell contents in position. Toluidine blue stains the cell nuclei and not the cytoplasm. Your teacher will make up a 0.5% solution.

3 Wash the root tips in cold water for about 5 minutes. Leave them to dry on some filter paper.

4   Place a few drops of distilled water into the watch glass. Place a few drops of 1 M dilute hydrochloric acid into the hollow glass block. (Using a hollow glass block is a safe way to work with acid. You can easily see which is acid and which is water. The hollow glass block is also stable and unlikely to spill.) Wear safety glasses.

5   Examine the garlic root tips. The last 2 mm of each root is a creamier colour than the rest. This is the part you want as it contains the tissue that has mitotic activity.

6   Using scissors, cut off the last 2 mm from about six roots and carefully place them in the acid.

7   Wait about 3 minutes. Using the forceps, carefully transfer all of the tips to the distilled water. They can stay in the water. Place the hollow glass block with acid safely out of the way.

8   Fold about six sheets of soft toilet tissue. Pick out two of the root tips from the water and place them at the centre of a clean microscope slide. Remove any water using a corner of toilet tissue to soak up the water.

9   Using the dropping pipette, add one drop of toluidine blue stain.

10 Place a cover slip over the stain. Tap it carefully with the forceps to separate out the cells. The root tip cells should separate out but the cells in the middle should appear white and unstained.

11 Carefully push the cover slip so that a corner sticks out at the edge of the slide. Lift it slightly with the forceps to allow the stain to wash over the cells. Repeat two or three times. (You want to stain the cells but not to over-stain or under-stain them.) Reposition the cover slip over the slide, and tap it gently again using the forceps.

12 Place the slide on the tissue paper, fold the tissue over the top and press gently to absorb excess stain. There will be a lot of air bubbles.

14 Examine your slide under a microscope at a medium magnification ($\times 100$). Search for cells that are box-shaped and in chains. The cytoplasm should appear nearly colourless and the nuclei dark blue.

15 Now change to high power ($\times 400$) and look for dividing cells.

If your preparation does not work (for example if the cells are over-stained or did not separate out), discard the slide in a beaker of water. Repeat the staining process using two more root tips from the water, adjusting your procedure to remedy the problem (for example if the cells were too crowded and overlapping, modify step 10; if the cells were over- or under-stained, modify step 11).

Questions
1 Why did you soak the roots in acetic alcohol?
2 What special tissue is found at the root tips?
3 What did you notice when you examined the cells under the microscope?
4 What conclusions did you reach from your observations?

Check your knowledge and skills
1 Using Figure 17.6 to help you, draw a flow diagram outlining the main stages in cell mitosis.

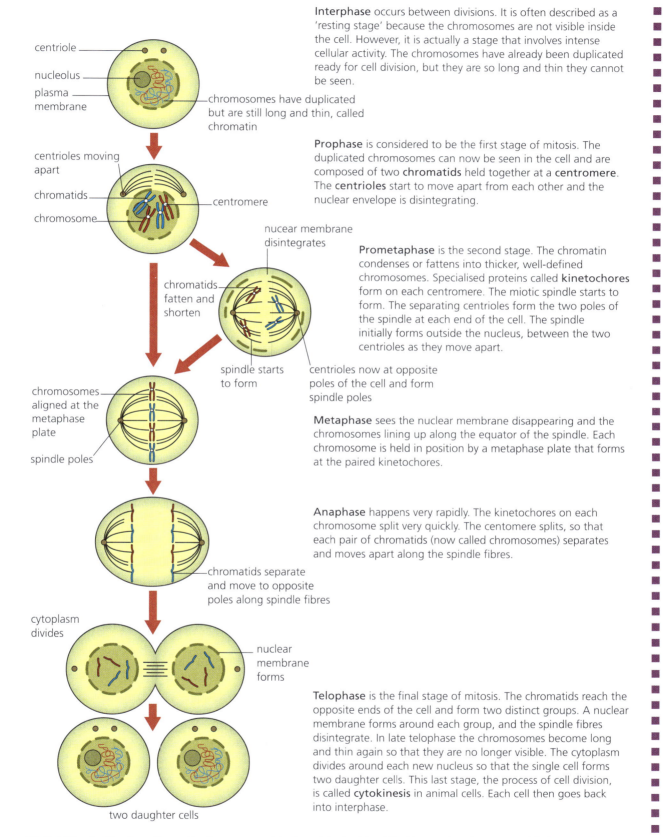

Interphase occurs between divisions. It is often described as a 'resting stage' because the chromosomes are not visible inside the cell. However, it is actually a stage that involves intense cellular activity. The chromosomes have already been duplicated ready for cell division, but they are so long and thin they cannot be seen.

centriole

nucleolus

plasma membrane

chromosomes have duplicated but are still long and thin, called chromatin

Prophase is considered to be the first stage of mitosis. The duplicated chromosomes can now be seen in the cell and are composed of two **chromatids** held together at a **centromere**. The **centrioles** start to move apart from each other and the nuclear envelope is disintegrating.

centrioles moving apart

chromatids

chromosome

centromere

nucear membrane disintegrates

Prometaphase is the second stage. The chromatin condenses or fattens into thicker, well-defined chromosomes. Specialised proteins called **kinetochores** form on each centromere. The miotic spindle starts to form. The separating centrioles form the two poles of the spindle at each end of the cell. The spindle initially forms outside the nucleus, between the two centrioles as they move apart.

chromatids fatten and shorten

spindle starts to form

centrioles now at opposite poles of the cell and form spindle poles

chromosomes aligned at the metaphase plate

spindle poles

Metaphase sees the nuclear membrane disappearing and the chromosomes lining up along the equator of the spindle. Each chromosome is held in position by a metaphase plate that forms at the paired kinetochores.

Anaphase happens very rapidly. The kinetochores on each chromosome split very quickly. The centomere splits, so that each pair of chromatids (now called chromosomes) separates and moves apart along the spindle fibres.

chromatids separate and move to opposite poles along spindle fibres

cytoplasm divides

nuclear membrane forms

Telophase is the final stage of mitosis. The chromatids reach the opposite ends of the cell and form two distinct groups. A nuclear membrane forms around each group, and the spindle fibres disintegrate. In late telophase the chromosomes become long and thin again so that they are no longer visible. The cytoplasm divides around each new nucleus so that the single cell forms two daughter cells. This last stage, the process of cell division, is called **cytokinesis** in animal cells. Each cell then goes back into interphase.

two daughter cells

Figure 17.6 Mitosis. To describe mitosis you need to be able to identify these structures: centrioles, nuclear envelope, chromosomes, chromatids, spindle.

Germination and growth in plants

In plants, only cells in particular places of the organism are able to divide by mitosis. These growth areas in the plant are called **meristems** or meristematic tissue. Meristematic tissue is usually found at the tips of shoots and buds, in the region behind the root tips, and in the cambium in vascular bundles. The location of the meristems, shown in Figure 17.7, causes the plant to grow in a branching shape. In animals, all the parts of the body can divide by mitosis, which leads to a more compact shape.

Growth in plants can be divided into two main types (see Figure 17.8).

- Primary growth – this refers to an increase in length and mainly involves the apical meristems (tips of roots and shoots).
- Secondary growth – this refers to an increase in width and involves the cambium in vascular bundles.

a Primary growth, in the early stage of the plant's life – the shoot mainly grows upwards, increasing in height.

b Primary and secondary growth – the stem thickens and lateral (side) stems begin to grow. As they thicken and harden with increased xylem tissue, they become known as branches.

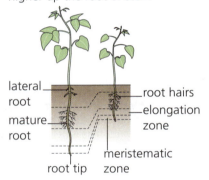

cell differentiation takes place higher up the root or stem

lateral root
mature root
root hairs
elongation zone
meristematic zone
root tip

cell division happens in the root cap

epidermis
meristematic tissue
root cap

cell enlargement or elongation takes place in the region behind the root cap

Figure 17.7 Mitosis in plants takes place only in meristematic tissue.

Did you know?

As a seed forms on a plant it dehydrates (dries out). Because a seed contains so little water, very little metabolic activity can take place inside the embryonic plant. It is in an inactive or dormant state. Dormancy is a useful way for seeds to survive for a long time in conditions that are not favourable for growth. The seed requires particular conditions in order to begin growing.

The structure of a seed

Many plants start as a seed which is an embryonic plant enclosed in a protective casing. The plant embryo consists of a **radicle** which grows downwards into a root and a **plumule** which develops towards the light as a shoot. The seed also contains some food for the plant in order to give it energy for the initial stages of growth.

c Secondary growth – the plant reaches its maximum height, and the branches and sterm get thicker.

Figure 17.8 Plant growth

A bean plant is an example of a **dicotyledon**. This is the name given to a very large group of flowering plants which sprout with two seed leaves when they first start growing. In the seeds of these plants, two **cotyledons** contain the starch and protein that feed the plant, as well as enzymes for the germination process. The cotyledons are surrounded by a tough coating called the **testa**, which protects the embryo from physical damage and from bacteria and fungi which could attack the seed. There is a tiny hole in the testa called the **micropyle**, next to the hilum, the scar which marks the place where the seed was joined to the pod.

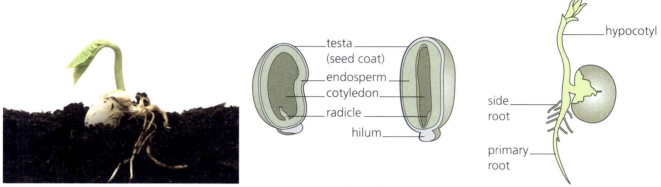

Figure 17.9 A bean seed is a dicotyledon. How can you tell this from the photograph? The diagram shows a dicotyledon seedling.

Germination

The growth of a plant begins with **germination**. Germination involves a series of changes within a seedling resulting in the development of seedlings (see Figure 17.10). The following conditions are required for germination to occur:

- **W**ater
- **O**xygen
- **W**armth (suitable temperature).

'WOW' is a good way of remembering the three essential conditions for germination. Light affects the quality of the plant's development but is normally not a requirement for the germination process.

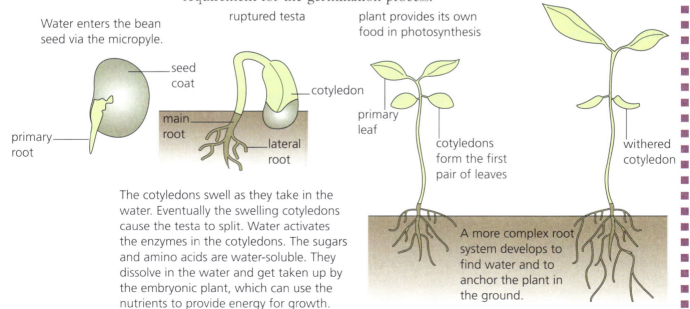

Figure 17.10 Stages in the germination of a bean seed, an example of a dicotyledon.

Effects of water on germination

Water is required in germination in the following ways.

- The uptake of water through the micropyle begins germination and ends dormancy in the seed.
- Water activates the enzymes that carry out metabolic reactions in the seed. Some of the enzymes activated as a result of the uptake of water are:
  - amylase, digests starch to maltose
  - maltase, digests maltose to glucose
  - lipase, digests lipids to fatty acids and glycerol
  - proteases, digest proteins to amino acids.

Effects of oxygen on germination

Oxygen is required for aerobic respiration as follows:

$$\text{glucose} \quad + \quad \text{oxygen} \quad \rightarrow \quad \text{carbon dioxide} \quad + \quad \text{water} \quad + \quad \text{energy}$$
$$C_6H_{12}O_6 \quad + \quad 6O_2 \quad \rightarrow \quad 6CO_2 \quad + \quad 6H_2O \quad + \quad \text{energy}$$

Aerobic respiration takes place in the cells in the seed and provides energy for metabolic reactions in the seed that lead to growth.

Effect of temperature on germination

A suitable temperature of approximately 5 °C to 40 °C is required so that enzymes have appropriate conditions in which to operate. Temperatures below 5 °C may result in inactivation of enzymes; whereas temperatures above 40 °C may cause denaturation of enzymes.

Effect of light on germination

Light is generally not needed for germination but if seeds germinate in its absence, the development of the seedlings is affected. **Etiolation** occurs in plants that are exposed to prolonged darkness and is characterized by:

- long, thin pale stems
- small, yellow leaves.

Did you know?

The yellow appearance of the etiolated leaves and paleness of stems occurs because chloroplasts fail to develop in the absence of light. Stems are abnormally long in an attempt to reach light. Have you ever turned over a pot or large stone to reveal etiolated plants beneath?

Key facts

The main events in germination are:

- uptake of water, which causes an initial increase in the wet mass of the seed
- digestion of food reserves stored in the cotyledons of dicotyledonous seeds (or in the endosperm of monocotyledonous seeds)
- break down of digested food (glucose) to provide energy using respiration; this causes an initial decrease in the mass of the seed as food reserves are used up
- translocation of digested food to the embryo where it is used for growth.

Activity 17.2 Investigating germination

Aim: To investigate germination in plants.

Equipment and apparatus

- 1.5 litre plastic soda bottles
- sharp scissors
- hot water
- Petri dishes
- plank of wood
- small screws
- screwdriver
- small seeds, for example mustard, radish, arugula, cress
- graph paper
- transparency paper (acetate)
- filter paper
- marker pen

a
cut this shape out using
sharp scissors

b
screw down these bottle
bases the correct length
apart to fit your bottle

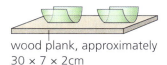

wood plank, approximately
30 × 7 × 2 cm

c

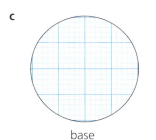

base

d

e

Petri dish

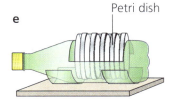

Figure 17.11 Method
for investigating
germination.

Procedure

1 Remove the label from the bottle. To do this, fill the bottle with hot water (but not boiling hot water, as the plastic will buckle). Screw the cap back on the bottle and soon the label should peel off easily. Empty the bottle and let it dry.
2 Create a reservoir to support the Petri dishes lying on their sides. To do this, using the scissors, carefully cut out a part of the side of bottle as shown in Figure 17.11a.
3 Now make a cradle for the bottle, using the coloured bases from two plastic bottles. Screw these to either end of a plank of wood as shown in Figure 17.11b.
4 On a piece of graph paper, draw a circle with a diameter of just under 9 cm (see Figure 17.11c). Make six of these circular grids. Photocopy the grids on to a transparency sheet. Carefully cut them out.
5 Place a grid into the lid of a plastic Petri dish. Place a piece of 9 cm diameter filter paper on top of the grid. Sprinkle a few drops of water to wet the filter paper all over. Make sure there are no air pockets between the filter paper and the grid. Place a row of seeds along a line on the grid.
6 Put the base of the Petri dish in place (see Figure 17.11d). Using the marker pen, label the lid with type of seed, number of seeds and the date. Leave it lying flat for a few minutes to allow the seeds to stick to the damp filter paper. Repeat steps 5 and 6 using different kinds of seeds.
7 Place the cut bottle into the cradle. Pour water into the bottle to a depth of about 2 cm. Gently slot the Petri dishes vertically in the cut bottle as shown in Figure 17.11e. Examine the Petri dishes at suitable intervals. Replace any seeds that fall off.

Questions

1 How many seeds germinated after:
   a 1 day   **b** 2 days   **c** 3 days   **d** 5 days   **e** 1 week?
2 **a** Calculate the percentage of seeds that germinated each day. Design a suitable way of showing the information on a graph.
   b Compare germination rates for your different seeds.
3 **a** How can you tell which part of the germinated seed is the radicle (root) and which part is the plumule (shoot)?
   b Which appears first, the radicle or plumule? Is it the same for other seeds?
4 **a** Measure the radicles and plumules as they grow. Design a suitable way to display the data you collect.
   b Do radicles and plumules grow at the same rate?

Unit 4 Measuring growth

There are several methods of measuring growth in plants and animals.

- Height or length – the height or length of an entire organism, or one part of it, can be measured. This is an easy way of measuring small- and medium-sized animals and plants. We can measure height in an organism's natural habitat or in laboratory conditions, and we can also measure different parts of the organism (for example wings). It is more difficult with very tall plants.
- Mass – measuring the mass of small- and medium-sized animals is the easiest. Weighing plants is more difficult, as it requires uprooting the plant, which may damage or kill the organism. It is possible to measure smaller pot plants this way, especially if the pot and soil are measured beforehand and their mass is subtracted from the measurement each time. It is also impractical to measure the mass of very large animals.
- Number of leaves – counting the number of leaves is a good indicator of plant growth and is easy to do, but it only measures one aspect of growth.
- Area or volume of body parts – this is another good indicator of plant growth, and can provide specialized data about the growth of an animal or plant. The drawback is that it is not always easy to calculate. It may also depend on the removal of the part to be measured, which is not always possible without killing or damaging the organism.
- Number of organisms – this is a useful indicator of population growth. It may sometimes be difficult to measure.

Wet and dry mass

The **wet mass** of an organism is its full mass, including all the water in its body. This mass fluctuates constantly as an organism takes in and loses water at different times of the day. However, the water we take in and lose daily does not become part of our biomass. For this reason, scientists sometimes measure the **dry mass** of an organism – that is, its mass with all the water removed.

For example, to work out the patterns of growth in a particular plant species, scientists take hundreds of seedlings and plant them under similar conditions. As the plants grow, the scientists take samples at each stage of growth and dry them gently in ovens. They then measure the mass of the remaining dry plant matter and process the results to find the patterns of growth for that species. Measuring dry mass gives a very accurate indication of real growth, but it kills the organism, so it is not a method of measurement that is practical for repeating over time.

In plants, counting the number of rings (in woody plants), counting the number of leaves, and measuring the surface area of leaves are additional methods of measuring growth.

Activity 17.3 Evaluating the different methods of measuring growth

SBA skills

Manipulation/
measurement (MM)

Aim: To evaluate several methods of measuring growth.

Questions

1 Draw up a table like the one below. Complete the table by listing the main advantages and disadvantages of each method of measuring organisms, and give examples of plants or animals we would measure in this way.

Method	Advantage	Disadvantage	Examples
Measuring length or height of organism			
Measuring wet mass of organism			
Measuring dry mass of organism			
Counting number of leaves on plant			
Measuring area or volume of body parts			
Counting number of organisms			

2 Which method would you use to measure growth in each of the following organisms? Give a reason for each of your answers.
   a banana tree
   b pet cat
   c tomato plants in a vegetable garden
   d young child.

3 Scientists plant 300 samples of a particular radish plant. Over 6 weeks, they take ten samples every two days, dry the samples, and measure the mass of dried matter. The table below shows their findings.

Day	0	2	4	6	8	10	12	14	16	18	20	22
Mass (g)	5	3	5	6	8	11	15	18	22	28	32	36
Day	24	26	28	30	32	34	36	38	40	42	44	46
Mass (g)	46	48	49	48	47	45	44	42	41	40	40	39

   a Why do you think the scientists measured dry rather than wet mass?
   b Explain briefly how you think they measured the dry mass.
   c Draw the data as a growth curve.
   d Label the four main stages of growth.
   e Why do you think the mass decreased in the first few days?
   f Give two different reasons to explain why the mass might have decreased in the last few days.

Taking accurate measurements

There is a saying that 'Measurement is an inexact science'. Each method of measurement involves a level of inaccuracy. This is a result of both instrument error – the inaccuracy introduced by a specific measuring instrument – and human error – inaccuracy on the part of the person who observes and records the measurement. When measuring quantities, we aim to keep the margin of error as small as possible.

Measuring mass

There are several kinds of instruments used to measure mass. A spring scale uses a spring mechanism, which stretches (as in the case of a hanging scale) or compresses (as in the case of a bathroom or kitchen scale) in proportion to gravity's pull on the object. These scales measure the displacement of the spring using a mechanism that estimates the gravitational force applied by the object. In electronic versions of the spring scale, the measurement is taken using a strain gauge.

Figure 17.12 Spring scales.

For all measurements:
- Decide which units of measurement would be most suitable for the object you are measuring.
- Select a suitable instrument. The instrument should be appropriate for the object's size.
- If possible, take the measurement twice or even three times. If the readings differ, you can take an average of the readings.
- Make sure you record the reading using the correct units of measurement. If the measuring instrument uses a different unit of measure, convert to the unit of measure you initially selected.

When you measure the mass of an object:
- Select the correct instrument for the size of object. For example, a kitchen scale is usually suitable for objects under 500 g; a bathroom scale is suitable for larger objects up to about 120 kg. It should also use appropriate units of measure.
- Make sure that the scale is calibrated to zero. On a balance scale, you do this by making sure the needle is pointing to zero steadily before you put anything on the scale. On an electronic scale, you do this by setting the display to zero.
- If you are measuring an item in a container (for example a plant in a pot) or with any additional mass (for example the soil around a pot plant) take the measurement of the container or the additional mass alone first.
- Measure the mass of the object slowly and carefully. Do not bounce or jolt the scale, as this can change the calibration.
- Wait for the scale to settle at the correct reading.

When you measure the length of an object:
- You would usually use a tape measure or ruler, although there are also other instruments for larger distances: trundle wheels and odometers for example.
- Select the instrument that has suitable units of measurement, for example mm, cm or m.
- Decide which extent you are going to measure. For example if you are measuring a bird, you might measure its wingspan (tip to tip), its legs, its standing height.
- You usually need to measure against a flat surface such as a wall, floor or table.
- For curved items, you can use a piece of string or rope, and then measure that length against a ruler or tape measure.

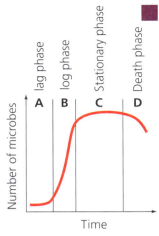

Figure 17.13 A growth curve.

Figure 17.14 Taking an accurate measure of the height of a seedling.

Plant growth curves

We can draw graphs to represent the changes that occur as plants grow. Such graphs usually take the form of a **sigmoid** (S-shaped) curve and are called growth curves. In Figure 17.13 the different phases of growth are labelled as follows:

A – Growth occurs slowly at first.
B – Once it begins, growth increases very rapidly.
C – Growth slows down and stabilizes.
D – Once the plant dies, growth is negative.

Activity 17.4 Measuring growth of a potted plant

Aim: To measure the growth of a pot plant.

Equipment and apparatus
- bean or pea seeds
- 3 pots
- potting soil
- ruler or tape measure
- water
- plastic container
- cotton or paper towel
- graph paper
- correction fluid
- marker pen

Procedure

1 Fill each pot two-thirds full with potting soil. Place two bean or pea seeds in each pot. Cover the seeds with 2 cm of potting soil. Water the pots and place them in a warm, light place.

2 Using the marker pen, label the pot plants 1 to 3.

3 After a few days, shoots should start appearing in each pot. Pull out the weaker plant from each pot so that there is only one plant per pot.

4 a Measure the height of each seedling daily, over a period of three weeks. Use a ruler or tape measure, and measure from the soil to the top of the main shoot. Record your observations in a table.

  b Also count the number of leaves on your plant every second day over the three-week period. Record your observations in a table.

5 a Choose three differently sized leaves on one of the plants. Mark the leaves you have chosen with white correction fluid. You can number them 1 to 3. Carefully place each leaf against a piece of graph paper and trace the outline of the leaf. Calculate the surface area of the leaf by adding up the numbers of squares (in mm) on the graph paper. Measure the surface area of each selected leaf every three days for your three-week investigation. Draw an appropriate table to record your results.

Questions

1 Draw growth curves of the data for the three plants.

2 What similarities and differences do you notice between the three growth curves?

3 Plot a graph to represent your results of leaf area.

4 Write a paragraph about what your investigation demonstrated about plant growth. Make sure you include a conclusion from each of your three methods of measurement (height, number of leaves and surface area of leaves).

5 Suggest two other ways to measure the growth of these plants.

Check your knowledge and skills

1 Identify the various phases of growth on the curve shown in Figure 17.15.

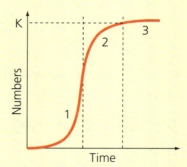

Figure 17.15 Growth curve for an annual plant.

2 Given that an annual plant completes its life cycle in one year and that a perennial plant persists from year to year, what would the growth curve of a perennial plant look like? Draw a sketch to show your answer.

Controlling growth and development

In Chapter 16 you learned how auxins affect tropisms. Auxins are a type of plant hormone produced by cells in the meristematic regions of plants, such as the root tips and shoot tips. The auxin diffuses downwards and causes the region just behind the tip to elongate. The more auxin that diffuses into an area, the faster the cells grow in that area (see Figure 17.16). If no auxin reaches an area of a plant, it does not grow.

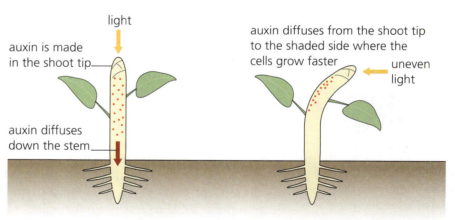

light

auxin is made in the shoot tip

auxin diffuses down the stem

auxin diffuses from the shoot tip to the shaded side where the cells grow faster

uneven light

When a plant receives light evenly around its growing tip, auxin diffuses evenly down the shoot. The cells behind the tip all grow at the same rate and the shoot grows straight up.

When a plant receives more light in one area than in another, auxin diffuses to the shaded side. This makes the cells on the shaded side grow faster and the shoot bends towards the light source.

Figure 17.16 A plant growing in unilateral light and in uneven light conditions.

Activity 17.5 Investigating the effect of an auxin on root growth in seedlings

SBA skills

Observation/recording/
reporting (ORR)

Manipulation/
measurement (MM)

Analysis and
interpretation (AI)

Aim: To find out the effect that different concentrations of auxin have on root growth in seedlings.

Equipment and apparatus

- 6 Petri dishes fitted with grids and filter paper (see page 189)
- 1.5 litre plastic bottles prepared to hold vertically stacked Petri dishes (see page 189)
- plastic $1\,cm^3$ dropper or $5\,cm^3$ syringe
- mustard seeds
- distilled water
- different concentrations of indoleacetic acid (IAA) solution
- blunt forceps
- hand lens
- paper towel
- marker pen

Indoleacetic acid (IAA)

Indoleacetic acid (IAA) is a powerful plant hormone that belongs to the auxin group. It is effective even at low concentrations. In order to dissolve IAA in water, first dissolve 0.1 g IAA in a few drops of 95% ethanol. Add 1 L of distilled water. This makes a solution of 100 ppm. You can store the stock solution in a refrigerator for up to four weeks.

- For 1 ppm IAA solution: mix $3\,cm^3$ stock solution with $297\,cm^3$ distilled water.
- For 10^{-1} ppm IAA solution: mix $30\,cm^3$ stock solution with $270\,cm^3$ distilled water.
- Similarly, use the stock solution in different ratios to make up dilutions of 10^{-2} ppm, 10^{-3} ppm and 10^{-4} ppm of IAA solution. Pour about $200\,cm^3$ of each solution into an appropriately labelled bottle, to a depth of about 1 cm.
- During the experiment, keep bottles with germinating seeds in the dark at 20–25 °C.

Procedure

Note: Get into groups of three or four. Each group will work with a specific solution, for example Group 1 works with 1 ppm IAA solution; Group 2 works with 10^{-1} ppm IAA solution, and so on. One group will work with distilled water – this is the control group.

1 Line six Petri dish lids with circular grids and filter papers (see steps 4 and 5) on page 189.
2 Using the marker pen, label each lid with the IAA concentration it contains or 'control'.
3 Soak the filter paper with $3–4\,cm^3$ of the appropriate solution.
4 Place about ten mustard seeds on the damp filter paper along a grid line on one side of each Petri dish.
5 Cover each Petri dish and slot it into the prepared bottle (see steps 1 to 3 on page 189 for the method). You will need one bottle per solution. Label the bottles. Each bottle should contain about $200\,cm^3$ of the solution indicated on its label.
6 Place all the bottles in a warm, dark place for two to three days.

7 After two to three days, collect your Petri dishes from the bottles. Use forceps to handle the seedlings. Use a hand lens to inspect the root hairs so that you can see where each root begins. Throw away any seedlings that differ significantly from the others in a dish: for example, throw away any that are exceptionally long, have not germinated, or have distorted growth.

8 Measure the length of the remaining roots by counting the millimetre boxes that each root covers on grid paper. If the root does not grow at right angles to the grid, use forceps to move it so that you can measure its length.

9 Record the results in a suitable table.

Questions

1 Calculate the average root length for each dish.

2 Calculate the % change for each IAA concentration, using this formula:

% change = (total average length of roots in IAA − total average length of control roots) × 100

If the change has a positive value, the roots were stimulated by the concentration of IAA. If the percentage change has a negative value, the roots were inhibited by the concentration of IAA.

3 Copy and complete the table below.

Concentration of IAA (ppm)	Class average root length	% change	Seedling root growth stimulated or inhibited
0			
10^{-4}			
10^{-3}			
10^{-2}			
10^{-1}			
1			

Human hormones and growth

All organisms grow and develop in stages. In a human life, growth and development refer to the processes of growing from a single fertilized cell (or zygote) to a mature adult. The main stages in human growth are as follows.

■ Prenatal – this is the stage of development before birth. It begins with the zygote, the fertilized egg. As the cells multiply in the first three weeks after fertilization, the zygote becomes a ball of cells called a blastocyst. Between the third and eighth weeks of a pregnancy, the human form becomes distinguishable as an embryo. From the eighth week up until birth, the unborn person is known as a foetus.

■ Child – this is the name given to an individual from birth to 18 years old. The child undergoes dramatic growth and development during this period. For the first 30 days, the child is known as a newborn or neonate. Next comes the infant stage (one month to one year). Between one and four years of age, the child becomes a toddler. Play age is the period between three and six years of age, followed by prepubescence (age four to 12 years), pubescence or adolescence (13 to 19 years).

- Adult – after the age of 19 years an individual is an adult. This stage too can be subdivided into early adulthood (20s and 30s), middle adulthood (40s to 50s), and late adulthood (60s upward).

Human growth

We can broadly divide human growth into three stages, but there are many many stages. Can you think of different ways of dividing the developmental stages of a human life?

Different parts of the body grow at different rates at each stage. This is because cell division takes place more quickly in some parts of the body than others. When you compare children's bodies with adults' bodies, you can see that a child's head is much bigger in proportion to the rest of its body. During adolescence, the arms and legs grow more quickly, and the proportions of the body change. Stages of human growth are shown in Figure 17.1b.

During the prenatal stage, an enormous amount of growth has to take place in order for the zygote to develop in complexity and become a fully formed foetus. Cell differentiation allows the cells to form all the body's specialized organs and tissues. During childhood, the body gradually increases in size. The next dramatic stage of development is adolescence. During this period, the body undergoes many changes to prepare it for reproduction.

Check your knowledge and skills

1 During which stage do you think the human body grows most rapidly?
2 During which stage does the least noticeable growth take place?
3 Is there any stage of life during which there is no growth? Explain your answer.
4 How do you know that the graph shown in Figure 17.1b is not to scale?

Key facts

- A gland is an organ that synthesizes a particular chemical and releases it into the body.
- Exocrine glands secrete their chemicals directly to a target area via a duct, for example salivary glands in the mouth and tear glands in the eye.
- Endocrine glands secrete chemicals that move by diffusion into the bloodstream, which carries them to the target area, for example the pancreas, thyroid, pituitary gland and gonads.
- The endocrine system produces hormones that control human growth and development.

The pituitary gland

The **pituitary gland** is sometimes called the 'master gland'. Although it is only the size of a pea, and weighs about 0.5 g, it controls most of the endocrine glands, and thus most of the hormonal activity in the body. The pituitary is situated in the brain, near to the **hypothalamus**, which manages the functioning of the pituitary. Table 17.1 on the next page lists the pituitary hormones and their functions.

The thyroid gland

The **thyroid gland** is situated near to the front of the neck. It produces thyroxine, a hormone which performs the following functions:

- regulates metabolic rate
- regulates growth and development.

Thyroxine and human growth hormone (HGH or somatrophin) stimulate protein synthesis in the cells. The increased formation of protein causes increased human growth. Under-production of thyroxine may lead to stunted growth and retarded mental development.

Part of pituitary gland	Hormone	Function
Posterior pituitary	Antidiuretic hormone (ADH)	Regulates the body's water content
	Oxytocin	Facilitates birth and breastfeeding. Functions associated with social recognition, trust, bonding and other behaviours
Intermediate lobe	Melanocyte-stimulating hormone (MSH)	Stimulates release of melanin in skin
Anterior pituitary	Pituitary growth hormone (GH)	Stimulates cell growth
	Follicle-stimulating hormone (FSH)	Regulates the growth and sexual maturing of the body
	Luteinising hormone	Triggers ovulation in females. Stimulates production of testosterone in males
	Thyroid stimulating hormone (TSH)	Stimulates thyroid to produce thyroxine
	Endorphins	Work as natural pain relievers during periods of exertion or stress
	Adrenocorticotropin (ACTH)	Stimulates production of androgens and cortisol
	Prolactin (PRL), also known as luteotropic hormone (LTH)	Stimulates lactation in women, orgasm in men and women

Table 17.1 Pituitary hormones and their functions.

The gonads

The gonads are the sex organs – the testes in the male and the ovaries in the female. These organs secrete hormones that influence growth and development as shown in Table 17.2.

Gonads	Functions
In males: the testes	Produce the hormone testosterone, which influences sexual development and causes development of sex organs. Produce gametes in the form of sperm for sexual reproduction. Stimulate the development of secondary sexual characteristics, including: – increased size of reproductive organs – ability to ejaculate – increased muscle development – deepening or 'dropping' of the voice – growth of body hair and pubic hair, as well as more hair on face and chest
In females: the ovaries	Secrete oestrogen, which stimulates development of secondary sexual characteristics, including: – enlargement of reproductive organs – growth of breasts – onset of menstruation – widening of hips – growth of pubic and body hair – increase in body fat Produce gametes in the form of ova (eggs) for sexual reproduction

Table 17.2 Functions of the gonads.

Chapter summary

Do you know?

If you are unsure of any of the facts in the list, refer to the page number in brackets.

■ Growth involves a permanent or irreversible increase in the size and mass of an organism. (page 180)

■ In complex organisms, the three main mechanisms of growth are cell division, cell elongation and cell differentiation. (page 181)

■ Growth requires both cell division and cell elongation. (page 181)

■ Mitosis is a type of division that takes place when a single cell makes a copy of itself and splits into two identical cells. (page 182)

■ Mitosis has five stages – prophase, prometaphase, metaphase, anaphase and telophase. At the beginning and end of each cycle, the cell is in interphase. (page 182)

■ In plants, growth takes place in the meristematic tissue. (page 186)

■ In flowering plants, the life cycle of the plant begins with germination. (page 187)

■ Water, oxygen and warmth ('WOW') are the key conditions required for germination to take place. (page 187)

■ There are several methods of measuring growth in plants and animals, including: measuring height or length; measuring mass; measuring the number of leaves; measuring the area or volume of body parts; measuring the number of organisms or specimens in a population. (page 190)

■ The wet mass of an organism is its full mass including the water in its body; whereas the dry mass is its mass with water removed. There are advantages and disadvantages to each method of measuring the mass of an organism. (page 190)

■ Plant growth is controlled by hormones including auxins. (page 194)

■ Animal growth is controlled by a variety of hormones. (page 197)

Are you able to?

If you have trouble in doing these things, refer to the page number in brackets.

■ Identify the main processes that facilitate growth and development. (page 181)

■ List and briefly describe the stages in mitosis. (page 185)

■ Identify the following structures in the cell: centrioles, nuclear membrane, chromosomes, chromatids, spindle. (page 185)

■ Locate the regions that contain meristematic tissue in a plant. (page 186)

■ Label the main parts of a dicotyledonous seed. (page 187)

■ Describe what happens during germination of a dicotyledonous seed. (page 187)

■ Explain the effects of water, oxygen, warmth and light on germination. (page 188)

■ Draw a sigmoid growth curve. (page 193)

■ Take accurate measurements of the growth of a plant or animal. (pages 191 and 193)

■ Interpret growth curves. (page 194)

■ State the function of hormones in controlling human growth and development. (page 198)

18 Asexual reproduction and sexual reproduction in plants

Continuing life

All living organisms have several processes in common. All need to carry out digestion, respiration, excretion and processes of homeostasis in order to survive. These processes ensure the survival of the individual organism. Reproduction is different: it is necessary for the survival of the species, rather than the survival of the individual.

In this chapter you will learn about the ways plants reproduce.

By the end of this chapter you should be able to:

■ Define and compare asexual and sexual reproduction

■ Understand asexual reproduction in simple organisms

■ Describe the main forms of asexual reproduction in plants

■ Describe the processes of pollination and fertilization

■ Draw and label the male and female reproductive parts of a plant

■ Compare the structures of wind-pollinated and insect-pollinated flowers

■ Determine the mode of dispersal of fruits and seeds

Unit 1 Asexual and sexual reproduction

Key fact

Sexual reproduction involves two parents, a male and a female, and has the advantage of producing an organism with an entirely original genetic make-up – a unique combination of genetic material from both parents.

Figure 18.1 a The children in this family have similar but different combinations of their parents' genetic material. **b** Identical twins have identical DNA because they are produced from a single fertilized egg that splits into two identical zygotes after the DNA has already duplicated.

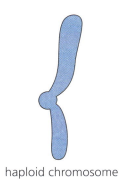

haploid chromosome

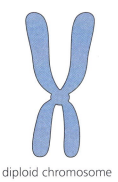

diploid chromosome

Figure 18.2 A haploid cell has a single set of unique chromosomes. A diploid cell has chromosomes in identical pairs.

The main purpose of cellular reproduction is the copying of cell contents, in particular the chromosomes. The chromosomes contain all of the cell's genetic information and determine its specific features, traits and capabilities. In eukaryotic cells, chromosomes contain DNA which needs to be copied in a very particular way. Not all cellular components are copied like DNA. Some structures can be synthesized from the new DNA after cell division; other structures (for example endoplasmic reticulum) are broken down during the cell cycle and then re-synthesized after cell division is complete.

In higher organisms, each cell usually contains two copies of each chromosome – one from the female parent and the other from the male parent. Together, these are called a **homologous pair** of chromosomes (and each member of the pair is called a homologue). A cell's **haploid number** is the total number of homologous pairs (or the number of unique chromosomes) in a cell. This number varies in different species. In humans, the haploid number is 23. The **diploid number** of a cell refers to the total number of chromosomes in a cell.

diploid number $= 2 \times$ haploid number

If the haploid number is thought of as n the diploid number would be $2n$. In humans the diploid number is 46.

In **asexual reproduction** a single parent organism produces offspring that are genetically identical to their parent.

Did you know?

All reproduction involves copying old cells into new ones. Most importantly, the new cells must be accurate enough copies that they can perform the same function as the original cell. An inaccurate copying process introduces mutations, or errors, in the cells of offspring. Some mutations do not result in any noticeable damage to the cell's functioning, but others may result in severe disorders or even death.

■ Methods of asexual reproduction

Some simple animals are able to reproduce asexually as well as sexually.

Budding

Hydras and other *cnidaria* can reproduce either sexually or asexually. In asexual reproduction, the new individual forms during a process called **budding** (see Figure 18.3). The parent develops a swelling on the side of its body. This eventually grows into a daughter bud, which will grow tentacles and begin to feed on small water organisms. Once it can feed itself, the daughter bud breaks off from the parent *Hydra* and floats in the water until it settles on a support and lives independently.

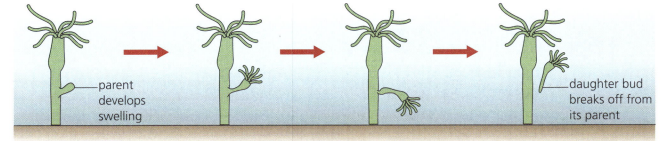

parent develops swelling

daughter bud breaks off from its parent

Figure 18.3 Budding in *Hydra* is an example of asexual reproduction.

Hydra

Hydra also reproduce sexually. The *cnidaria* life cycle shown in Figure 18.4 shows the stages of budding, as well as the sexually reproductive stage.

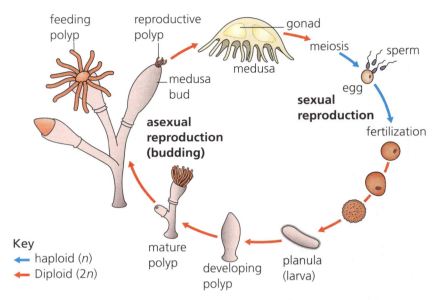

feeding polyp
reproductive polyp
gonad
meiosis
sperm
medusa
medusa bud
egg
sexual reproduction
fertilization
asexual reproduction (budding)

Key
← haploid (*n*)
← Diploid (2*n*)

mature polyp
developing polyp
planula (larva)

Figure 18.4 *Cnidaria*, for example *Hydra*, have asexual and sexual reproductive stages in their life cycles.

Regeneration

In some animals asexual reproduction takes the form of **fragmentation and regeneration**. The animal becomes fragmented into parts, and then each part regenerates to form a new individual. Examples of animals that fragment and regenerate include flatworms, sea sponges and starfish.

a A flatworm reproduces asexually by breaking into two pieces, each of which regenerates into a new worm.

b A starfish can regenerate its entire body from one chopped-off part.

Figure 18.5 Fragmentation and regeneration are examples of asexual reproduction.

Parthenogenesis

Parthenogenesis is a form of asexual reproduction in which the female produces eggs that develop without fertilization (see Figure 18.6). The unfertilized egg develops into an individual identical to the parent individual. Parthenogenesis occurs in some plants as well as animals including water fleas and aphids. It even occurs in some higher animals including bees, lizards and turkeys. In honeybees, the queen bee produces eggs. She either fertilizes these as she lays them (these develop into worker bees), or she lays them unfertilized. The unfertilized eggs develop into haploid males called drones (see Figure 18.7).

Figure 18.7 Honeybees reproduce asexually (using parthenogenesis) to produce drones (top) and sexually to produce worker bees (bottom).

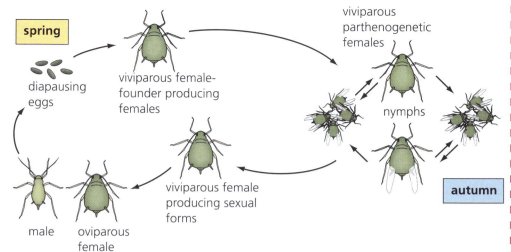

Figure 18.6 Parthenogenesis in beetles.

Sporification

Fungi can reproduce asexually by producing and dispersing spores. A typical spore-producing fungus is *Penicillium*, which grows on decaying food. The fungus has a threadlike mycelium that grows over the surface of the food, absorbing nutrients and digesting them for energy. From the mycelium, the fungus develops vertical structures called hyphae which produce tiny spores at their tips. These spores are dispersed usually in air or water. Once they settle in a new place, they may grow into a new organism. Mushrooms are another example of a fungus that reproduces using **sporification**.

Figure 18.8 *Penicillium* growing on bread.

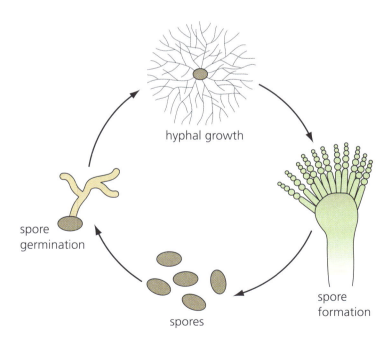

Figure 18.9 Sporification in funghi.

Stolons and rhizomes

Asexual reproduction in plants is also known as vegetative propagation. When a plant has plentiful resources, it may grow a bud on one of its stems. This bud develops into a complete new plant with roots, stems and leaves. The old plant may die and the new plant survives independently.

Strawberry plants have short flowering shoots. Their stems are known as **rootstocks**. Once the main shoot has flowered, the rootstock produces buds down its sides. These are called **stolons** or runners, and they grow outwards horizontally over the ground. At the end of the runner a node forms, which produces a bud that develops roots and shoots. The stolon provides enough nutrients and water to the new plant to sustain it until it settles into a new place and begins to grow and sustain itself. A new runner may continue to grow from the new plant, so that a stolon may produce several independent new plants.

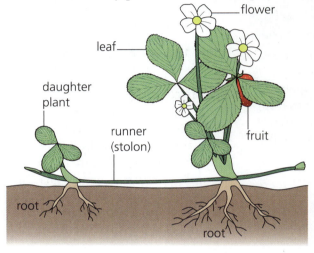

Figure 18.10 Strawberries reproduce asexually by producing runners.

Figure 18.11 Couch grass reproduces asexually by developing rhizomes.

Another form of vegetative propagation is the development of **rhizomes**. A rhizome is an underground stem. For example, couch grass sends horizontal shoots from lateral buds near its stem base. These grow horizontally underground. At the end of each rhizome is a node with a growing point that can develop to form shoots and roots. Many grasses propagate in this way.

Comparing asexual and sexual reproduction

- **Asexual reproduction** – this generates new individuals without any sharing or combining of genetic material between two parents. The daughter organism is genetically identical to the parent organism.
- **Sexual reproduction** – this enables two organisms to share some of their genetic material so that their offspring have half of their genetic material from each parent. This mode of reproduction is used by higher organisms, both plants and animals.

Table 18.1 shows the advantages and disadvantages of sexual and asexual reproduction.

Check your knowledge and skills

1 Define the following terms:
   a asexual reproduction   b diploid number   c rhizome   d rootstock.
2 List three advantages and three disadvantages of asexual reproduction in plants.
3 Why do you think humans cannot reproduce asexually?

Method of reproduction	Advantages	Disadvantages
Asexual reproduction	The species can continue even if the individual organism has no suitable mate nearby The parent can produce a large number of offspring in a short time Offspring are produced quickly The parent does not have to spend time and energy on finding a mate Asexual reproduction does not require energy for the metabolic processes involved in sexual reproduction	Little or no genetic variation is introduced as the offspring are identical to the parent. Any genetic disadvantages are passed on to the offspring. Variations often introduce adaptations that help the organism to survive If there is a change in the environment that kills off the parent organism, the offspring will be vulnerable in the same ways. Short-term changes in the environment (for example floods or droughts) can wipe out the parent and offspring Overcrowding can result in limited resources for the offspring, since offspring are produced in such large numbers and near to the parent organism
Sexual reproduction	The process of **meiosis** introduces genetic variation, which helps to introduce adaptations It is possible for the offspring to be stronger and more disease-resistant than their parents If the environment changes, the offspring are more likely to be able to adapt to it	Meiosis can also produce disadvantageous variation; it is possible to produce offspring that are weaker or more susceptible to disease than their parents It is a very slow process A lot of time and resources are spent finding a mate Few offspring may be produced

Table 18.1 Advantages and disadvantages of sexual and asexual reproduction.

Unit 2 Sexual reproduction in plants

Gametes, or sex cells, are cells that contain *half* the genetic material of a normal cell in an organism. Gametes are specifically produced for sexual reproduction. Flowering plants produce male gametes in the form of pollen and female gametes in the form of ovules. In humans, male gametes are known as sperm, and female gametes are called ova (singular: ovum).

Sexual reproduction requires one male gamete and one female gamete. In the case of plants, at least one pollen grain must fertilize an ovule in order for the plant to reproduce sexually. In humans, a sperm cell must fertilize an ovum in order for sexual reproduction to take place.

In plants, as in animals, sexual reproduction involves the fusing of male and female gametes. Sexual reproduction takes place in flowering plants. The flower forms the reproductive organ of the plant (see Figure 18.12, overleaf). Unlike humans, plants can usually develop both the male and female sexual reproductive systems.

■ Parts of a flower

The female part of the flower is called the **pistil**. It consists of the style, **stigma** and the ovary and the ovum which contains the ovules (female gametes). The male part of the flower is called the **stamen**. It consists of the filament and the **anther** which contains the pollen grains which house the male gametes.

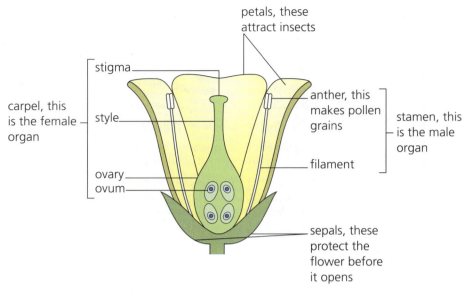

Figure 18.12 Structure of a typical flower.

■ Pollination

Pollination is the process that brings together male and female gametes in plants. It takes place when **pollen** leaves the anther and lands on the stigma of a flower. There are two types of pollination:

■ self-pollination – this takes place when pollen falls on to the stigma of either the same flower or on a flower on the same plant.
■ cross-pollination – this takes place when pollen lands on the stigma of a flower of the same species that is on a different plant.

Pollination is aided by wind, by insects, birds and animals. Flower adaptations tend to make each plant more likely to rely on either wind or animals.

■ Wind-pollinated flowers – these have pollen adapted to float easily in the air and stamens adapted to catch tiny floating pollen granules.
■ Insect-pollinated flowers – these have bigger, stickier pollen grains. They attract insects with their scent and flowers. The insect lands on the flower to suck the nectar at the flower's base. As the insect's body brushes against the anther, the pollen sticks to its body. As the insect moves past the stigma, pollen grains fall off and stick there.

Figure 18.13 a Grasses, sweetcorn and sugar cane are examples of wind-pollinated plants. **b** Plants with bright flowers such as Hibiscus are usually pollinated with the assistance of birds or insects.

Figure 18.14 Structure of a wind-pollinated flower.

Adaptation	Wind-pollinated flowers	Insect-pollinated flowers
Pollen	Pollen adapted to detach from plant and blow easily in the wind; lightweight grains; produced in massive quantities	Pollen adapted to stick easily to insects' bodies; large, spiky grains; produced in smaller quantities
Flower	May be dull in colour; may have many small flowers	Large, prominent, attractive to birds and insects
Petals	Small, green or brown; no noticeable scent; no nectaries present	Large, brightly coloured; scented nectaries at the base
Stamen	Long thin filaments; anthers that hang outside the flower	Short; anthers inside the flower
Stigma	Large, branched, feathery stigma	Sticky stigma inside the flower
Pollen granules	Tiny, lightweight, produced in massive quantities	Large, sticky or spiky, produced in smaller quantities

Table 18.2 Adaptations of wind-pollinated and insect-pollinated flowers.

Now compare the structure of the insect-pollinated flower shown in Figure 18.12 with that of the wind-pollinated flower in Figure 18.14. What are the differences?

Fertilization

Like sexual intercourse in humans, plant pollination brings the male and female gametes into close contact. Once this has happened, the male and female gametes must fuse to form a diploid cell. Figure 18.15 shows how the male and female gametes come together for **fertilization** to occur.

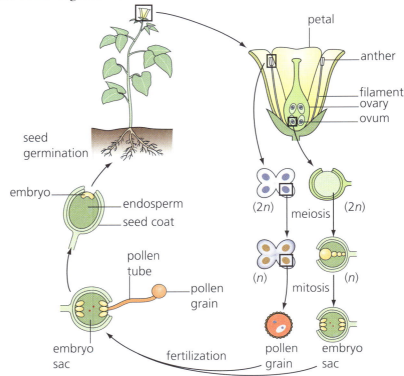

Figure 18.15 Fertilization in flowering plants.

Fruit and seed dispersal

For seeds to form new plants, they need to leave the parent plant. If all the seeds landed around the parent, they would be overcrowded, and the parent plant would not leave them enough light, water or nutrients to develop properly.

For this reason, plants rely on several methods of seed **dispersal**. Seed dispersal refers to any process by which plants spread their seeds to habitats for new germination. Plants disperse seeds by:

- gravity – for example by falling or rolling
- ingestion by animals – animals eat the fruits and discard or excrete the seeds
- wind dispersal – the seeds blow on air currents
- water dispersal – the seeds float away on water
- mechanical dispersal – relatively uncommon; when the fruit comes to maturity it explodes and dispels the seeds out of it
- fire – in some areas, natural fires cause seed pods to open and release seeds.

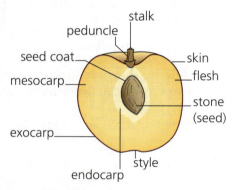

Figure 18.16 Succulent fruits, such as peaches, are adapted to attract animals: they are colourful, scented and have sweet, soft flesh that animals like to eat.

Animal seed dispersal

In deserts in North Africa, elephants eat fruits fallen from trees. The elephants then travel several miles away from the parent plant before releasing the seeds in their dung. The other components of dung provide an environment that helps the seed to germinate. In jungle habitats, monkeys and birds eat fruits and nuts, also dropping the seeds away from the parent plant in their faeces. Some animals collect seeds and bury them for later use, or discard the seeds as they eat them.

Animals can also aid dispersal of plants that do not produce edible fruits. Some plants have fruits or seed pods that have spiky, hooked or prickly structures (see Figure 18.17). They stick to the animal's fur or feathers, and then drop off the animal days later.

Water seed dispersal

Plants that grow near lakes, rivers, streams or seas can rely on water to spread their seeds. These plants tend to grow lightweight seeds that float easily. They may also have external fluffy structures that help the seeds to float on water. Trees that grow in coastal habitats often have woody, waterproof seed coats so that the seeds can float in salty seawater for long periods without damage to the plant embryo. Seeds that spread in this way include willows, coconut, sea beans and water mint (see Figure 18.18).

Wind seed dispersal

Some seeds are not only lightweight but are also shaped in a manner that will enable them to be carried away by the wind. Dandelions, cotton seeds and mahogany seeds are well adapted for wind dispersal (see Figure 18.19).

Figure 18.17 The seeds of the conker tree have spiky seed coats.

Figure 18.18 The seeds of the coconut tree are lightweight and have waterproof seed coats.

Figure 18.19 The seeds of the mahogany tree are shaped so that they are easily carried by the wind.

Check your knowledge and skills

1 Describe, giving examples, three examples of asexual reproduction in plants.
2 A strawberry plant reproduces asexually and sexually.
   a Sketch a strawberry plant and annotate your sketch to explain where sexual and asexual reproduction take place.
   b Write a paragraph explaining how each form of reproduction advantages the plant.
3 State the functions of the petals and sepals on a flower.
4 Describe how fertilization occurs after pollination.

Chapter summary

Do you know?

If you are unsure of any of the facts in the list, refer to the page number in brackets.

- All living organisms need to reproduce in order for the species to survive. (page 200)
- Sexual reproduction requires two parents of each sex and asexual reproduction involves a single parent. (page 200)
- Some methods of asexual reproduction include budding, fragmentation and regeneration, parthenogenesis, sporification and the production of stolons and rhizomes. (page 201)
- Meiosis is a type of cell division that produces gametes for sexual reproduction. (page 205)
- Gametes (or sex cells) contain half the genetic information of a normal cell. (page 205)
- Sexual reproduction involves the formation and fusion of gametes to produce new offspring. (page 205)
- Pollination is the way by which pollen grains reach the stigma. (page 206)
- The embryo remains in the seed, protected until the seed germinates and the new plant emerges. (page 207)

Are you able to?

If you have trouble in doing these things, refer to the page number in brackets.

- Describe the process of sexual reproduction. (page 200)
- Give examples of asexual reproduction in animals and plants. (pages 201 to 204)
- Identify advantages and disadvantages of sexual and asexual forms of reproduction. (page 205)
- Identify the reproductive organs of flowering plants. (page 206)
- Describe adaptations in wind- and insect-pollinated plants. (page 207)

Sexual reproduction in humans

Reproductive success

Humans, like most animals, are not capable of asexual reproduction. The human body has specialized sex organs that enable it to reproduce and thus maintain its population. The development of the human reproductive system starts before birth; although the sex organs only mature around the age of puberty.

In this chapter you will learn about how the human body reproduces and about ways in which pregnancy is prevented – contraception.

By the end of this chapter you should be able to:

■ Describe the process of meiosis, and identify the stages in meiosis I and II

■ Label the reproductive systems of a human male and a female

■ Identify stages in the menstrual cycle

■ Explain how fertilization takes place

■ Discuss the roles of hormones in human reproduction

■ List the main functions of the placenta in pregnancy and birth

■ Explain how various contraceptive methods work

Unit **1**

Meiosis

In most animals that reproduce sexually, each organism has primary sex organs which produce **gametes** (sex cells) as well as other organs which store and transport the sex cells. In humans, the primary sex organs are called the **gonads**. Males develop gonads called the **testes** which produce sperm. Females develop gonads called **ovaries** which produce eggs or ova.

The gametes, sperm and ova, are produced by a process called **meiosis**. Unlike the process of mitosis (see page 182), meiosis is a type of cellular division that produces non-identical cells. Each gamete contains only half of the genetic material found in other cells of the organism. Whereas the normal cells of the body are diploid, the gametes are haploid. Meiosis produces haploid cells.

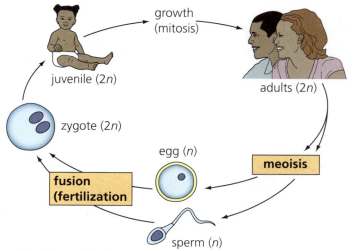

Figure 19.1 In the life cycle of diploid (2*n*) animals, each parent organism contributes a haploid (*n*) gamete in order to reproduce sexually.

Making gametes

Key facts
- Meiosis is a process of cell division which results in the production of four daughter cells from a single parent cell.
- It is the process by which gametes (sperm/eggs/pollen) are made. Therefore this process only occurs in organisms that reproduce sexually.
- The daughter cells are not identical to one another or to the original parent cell. Meiosis converts a diploid cell to haploid gametes.
- Meiosis involves two sets of divisions. We call these meiosis I and meiosis II.
- Meiosis increases diversity in the offspring.

By the time a female human is born, her reproductive organs are fully formed and all her eggs have been produced. At birth, these eggs are still immature. At puberty the eggs begin to mature, at a rate of one each month. In males the production of sperm only starts at the onset of puberty.

In some cases, two eggs may mature each month. Non-identical twins are born when both eggs are fertilized and the embryos develop and grow. Sometimes, both eggs may be fertilized, but only one embryo survives the pregnancy.

Key fact
The cell in Figure 19.2 has four pairs of chromosomes. Remember, humans have 23 pairs.

Phases of meiosis in humans
Meiosis has two stages – meiosis I and meiosis II (see Figure 19.2, overleaf). Each stage has five phases – prophase, metaphase, anaphase and telophase, with interkinesis forming the resting phase between the stages. During interkinesis, chromosomes shorten and condense into chromatids. There is also a development between prophase and metaphase, known as prometaphase, which is sometimes considered an additional phase. Look back at Figure 17.6 on page 185 to compare meiosis and mitosis.

Key facts
- **Chromatids** are strands of newly duplicated chromosome still joined together. One duplicated chromosome forms two chromatids.
- A **homologous chromosome** is a pair of chromosomes of the same type. (One from the mother and one from the father.) For example, a pair of chromosomes 'number 6' in the same cell is called homologous.
- The **centromere** is the structure that joins the chromatids.

Unit 2
Sexual reproduction in humans

Males and females each have a very complex set of reproductive organs specifically adapted for sexual reproduction.

The male reproductive system
- The male sexual organs are called the testes (singular testis) and they are located within the scrotum which hangs outside the body in the genital area.
- The testes have two main functions – to produce male gametes and to deliver these gametes to the female reproductive system. The structure of the testes is well suited for both these jobs.
- In male animals and humans the gametes are called spermatozoa (singular spermatozoan) or sperm.

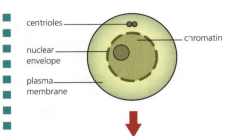

Interphase I occurs between divisions. It is often described as a 'resting stage' because the chromosomes are not visible inside the cell but it is actually a stage that involves intense cellular activity. The chromosomes have already been duplicated ready for cell division, but they are so long and thin they cannot be seen.

Prophase I this is where the already duplicated homologous chromosomes pair up and strands cross over, forming **chiasmata**. The chromosomes condense making them visible using a microscope. The nucleus starts to disappear. The cell contains 46 chromosomes (or 92 chromatids) at this point.

Metaphase I here the homologous chromosomes align at the equator (midline) of the cell. The alignment is random, with each parental homologue on one side.

Anaphase I now the crossed-over sections separate. The chromosomes move to opposite sides of the cell. The cells still contains 46 chromosomes – 23 at each side.

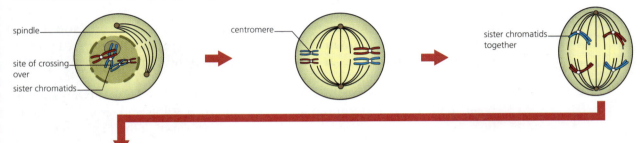

Telophase I the nuclear enevelopes form around each set of chromosomes forming two cells in preparation for meiosis II. There are 23 chromosomes in each daughter cell. **Cytokinesis** occurs, with each chromosome still consisting of two chromatids.

Prophase II here chromosomses in each cell shorten and become visible under the microscope. The nuclear envelope disappears. Each new cell contains 23 chromosomes (46 chromatids).

Metaphase II now the chromosomes (still connected by a centromere) line up on the middle of the cell. Spindle fibres coming from the centrioles connect to the centromeres from either side.

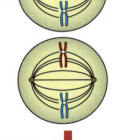

Anaphase II the centromeres split, allowing the sister chromatids to separate and move to opposite sides of the cell until they have formed two groups near the poles of the spindle. By late anaphase, the daughter chromosomes are no longer chromatids.

Telophase II the chromosomes, previously distinct and condensed, begin to disperse and become less visible under the microscope. A nuclear envelope re-forms around each cluster of chromosomes. The spindle begins to break apart and a nucleolus becomes visible in each daughter nucleus. Each of the four daughter cells contains 23 chromosomes

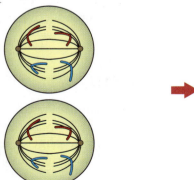

Figure 19.2 Meiosis.

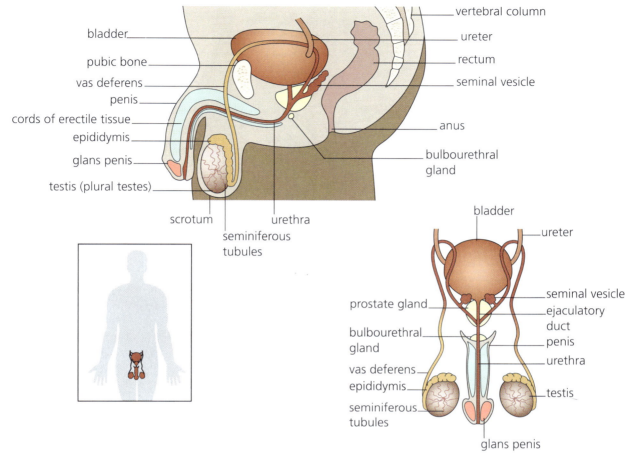

Figure 19.3 The male reproductive system.

Did you know?

The scrotum, that houses the testes, hangs outside the body because sperm cells survive better at temperatures which are lower than body temperature.

- There are usually two testes on a human male. Each contains coiled tubules called seminiferous tubules. The sperm are made inside these tubules. Once formed, the sperm move to the epididymis which is the first part of the vas deferens. In the epididymis the sperm mature further and are stored until ejaculation (release of sperm from the penis inside the vagina). A typical human male will make more than 100 million sperm each day.
- The testes are endocrine glands (see page 166) which produce a hormone called **testosterone** which is responsible for the development of secondary sexual characteristics in males.
- Male secondary sexual characteristics develop around puberty. They include a deepening of the voice, growth of facial and pubic hair, increased penis size and a more muscular body shape.

The female reproductive system

The female reproductive system has several functions:

- to produce female gametes (eggs)
- to receive sperm from the male
- to maintain an environment suitable for the implantation of an embryo
- to support the development of the foetus.

The female reproductive organs are situated inside the body in the pelvic region (see Figure 19.4, overleaf). The vagina is a muscular channel about 4 cm in length

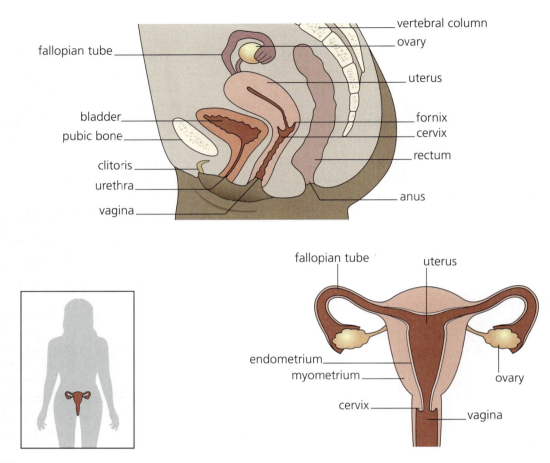

Figure 19.4 The female reproductive system.

that forms the entrance to the inner parts of the female reproductive system. During sexual intercourse the penis enters the vagina and deposits sperm at ejaculation. The vagina leads to the cervix (sometimes called the neck of the womb). The **uterus** (or womb) is a hollow pear-shaped organ. This is where implantation of the embryo takes place after fertilization. This is also where the foetus develops until birth.

Each month the ovaries release an unfertilized ovum (egg). The **fallopian tubes** convey the egg from the ovaries to the uterus. The ovaries are also where hormones such as **oestrogen** are made. This hormone is responsible for the secondary sexual characteristics in females. These characteristics include enlarging of the breasts, growth of hair under the arms and in the pubic areas, and a broadening of the hips.

Key facts

The ovaries are responsible for:

- producing ova (this happens before birth)
- maturing and releasing ova (from puberty until menopause)
- producing hormones such as oestrogen.

Menstruation

Upon reaching **puberty**, at around 12 or 13 years old, a female begins to ovulate. In other words, she begins to release a matured ovum each month from her ovaries. Ovulation is part of the female **menstrual cycle** which begins at puberty and ends at **menopause** (which takes place in most women between the ages of 45 and 50). The length of the menstrual cycle varies in different women but it is usually a cycle of around 28 days involving the following stages.

- **Menstruation** or menses – during days 0–7 of the cycle, the lining of the uterus disintegrates. Blood and tissue are expelled through the vagina.
- Development of the follicle – over the first 10 days of the cycle, a follicle matures in one of the ovaries. The follicle is a group of cells which contains a single immature ovum or egg. As the follicle matures, it releases oestrogen.
- Thickening of the uterus lining – the oestrogen released by the follicle causes the endometrium (lining of the uterus) to thicken with blood vessels. The endometrium provides an environment that is able to sustain a growing foetus.
- **Ovulation** – at day 14 of the cycle, luteinising hormone is released by the pituitary gland (controlled by gonadotropin-releasing hormone, or GnRH, from the hypothalamus). This triggers the release of the mature egg from the follicle.
- Development of the corpeus luteum – the empty follicle left behind in the ovary, now called the corpus luteum, begins to release another hormone called **progesterone**. This hormone develops the uterus lining in preparation for a possible pregnancy. If fertilization does not occur, the release of progesterone is inhibited, and the lining of the uterus is shed. Thus menstruation begins again.

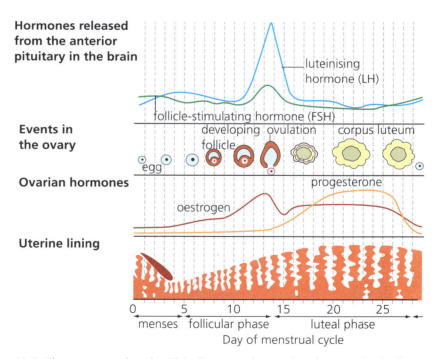

Figure 19.5 The menstrual cycle. This diagram shows the changes in the hormones, in the ovaries, and the changes to the uterus that take place over each cycle. Note that the hormones in the menstrual cycle create a negative feedback system (see page 144).

Menopause

The menstrual cycle continues until age 45–50 years, at which time the menopause begins. The menopause is the natural cessation of the menstrual cycle, signalling the end of a woman's reproductive age. In other words, she will no longer be able to fall pregnant. Ovulation and menstruation cease around this time.

Courtship and pair bonding

In most mammals, including humans, males and females engage in courtship behaviour before sexual intercourse. Courtship is any behaviour that leads to mating and reproduction. Males may try and impress females in various ways, and usually a female chooses the male that appears most likely to father and provide for her young.

Courtship also often develops a pair bond between the male and female partners that may help them to care for their offspring together. Pair bonding may be short-term (for the purposes of mating only), long-term (lasting a significant period of the animal's life cycle), or even lifelong (lasting until one or both animals die). Humans are not the only animals that demonstrate pair bonding behaviour; many birds and mammals demonstrate this too. Successful courtship leads to copulation (sexual intercourse) and fertilization.

Sexual intercourse

During sexual intercourse, sperm leave the epididymis and flow through the tubular structure called the vas deferens towards the penis. Before reaching the penis the sperm pass through a gland called the **prostate gland** and a vesicle called the seminal vesicle. These two structures produce fluids which bathe the sperm as they move towards the penis. The fluids from the prostate and seminal vesicle provide a medium for the sperm to swim in after they are ejaculated in the vagina. The mixture of sperm and seminal fluid is called semen.

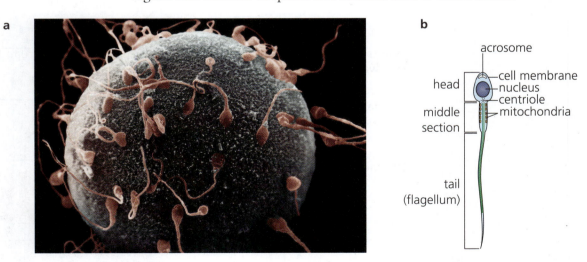

Figure 19.6 a Human sperm cells approaching the ovum for fertilization; **b** the structure of a human sperm cell.

Each spermatozoan is shaped like a tadpole (see Figure 19.6b). The head area is where the haploid nucleus is located. The tip of the head contains a small sac called the acrosome. This contains hydrolytic enzymes responsible for burrowing through the exterior of the egg cell during fertilization. The 'neck area' contains lots of mitochondria, which indicates that a lot of energy is generated here. The tail is for the purpose of allowing the sperm to be motile (to be able to swim up the vagina). Sperm can survive for up to four days in the body of the female after ejaculation.

Fertilization and pregnancy

In biological terms, sexual intercourse (also known as coitus or copulation) is the act of bringing the male reproductive organ into the female reproductive tract for the purpose of fertilization.

Usually intercourse begins with sexual arousal. In the male, arousal causes the erectile tissue inside the penis to fill with blood, causing the penis to lengthen and harden. In the female, arousal causes lubrication of the vagina. The male inserts his penis into the female's vagina, and the act of intercourse gradually increases the sense of sexual arousal until the point of ejaculation, when semen is released into the vagina. (Small droplets of semen escape from the penis during intercourse, so even without ejaculation, intercourse still brings the gametes into very close contact.) At the point of ejaculation, 5 ml of semen is released – containing around 500 million spermatozoa. The sperm, which are highly motile, use their long tails to propel themselves towards the unfertilized ovum.

If a viable egg is present, many sperm surround it and try to penetrate the cell membrane. If one sperm successfully enters the egg, a hard exterior forms around the egg preventing any more sperm from entering. The haploid nucleus in the sperm that enters fuses with the haploid nucleus of the egg cell. This is called fertilization. This creates a single diploid cell called a zygote. The fertilized egg continues to duplicate using mitosis. Eventually it becomes an **embryo**. About a week after fertilization the embryo embeds into the thickened uterus wall. At this point the woman is said to be pregnant. After about eight weeks, the embryo has formed a **foetus**.

> ## Did you know?
> It is during the embryonic stage of development that the growing embryo is most vulnerable to environmental factors such as alcohol and drugs.

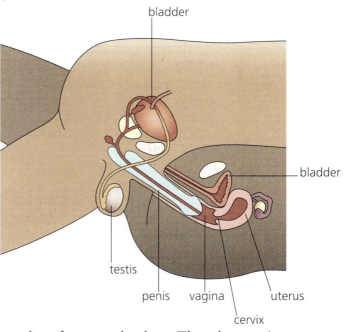

bladder

bladder

testis

penis   vagina   uterus

cervix

Figure 19.7 Sexual intercourse brings the gametes into close contact, increasing the chances of fertilization.

After implantation the **placenta** develops. The placenta is an organ that is engorged with blood vessels and is responsible for several functions:

- it delivers oxygen to the foetus via the umbilical cord
- it protects the foetus from disease and infection
- it produces hormones that support pregnancy

- it prepares the mother's body for lactation (producing milk)
- it stimulates the uterus walls during labour to allow birth to take place.

The placenta allows for exchange of substances between the mother's blood and foetal blood without the two bloods coming into direct contact with each other.

Check your knowledge and skills

1 What characteristic must the membrane of the placenta have so that it can carry out its job efficiently?

2 What are the major structural differences between the egg and sperm?

Did you know?

Each spermatozoan carries a single sex chromosome that determines whether the embryo is male or female. Female sperm swim more slowly towards the ovum, but can survive longer than male sperm.

Contraceptive methods

The human body has evolved so that sexual intercourse frequently and successfully leads to pregnancy. However, many couples wish to have sexual intercourse without pregnancy. **Contraception** is the name for any method of preventing fertilization. Some contraceptive devices offer protection against diseases such as **sexually transmitted diseases (STDs)**.

Natural contraception methods

The rhythm method

The rhythm method is also known as natural family planning (NFP). It involves keeping a detailed record of the woman's menstrual cycle (see Figure 19.8). The couple abstains around the time of ovulation. The woman may keep a regular calendar to watch her menstrual cycle, and track her body temperature around the time of expected ovulation (body temperature rises very slightly at ovulation). Because women's cycles vary widely, it is very difficult to follow this method reliably.

Withdrawal

Withdrawal refers to the practice of removing the penis from the vagina before ejaculation takes place. Many people would not even consider this to be a form of contraception, as most sexual intercourse involves the release of some semen during penetration, before ejaculation takes place. Even 1 ml of semen contains hundreds of thousands of sperm, so fertilization is still highly likely.

Check your knowledge and skills

1 a Using the information in the text, can you work out why some days on the menstrual calendar are marked 'good for conceiving a boy' and others are marked 'good for conceiving a girl'?

  b How do you think this was worked out?

2 Give three reasons that the rhythm method is not a safe method of contraception for couples that seriously do not want to conceive.

Figure 19.8 The rhythm method involves the woman charting her menstrual cycle.

■ Surgical contraception methods

Surgical methods involve cutting and tying off the passageways along which the gametes travel to the reproductive organs. In men, this is called a vasectomy. A doctor cuts and ties off the vas deferens. In women, a doctor cuts and ties off the oviducts. This is called tubal ligation.

Vasectomies and tubal ligations are irreversible forms of surgery, so they are most suitable for older couples who are certain that they do not want more children.

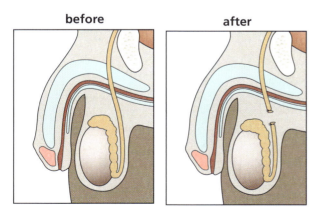

Figure 19.9 A vasectomy – a surgical contraceptive method in the male.

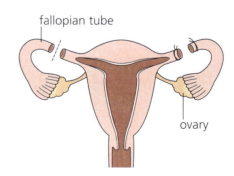

Figure 19.10 Tubal ligation – a surgical contraceptive method in the female.

Barrier contraception methods

Barrier methods include condoms and diaphragms. They are usually made from latex rubber. A condom is a thin latex rubber sheath, worn over the erect penis during intercourse (see Figure 19.12). It prevents the sperm from coming into contact with the egg. The female condom works in a similar way, but it is inserted into the upper end of the vagina before intercourse. Condoms are single-use devices which must be thrown away after use.

The diaphragm is a re-usable rubber dome which is inserted into the upper end of the vagina (see Figure 19.13). Like a condom, it prevents the sperm from coming into contact with the egg. All barrier devices are most effective when used with spermicidal jellies.

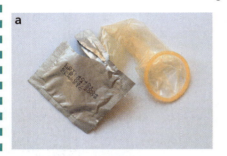  

Figure 19.11 Barrier methods of contraception: **a** male condom; **b** female condom; **c** diaphragm.

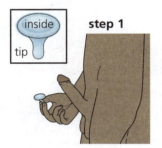

step 1

First, the penis must be erect. Check the condom package is unbroken before use. Carefully tear the corner of the packet and squeeze out the condom.

step 2

Place the condom over the tip of the penis.

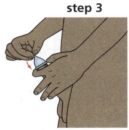

step 3

Roll the condom downwards towards the base of the penis.

step 4

Ensure the penis is entirely covered before penetration.

Figure 19.12 Putting a condom on the penis.

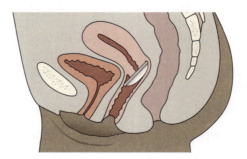

Figure 19.13 The position of the diaphragm in the female.

Hormonal contraception methods

Hormonal methods include the contraceptive pill, ring, patch or injection. Each of these methods delivers a combination of hormones (progesterone and oestrogen) that suppresses and thickens the cervical mucus in the woman.

- Contraceptive pill – this the most common form of hormonal contraception. It must be taken daily in order to be effective.
- Hormone injections – these are administered every three months by a doctor.
- Hormonal ring – this is a small, flexible ring that the woman inserts into her vagina for three out of every four weeks; it releases hormones.
- Patch – a sticker worn by the woman for three weeks at a time; it releases hormones into the skin that are absorbed by the body.
- Hormonal implant – this is a small plastic device, the size of a matchstick, which is surgically implanted under the skin of the arm; it is effective for about three years.

a
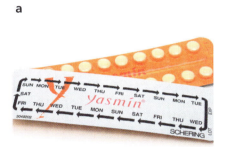

b
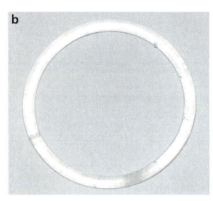

c

Figure 19.14 Hormonal methods of contraception: **a** the pill; **b** hormonal ring; **c** hormonal patch.

The 'morning after' pill

Do not confuse the oral contraceptive pill with the 'morning after' pill. The 'morning after' pill is a type of emergency contraception that can be taken after having unprotected sex. This preparation is usually used to block implantation if fertilization has occurred. One preparation, known as RU486, can also get rid of the foetus even if implantation has occurred. This has caused lots of controversy since it is looked on as a form of abortion.

Other contraceptive methods

The **intrauterine device (IUD)** is a T-shaped plastic device that is inserted into the uterus (see Figure 19.15, overleaf). There are slightly different versions of this device available. The original 'copper T' has a coil of copper around it. The device irritates the lining of the uterus, making implantation of a fertilized egg difficult. White blood cells move to the lining of the uterus and help to destroy the sperm. The Mirena, a more recent version of the IUD, acts in the same way, but releases small amounts of hormones which thicken the cervical mucus; this acts as an additional barrier to sperm.

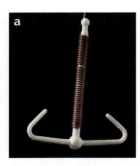
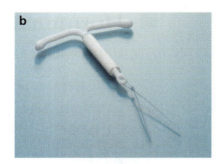

Figure 19.15 Intrauterine devices: **a** original 'copper **T**'; **b** Mirena.

Abstinence

Abstinence refers to the practice of refraining from sexual intercourse entirely. Abstinence is only effective for as long as both parties abstain, however. As soon as the couple breaks its resolve to remain abstinent, they may be at high risk for unwanted pregnancy and STDs if they do not use contraception.

Biology at work

Chapter 19 – Does abstinence work as a contraceptive?

Type	Method	Advantages	Disadvantages
Natural	Rhythm (also called natural family planning or NFP)	No devices or drugs necessary; no cost involved	Very unreliable as sperm can survive for several days after ejaculation; requires strict planning
Natural	Withdrawal		Very unreliable as some sperm are released before ejaculation
Surgical	Tubal ligation	100% effective; no further contraceptive methods required	Irreversible, so only suitable for people who are certain they will never want more children
Surgical	Vasectomy		
Barrier	Male condom	99% reliable when used properly; protects against STDs; male condom widely available; female condom can be inserted some hours in advance	May interrupt arousal or reduce penis sensitivity; some people complain about the sound of the latex; female condoms not widely available
Barrier	Female condom or diaphragm		
Hormonal	Contraceptive pill	Highly reliable	Need to remember to take it regularly
Hormonal	Contraceptive injection	Highly reliable; no need to remember to take pills	Needs to be repeated every 13 weeks; no protection against STDs
Hormonal	Spermicidal jellies	Highly effective when used with barrier devices	Ineffective alone
Hormonal	Ring or patch	Effective when used properly	The ring may not be placed properly, or may come loose; patch ineffective if it becomes loose or falls off
Hormonal	Implant	Highly effective; no further effort or cost after initial implantation; only has to be replaced after 3 years	Not widely available; must be removed after 3 years as it can interfere with menstrual cycle
Other	Intrauterine devices (IUDs)	99% reliable	Can be painful during menstruation; no protection against STDs
Other	Abstinence	Protects against STDs	Sudden change of mind can lead to a highly risky position for unwanted pregnancy and STDs

Table 19.1 A comparison of different contraceptive methods.

Check your knowledge and skills

1 Identify the following stages in meiosis.

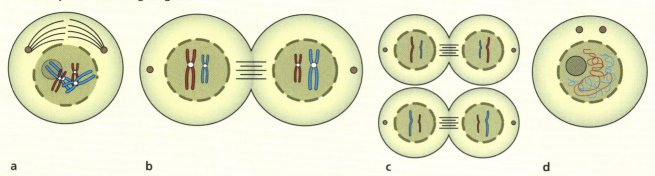

a   b   c   d

Figure 19.16

2 Make a sketch of the following diagrams of the male and female reproductive organs and add labels.

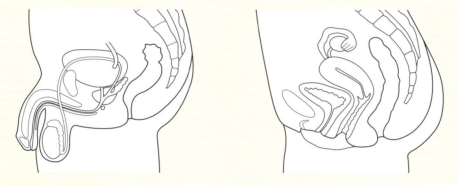

Figure 19.17

3 Where on your diagrams do the following take place?
  a sperm is deposited
  b egg matures
  c fertilization
  d thickening of the endometrium
  e implantation of the embryo.
4 Explain the role of hormones in controlling the female menstrual cycle.
5 An egg is released from a woman's ovary on 8 July.
  a Over which days is fertilization most likely to occur?
  b When is the next 'safe period' for intercourse to occur without conceiving?
  c If fertilization does not occur, state when the woman is most likely to start menstruating.

Chapter summary

Do you know?

If you are unsure of any of the facts in the list, refer to the page number in brackets.

- Meiosis is the type of cell division that is involved in the making of gametes for sexual reproduction. (page 210)
- In humans, the male gametes are known as sperm and the female gametes are known as ova. (page 210)
- Sperm are produced in the testes of the male. The prostate and seminal vesicle provide the fluids for the sperm to swim in when they are ejaculated from the penis. (page 211)
- Ova are produced in the ovaries of the female. (page 213)
- Sexual intercourse in humans, like pollination in plants, is the process that brings male and female gametes into close contact. (page 216)
- Sexual reproduction requires two parents, one of each sex. (page 217)
- Fertilization is the process of fusing the male and female gametes. (page 217)
- Contraception is any method that prevents fertilization. (page 218)
- Condoms are the only contraceptive method that also help to prevent the transmission of sexually transmitted diseases. (page 222)

Are you able to?

If you have trouble in doing these things, refer to the page number in brackets.

- Identify the stages in meiosis I and II. (page 212)
- Label diagrams of male and female reproductive systems. (pages 213 and 214)
- Identify the main hormones involved in human reproduction, and their functions. (pages 213 to 215)
- Describe the stages in the menstrual cycle. (page 215)
- Describe the process of fertilization. (page 217)
- Explain the functioning of contraceptive methods. (pages 218 to 222)

How characteristics are passed on and changed

Take a look at your classmates. Even though we are all part of the same species, no two people look quite the same. Biologists call this **genetic diversity**.

In this chapter you will learn why there are so many differences between people and how they come to have them.

By the end of this chapter you should be able to:

■ Define the terms chromosome, gene, allele, genotype and phenotype

■ Discuss dominant and recessive genes, incomplete dominance, co-dominance

■ Use Punnett squares to explain inheritance of single pairs of characteristics

■ Distinguish between continuous and discontinuous variation

■ Explain the importance of genetic variation

■ Predict the results of crosses

■ Describe how sex in humans is determined

Unit 1 Mendel and genetics

In the late 1800s a monk called Gregor Mendel conducted experiments using pea plants. His findings formed the basis of our current understanding of inheritance and how characteristics get passed on from one generation to the next.

Mendel chose to work on pea plants because they have characteristics that occur in a very simple form (today known as Mendelian traits). He worked in a very systematic way. He was particularly careful that his plants should not accidentally receive any foreign pollen, as this would invalidate his results. During his research, he made nearly 300 crosses between about 70 different pure-bred plants – he worked with approximately 28 000 pea plants in total!

Table 20.1 overleaf shows the traits (characteristics) of pea plants that are inherited. The trait can either be 'dominant' or 'recessive' – a dominant trait is expressed in a pea plant and a recessive trait is not.

Figure 20.1 Mendel is known as the father of modern genetics because of his work on pea plants.

What Mendel noticed

Mendel observed that when he bred two pea plants, a particular trait would disappear from the first filial generation (F_1 generation). However, in the next generation, (the second filial generation or F_2 generation), the trait would reappear.

For example, let's say he had a pea plant that, when self-pollinated, would only produce green peas. We can call this plant pure-bred to green. He then cross-bred this with a pure-bred yellow pea. If the results are that all the offspring have yellow peas, it would initially seem that all the genetic information for green peas was lost. However, if in the next generation the green peas reappear, we can conclude that the green variation is 'masked' by the yellow. Mendel used this kind of experiment to conclude that the gene for yellow peas is a **dominant gene** (usually expressed whenever it is present), and the gene for green peas is a **recessive gene** (only expressed when the dominant gene is not present).

Trait	Dominant expression	Recessive expression
Form of ripe seed	Smooth seed coat	Wrinkled seed coat
Colour of seed albumen	Yellow	Green
Colour of seed coat	Grey	White
Form of ripe pods	'Inflated'	'Constricted'
Colour of unripe pods	Green	Yellow
Position of flowers	Axial	Terminal
Length of stem	Tall	Dwarf

Table 20.1 Inherited traits in pea plants.

Labelling for dominant and recessive genes

The usual notation for labelling genetic variations is to use a letter that represents the trait. The capital letter represents the dominant trait and the lower case letter represents the recessive trait. For example, we might use the letter 'g' to represent pea colour.

The combination of genes on an organism's chromosomal DNA is the **genotype**. For pea colour, there are three possible genotypes: GG, Gg and gg. The genotype determines what trait is expressed – the **phenotype**. There are two possible phenotypes for pea colour: yellow peas or green peas. The genotypes GG and Gg would both result in yellow peas. Yellow is dominant over green. The genotype gg would result in green peas.

How to draw a Punnett square

A **Punnett square** is a table that shows the possible outcomes of gene combinations for a particular trait. The Punnett square is named after the geneticist Reginald Punnett who discovered some of the principles of genetics, including sex determination. The following steps are followed when drawing and using a Punnett square.

■ Assign a letter to represent the trait. For example, in the squares below, the letter 'g' represents the trait of pea colour.

■ Use a capital letter to represent the dominant form of the trait (in this case 'G' represents yellow) and a lower case letter to represent the recessive form of the trait (in this case 'g' represents green).

■ Draw a small square, divided into four blocks.

■ Along the top two blocks, write the genotype of the first parent, and down the left-hand side, write the genotype of the second parent. (Remember, the genotype is the combination of genes; the phenotype is the expressed characteristic.) The Punnett square in diagram 1 represents a first generation cross of a pure-bred yellow (GG) plant with a pure-bred green plant (gg).

■ Now write the genotypes in each of the empty blocks. (The capital letter is written before the lower case letter.)

■ Then write the phenotypes under each resulting genotype (see diagram 2).

①

	g	g
G		
G		

②

	g	g
G	Gg yellow	Gg yellow
G	Gg yellow	Gg yellow

③

	G	g
G	GG yellow	Gg yellow
g	Gg yellow	gg green

As you can see, all the offspring have one dominant and one recessive gene for pea colour. All the offspring in diagram 2 express the dominant trait.

The Punnett square in diagram 3 shows what happens when two of the Gg offspring are crossed to produce the second generation. In the second generation, one plant has two dominant genes for the colour (**homozygous dominant**), two have a mix of information (**heterozygous**), and one has two recessive genes (**homozygous recessive**). One in four pea plants express the recessive trait (green peas) and three in four express the dominant trait (yellow peas).

The structure of chromosomes

Some characteristics or traits from parents are passed down to the next generation – the study of how this is done is called **heredity**. In the early 19th century, biologists spent much time debating what caused our inherited characteristics. They discovered that each complete cell in a living organism contains a nucleus that contains all the information that determines the organism's characteristics.

Our **chromosomes** are composed of **deoxyribonucleic acid (DNA)** which contains individual pieces of genetic information called **genes**. A molecule of DNA takes the form of a double helix, which looks a bit like a spiral staircase (see Figure 20.2, overleaf). The building blocks of DNA are called nucleotides. Each nucleotide consists of three molecules:

■ a phosphate
■ a sugar, and
■ a base.

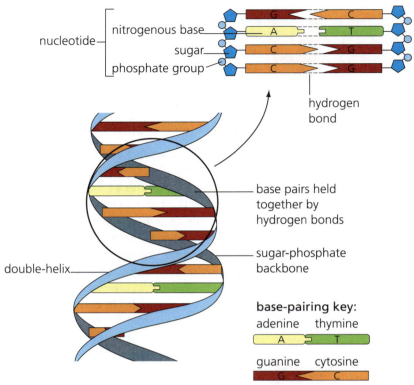

nucleotide ⎯ nitrogenous base
sugar
phosphate group

hydrogen bond

base pairs held together by hydrogen bonds

sugar-phosphate backbone

double-helix

base-pairing key:

adenine   thymine

A   T

guanine   cytosine

G   C

Figure 20.2 The structure of DNA.

Term	Definition	How we use the term
DNA	Deoxyribonucleic acid; a double-stranded chain of nucleotides; the arrangement of base pairs in the DNA determines genetic information	DNA is the substance that makes up all the chromosomes
Chromosomes	The self-replicating genetic structure in cells containing the cellular DNA; human beings have 23 pairs of chromosomes	A gene is always located at the same position on a given chromosome
Gene	A segment of DNA that determines a particular characteristic or set of characteristics; each chromosome contains around 100 000 genes	If a person has a dominant tongue-roller gene, they are able to roll their tongue
Allele	A particular form of a gene; one member of a pair of genes	There are two alleles of the tongue rolling gene – tongue-roller (R) and non-tongue-roller (r)
Dominant	An allele that is usually expressed, even if only one copy is present	Tongue-roller (R)
Recessive	A gene which is expressed only if there are two copies	Non-tongue-roller (r)
Homozygous	Having two identical alleles of a gene; they may be dominant or recessive	rr or RR
Heterozygous	Having two different alleles for a particular trait in the cells of the organism	Rr
Phenotype	The physical characteristics of an organism	Person can roll their tongue or not roll their tongue
Genotype	The genetic constitution of an organism, the combination of alleles for a particular trait	Rr, rr, RR

Table 20.2 Inheritance terms – definitions and examples of their use.

Each pair of chromosomes carries a matching set of genes (one inherited from the mother and one from the father). An **allele** is the term given to each of the gene sets in a pair of genes. For example, the gene for eye colour has an allele for blue eye colour and an allele for brown eye colour. Your eye colour depends on the combination of alleles you have inherited from your parents. In a pair of chromosomes, you may have two identical alleles (a homozygous gene pair) or two different alleles (a heterozygous gene pair) for a particular characteristic. Table 20.2 summarizes the terms used in inheritance studies.

Unit 2

Genetic variation

One of the main features of sexual reproduction is that the offspring produced are never identical to either their parents or their siblings (except in identical twins who are genetically identical to each other). Aside from identical twins, however, every individual in a species has its own genetic make-up. Some examples of genetically determined characteristics are:

- height
- eye colour and shape
- voice tone
- skin tone.

Most of our variations are inherited. In other words, they are determined by the genetic material we receive from our parents. However, some variations are determined by our environment. Variations caused by environmental factors cannot be passed on to the next generation unless their effect causes a change in the DNA.

Continuous and discontinuous variation

Consider the range of heights in your class. Let's say you took three people of different heights. You might describe their heights as short, medium and tall (see Figure 20.3, overleaf).

If you collected more examples of people from a wider range of ages, you might find that all the people in your original sample fell into the average category.

Continuous variation is influenced by genes and the environment. Genes give a person the ability to grow by providing the blueprint for the growth hormones. However, environment also influences their height. Environmental factors, such as nutrition, can affect the person's growth. In the case of plants the availability of light, water and fertile soil can affect the growth rate.

Some characteristics fall into more distinct groups. For example, there are four blood types, and each person's blood type falls into one of these groups. Similarly, there are two sexes – a person is either male or female. Even though there are many different colours of human eyes, eye colour is another variation that falls into several distinct groups. These kinds of variations are called **discontinuous variation**. Discontinuous variation is usually determined by our genes. For example, a change in the environment will not affect your blood type or your sex.

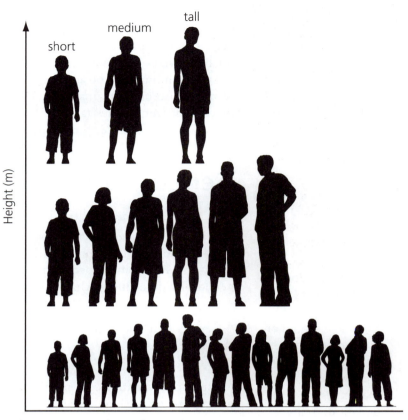

a Three teenagers of different heights. If you collected three more examples, you might find that your descriptions would change.

b Now the girl labelled 'tall' has fallen into our medium category in comparison to the other people we have measured.

c Height is an example of continuous variation; there is a continuous gradation from one extreme to the other. Other examples are skin colour, hair texture and foot length.

Height (m)

Figure 20.3: How continuous variation differs from non-continuous variation.

Activity 20.1 Inherited versus environmental characteristics

Questions

1 For each of the characteristics listed below, discuss whether it is inherited or environmentally influenced. If it is inherited, determine whether it is continuous or discontinuous.
   a blood group – there are four blood groups, A, B, AB and O
   b hand-clasping – when you clasp your hands together, some people naturally place the right thumb on top, and others place the left on top
   c freckles – some people have no freckles; others have few to many
   d eye colour – eye colours are brown, blue, green, gray or hazel
   e body weight – varies from light to heavy
   f hair colour
   g speaking accent
   h foot length
   i finger print pattern
   j ability to roll tongue
   k whether earlobes are attached or free
   l hairline – rounded or pointed
   m hair along the middle of the back of the finger

Figure 20.4 This athlete needs to be fast, strong and agile. Her athletic skills are in part a result of her genes and in part a result of her environment. Many hours of training, a proper diet and a suitable lifestyle help to improve her race times.

handclasping (right or left thumb on top)

tongue rolling (can or can't)

earlobes (attached or unattached)

hairline (widows peak or straight)

Figure 20.5 Examples of discontinuous traits. This type of trait is not influenced by environment.

Activity 20.2 Survey of types of variations

Equipment and apparatus
- notebook
- pen or pencil
- meter rule

Procedure
1 Select 25 to 50 subjects all of the same age and gender.
2 Measure the following continuous characteristics:
  a height
  b length of index finger or length of foot
3 Determine the following discontinuous characteristics for each person:
  – eye colour
  – ability to roll tongue
  – attached or free earlobes.
4 For the continuous variations surveyed, construct a histogram of the results. For the discontinuous variations surveyed, construct a bar chart of the results.

Unit 3

Patterns of inheritance

In the first unit of the chapter we looked at Mendel's discoveries about dominant and recessive traits in peas. During Mendel's lifetime, his research went unrecognized. It was about 15 years after his death that scientists started to realize the significance of his discoveries. In this unit you will learn about how genes get passed from one generation to the next.

Dominant and recessive genes

Even though an organism produced as a result of sexual reproduction receives equal amounts of DNA from each parent, it does not express its physical characteristics in equal proportions. For example, if your mother has blue eyes and your father has brown eyes, it is very unlikely that you have one blue eye and one brown eye. This is because some genes override others in the expression of inherited characteristics. In most gene pairs, each gene tends to be either recessive or dominant.

If a dominant and a recessive gene occur together (a heterozygous pair), the dominant gene is usually expressed. Recessive genes tend only to be expressed when two recessive genes are both present (homozygous pair).

Key facts

■ Sexually reproduced organisms receive two sets of genes for each characteristic – one set from the female parent and one from the male parent. There are two copies of each gene in a typical cell.

■ Each pair of genes is made up of two alleles. Each pair of chromosomes carries one allele, at corresponding positions on the chromosome, for a particular trait.

■ A physical characteristic (for example brown eyes) is known as a phenotype; the genetic composition of a gene pair is called its genotype. For eye colour, if B represents the dominant allele for brown eyes and b represents the recessive allele for blue eyes, the possible genotypes are BB, bB, Bb and bb.

■ If both the dominant allele and the recessive allele are inherited for the same characteristic, only the dominant allele is expressed.

■ When an individual expresses a dominant trait, there are several possible genotypes (for example BB, bB, Bb), but when an individual expresses a recessive trait, there is only one possible genotype (bb).

■ Humans have 46 chromosomes (23 pairs) in each cell, except for the gametes (sperm/egg) and red blood cells.

■ Gametes have half the usual number (23 pairs). Red blood cells do not have a nucleus and therefore have no chromosomes.

Genetic crosses

Suppose that eye colour is controlled by one gene with one pair of alleles. Let 'B' represent brown (dominant) and 'b' represent blue (recessive). Figure 20.6 shows a genetic cross diagram for the inheritance of eye colour involving two heterozygous parents (each is 'Bb' for eye colour).

The genetic cross in Figure 20.6 suggests that these parents have a 1 in 4 chance of having a baby with blue eyes and a 3 in 4 chance of having a brown-eyed baby.

Now consider the possible offspring of different set of parents – a heterozygous brown-eyed parent and a homozygous blue-eyed parent as shown in Figure 20.7.

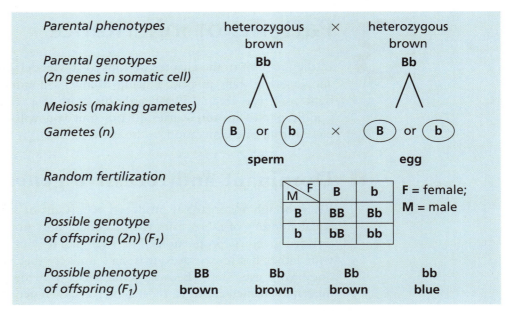

Figure 20.6 Genetic cross diagram for eye colour inheritance.

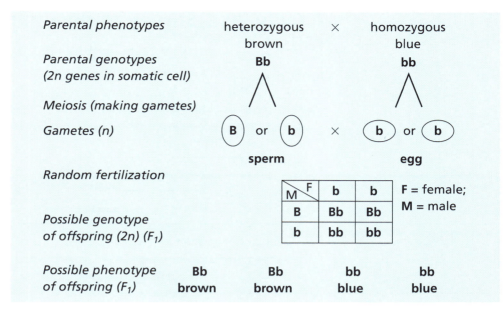

M \ F	b	b
B	**Bb**	**Bb**
b	**bb**	**bb**

F = female;
M = male

Figure 20.7 Genetic cross diagram for eye colour inheritance.

Test crosses

A **test cross** determines whether a particular characteristic of a plant or an animal is homozygous dominant (pure bred) or heterozygous dominant (hybrid). This is done by cross-breeding the animal or plant with a known pure bred with the recessive phenotype.

Activity 20.3 Test crosses

This cross suggests that these parents have a 50% (or 1 in 2) chance of having a baby with blue eyes or brown eyes.

Questions

1 Think of real-life situations in which we would need to:
   a carry out a test cross of an animal
   b carry out a test cross of a plant.
2 In dogs, the recessive gene 'd' causes deafness. A dog breeder has a male show dog that she wants to use for breeding. The show dog can hear, but the breeder only wants to use him for breeding if he does not carry the recessive gene.
   a A deaf female dog has the genotype 'dd'. What kind of gametes can she produce?
   b What are the possible genotypes of the male show dog?
   c What kind of test cross must the breeder do in order to work out the show dog's genotype?
   d Draw two Punnett squares to show the possible outcomes for each genotype.

Activity 20.4 Survey of dominant and recessive traits

Questions

1 Table 20.3 lists several human traits that are determined by simple dominant/recessive mode of inheritance. Make a copy of the table and record the phenotypes and the percentage of the phenotypes among students in your school.

Trait	Phenotype	Tally	Number of students with phenotype	
			number	percentage
Handedness	Right-handed*			
	Left-handed			
Ear lobes	Free*			
	Attached			
Hairline	Widow's peak*			
	Straight			
Little finger	Bent*			
	Straight			
Tongue rolling	Yes*			
	No			
Rh Factor	Rh⁺*			
	Rh⁻			

Table 20.3 Human traits determined by a single gene.
* indicates dominant allele

Incomplete dominance

For some phenotypes, neither of the alleles is dominant or recessive. Instead, each allele is partially expressed. One example of this is the gene for flower colour. A snapdragon, for example, has two alleles for flower colour. One produces white flowers and the other produces red flowers. However, when a plant with red flowers is cross-bred with a plant with white flowers, some of the resulting plants have pink flowers. This is called partial dominance or **incomplete dominance**.

Figure 20.8 shows the cross for homozygous red (RR) and homozygous white (R*R*) snapdragons.

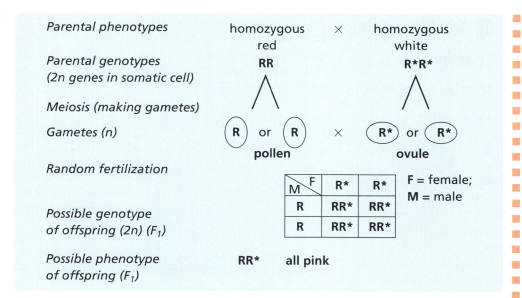

Figure 20.8 Genetic cross diagram for inheritance of flower colour in snapdragons.

From this cross it can be seen that all offspring are heterozygous and exhibit a pink phenotype, i.e they have pink flowers.

Figure 20.9 shows the cross for two heterozygous pink (RR*****) snapdragons.

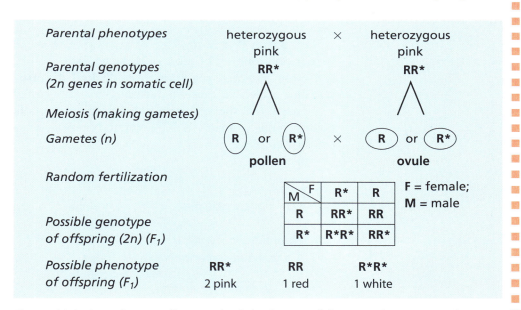

Figure 20.9 Genetic cross diagram for inheritance of flower colour in snapdragons.

■ Sex determination

A normal human cell contains 23 pairs of chromosomes (see Figure 20.10, overleaf). 22 of these pairs are **autosomes** (any genetic material that does not determine sex). The 23rd pair is the sex chromosomes. In females, both sex chromosomes are similar and relatively long. They are called X chromosomes. In males, the sex chromosomes are different. There is one X chromosome and one Y chromosome. The presence of the Y chromosome determines the development of male sexual characteristics.

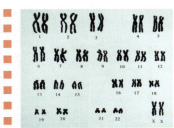

Figure 20.10 A full set of human chromosomes as seen under a microscope. The chromosomes have been coloured to make them easier to see. The 23rd pair of chromosomes shows the sex chromosomes. Note that the X chromosome is longer than the Y chromosome.

In women, meiosis produces gametes in the form of ova that carry X chromosomes. In men, meiosis produces gametes in the form of sperm. Each sperm cell carries either an X or a Y chromosome. Thus, at fertilization, the sex chromosome of the sperm cell determines whether the offspring will be male or female as shown in Figure 20.11.

female parent

		X	X
male parent	X	XX female	XX female
	Y	XY male	XY male

Figure 20.11 Sex determination of offspring during fertilization in humans.

Sex–linked traits

Any genes situated on the 23rd chromosome are known as **sex-linked genes**. In all other chromosomes, the gene on one chromosome has a corresponding gene on the other homologue in the pair. However, as you can see in Figure 20.10, this is not always true of the sex chromosomes, because the X chromosome is longer than the Y chromosome.

This means that in the case of genes carried on the X chromosome, the male carries only one set of most of these genes.

Colour blindness is a known **sex-linked disease** in humans. It is caused by a recessive allele that causes a person to have abnormal colour vision.

Figure 20.12 shows the possible offspring of a colour-blind man ('X^nY') and a heterozygous woman ('X^nX^N'), where N represents normal sight and n represents colour blindness.

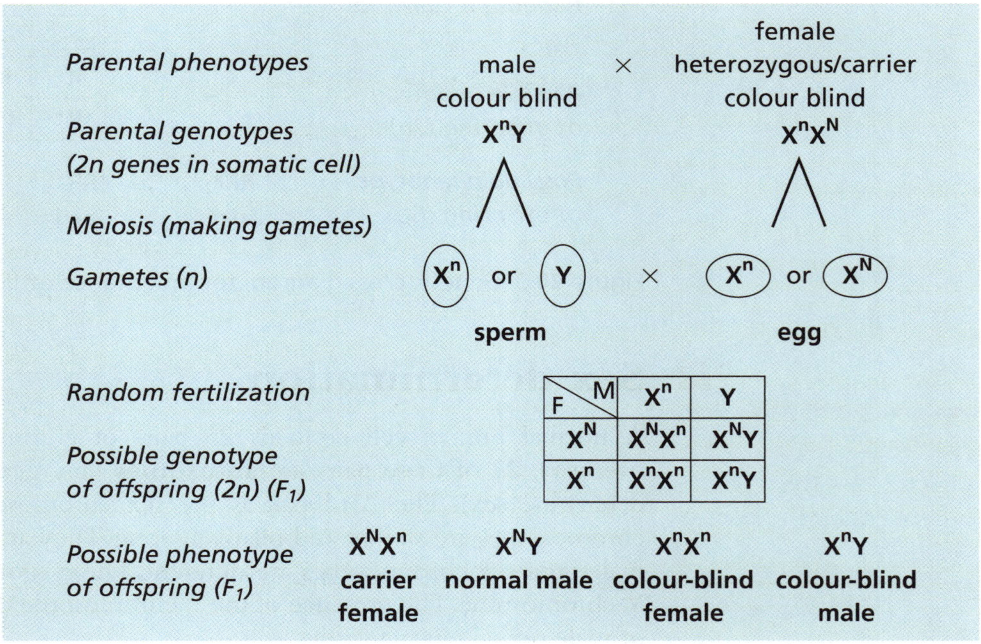

Parental phenotypes	male colour blind $\times$	female heterozygous/carrier colour blind
Parental genotypes (2n genes in somatic cell)	X^nY	X^nX^N
Meiosis (making gametes)		
Gametes (n)	X^n or Y $\times$	X^n or X^N
	sperm	**egg**

Random fertilization

Possible genotype of offspring (2n) (F_1)

F \ M	X^n	Y
X^N	X^NX^n	X^NY
X^n	X^nX^n	X^nY

Possible phenotype of offspring (F_1)

X^NX^n	X^NY	X^nX^n	X^nY
carrier female	**normal male**	**colour-blind female**	**colour-blind male**

Figure 20.12 Genetic cross diagram for sex-linked inheritance of colour-blindness.

Co-dominance

Blood type	
phenotype	**possible genotype**
A	$I^A I^A$ or $I^A I^O$
B	$I^B I^B$ or $I^B I^O$
AB	$I^A I^B$
O	$I^O I^O$

Table 20.4 Blood types: phenotypes and possible genotypes. 'I' indicates the gene for blood type. The superscript represents the allele on the gene.

You have learnt that incomplete dominance means that both alleles get partially expressed. **Co-dominance** means that both alleles are fully expressed. We can see an example of this in human blood types. Each person has two of three alleles – A, B and O. A and B are co-dominant and O is recessive.

The blood type gene controls the placement of antigen markers on cells. A person with blood type A has one kind of antigen marker, while a person with blood type B has a slightly different kind of antigen marker. Someone with blood type O has no antigen marker. A person with blood type AB has both A and B antigen markers.

There is no 'in-between' antigen as would be expected if the alleles showed incomplete dominance. Both of the alleles are completely expressed and the person has both blood types at the same time – the alleles are co-dominant. Table 20.4 shows the possible blood types.

Figure 20.13 shows a cross of two parents – one heterozygous for blood type A and the other heterozygous for blood type B. The figure shows that there is a 1 : 1 : 1 : 1 chance that the parents may produce offspring that have any of the four blood types.

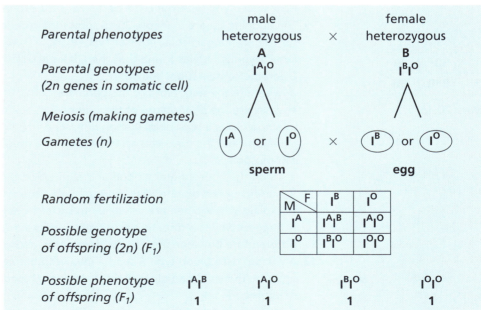

Figure 20.13 Genetic cross diagram for inheritance of blood type.

Check your knowledge and skills

1 Choose the correct term to complete each sentence.
   a Sperm and eggs are known as _____ (sex chromosomes/sex cells).
   b The expression of a trait is known as the _____ (phenotype/genotype).
   c The expression of a trait is determined by the _____ (phenotype/genotype).
   d Different versions of the same genes are called _____(alleles/chromosomes/phenotypes).

2 John can roll his tongue. He has the alleles R and r for the trait. Choose the correct term to complete each sentence.
   a Rr is John's _____ (genotype/chromosome).
   b John is said to be _____ (homozygous/heterozygous) for the trait.

3 a In terms of genotype, describe when a recessive phenotype can be expressed.
   b Give an example.

4 In genetic terms, name the two types of homozygous situations.

5 In a typical dominant/recessive situation, describe the phenotype of the heterozygote. Illustrate further with an example.

6 a In dogs, long hair is dominant over short hair. If you are a dog breeder and you were sold a long-haired dog that you were told is pure breeding, what could you cross it with to determine if this is true?
   b Give the result of the cross if your dog was pure breeding.
   c Give the result of the cross if your dog was not pure breeding. You can use Punnett squares to illustrate your answers.

7 The Punnett square in Figure 20.14 shows the possible combinations of alleles for fur colour in a particular litter of rabbits. Black fur, B, is dominant over white fur, b. Given the combinations shown:
   a what are the genotypes of the parents?
   b what are the phenotypes of the parents?

8 How would you determine if a particular trait exhibits incomplete dominance?

9 The Punnett square in Figure 20.15 shows the combinations of alleles for fur colour in a particular litter of rats. (Black = B, White = B* and Grey = heterozygous.) It was found that the alleles exhibited incomplete dominance. Given the combinations shown:
   a what are the genotypes of the parents?
   b what are the phenotypes of the parents?

10 Haemophilia is a sex-linked human disease. A haemophiliac's blood lacks the ability to clot. The disease is an expression of a mutated gene carried on the X chromosome.
   a Using the appropriate symbols and a genetic cross diagram, explain how two parents of the disease could end up with a son with haemophilia.
   b Explain why women are less likely to suffer from haemophilia.
   c Suggest why this disease is especially dangerous for females to have.

11 Differentiate between incomplete dominance and co-dominance.

12 A woman with blood type A sued a man with blood type B ($I^B I^B$) for child support. The child had blood type O. A simple blood test was done to determine paternity.
   a Using a genetic diagram, such as a Punnett square, determine if the sued man could possibly be the child's father.
   b If he is not the father, determine the possible genotype of the true father.
   c Determine the genotype of the mother.

M\F	?	?
?	Bb	Bb
?	Bb	Bb

Figure 20.14

M\F	?	?
?	BB*	BB
?	B*B	BB

Figure 20.15

Chapter summary

Do you know?

If you are unsure of any of the facts in the list, refer to the page number given in brackets.

- Some traits can be dominant and others can be recessive. (page 225)
- Genes are sections of DNA that code for a trait or characteristics. (pages 227–8)
- Sexually reproduced offspring are genetically unique (with the exception of identical twins). (page 229)
- Genetic variation is the variety of genes in a species or population. (page 229)
- Genetic diversity refers to the various phenotypes that make up a species. (page 229)
- Continuous variation involves a continual and gradually differing range of traits (for example from very short to very tall animals). (page 229)
- Discontinuous variation involves a range of traits that fall into distinct groups (for example different eye colour). (page 229)
- Some traits are incompletely dominant over others and are inherited only when the genotype is in a heterozygous state. (page 234)

Are you able to?

If you have trouble doing this, refer to the page numbers in brackets.

- Use a Punnett square to show a monohybrid cross. (page 226)
- Describe the structure of a chromosome. (page 227)
- Define the terms: gene, allele, genotype and phenotype. (page 228)
- Distinguish between continuous and discontinuous variation. (page 229)
- Describe the effect of the environment on gene expression. (page 230)
- Predict the ratio of phenotypes of the offspring given the genotype of the parents. (page 232)
- Construct and use genetic diagrams to explain inheritance of single pairs of characteristics. (page 232)
- Describe and give examples of dominant, recessive and incomplete dominance. (pages 231–235)

Choosing characteristics for survival

As environmental changes occur, organisms have to do something to compensate. Some adapt to the changes which allow them to survive. Some of these adaptations are passed onto the next generation, allowing the new generation to thrive in the environment until it changes again. This ability of organisms to adapt to changes so that their species can survive is called natural selection. This forms the basis for the theory of evolution.

By the end of this chapter you should be able to:

■ Explain the principles of common descent and evolution
■ Define natural selection
■ Distinguish between natural and artificial selection
■ Understand that genetic engineering can change the traits of an organism
■ Discuss the advantages and disadvantages of genetic engineering

Unit 1

The principle of common descent

Figure 21.1 Charles Darwin (1809–1882).

Charles Darwin was a British naturalist whose work presented evidence that all life on Earth descended from a common ancestor. In 1831 Darwin embarked on a voyage on a ship called the *HMS Beagle*. His journey took him from England around the east and west coasts of South America, to the Galapagos Islands and around the southern coasts of Australia and Africa. On his travels, Darwin collected hundreds of specimens and made detailed observations about genetic variations and adaptations. When he returned to England, he developed his **theory of evolution** which was based on the following arguments.

■ Species have great fertility. Organisms usually produce more offspring than are necessary to replace them, and more offspring than can successfully survive to adulthood. Yet, despite the overproduction of offspring, most populations remain stable in size.
■ Food resources are limited, but populations remain relatively constant.
■ From the two points above, Darwin inferred that organisms compete for limited resources.
■ Members of the same species vary from one another.
■ Many variations are heritable.
■ The two points above suggest that organisms with the most beneficial traits for survival are more likely to survive and reproduce, a principle he called 'survival of the fittest'.

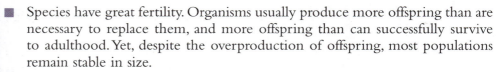

Survival of the fittest

Sometimes the phrase 'survival of the fittest' leads to misunderstandings:

■ Sometimes no organism is 'fitter' or more likely to survive than another. Sometimes survival is random.
■ Survival, in the context of natural selection, only refers to survival up to the reproductive age and the ability to reproduce.

- Different organisms may be well suited to different types of environment, so there is no way of comparing whether one species or group of organisms is 'fitter' than another.
- 'Fitness' only suggests traits for survival in the given environment. As the conditions of the environment change, the criteria for the fittest may change too. An organism that was previously the most fit for the environment may therefore find itself at a disadvantage.

The characteristics of living organisms change with each successive generation. Some phenotypes get passed along generations, others change from one generation to the next, and others disappear. Over a long period of time, this leads to gradual changes to each species. This process is known as **evolution**, which can be defined simply as 'descent with modification'.

However, many things change over time. For example, a snake might shed its skin several times during its lifetime. A tree might lose its leaves each season. A mountain may be eroded. These are not examples of biological evolution because the modifications are not caused by descent through genetic inheritance.

Activity 21.1

Questions

1 For each of the following changes, say whether it is an example of biological evolution or not. Give reasons for your answers.
   a Gradually rising sea levels.
   b Appearance of bacteria with an increased resistance to antibiotics.
   c Increase in the average height of the Chinese population over the last 150 years.
   d Prevalence of thick fur on animals that live in cold climates.
2 Why can 'survival of the fittest' be a misleading term?

Unit 2 Evolution and natural selection

In order to understand how evolution takes place, we need to understand the principle of descent. Look at the example in Figure 21.2.

These worms live in an area with many plants. 20% of the worms have the gene for brown colouring and 80% have the gene for green colouring.

During a three-year drought, many plants die. The worms survive, but do not grow as much.

After three years, the gene frequency in the population has changed. 80% of the worms have the gene for brown colouring and 20% have the gene for green colouring.

Figure 21.2 How gene frequency changes in a population.

Random mechanisms of change

Random mechanisms can be divided into:

- mutations
- gene flow
- genetic drift.

Mutations

Each time an animal or plant reproduces sexually, small errors may creep into the new chromosomes as they get copied. These errors are known as mutations. Some of these mutations might have no effect on the animal's phenotype. Others may introduce changes, which might be advantageous, disadvantageous or make no difference at all. In the worm example, it is possible that mutations could cause parents with the genes for green colouring to have offspring with brown colouring.

An organism's DNA provides a blueprint for much of that organism's development and behaviour. Changes to the DNA can cause all sorts of changes to the organism's life. There are two main causes of mutations:

- DNA copies inaccurately.
- External influences cause mutations. For example, exposure to radiation makes DNA more likely to copy incorrectly.

It is important to realize that although mutations may introduce advantages or disadvantages for an organism, they occur randomly. In other words, the mutation happens as a matter of chance. Neither the organism nor the DNA itself can decide or try to cause mutations. Mutations happen randomly and their occurrence is unrelated to any benefit or disadvantage they may bring to the individual.

Also, not all mutations get passed onto the next generation. Somatic mutations – mutations in non-reproductive cells, such as skin tissue or liver tissue – do not get passed on to the next generation. Only the mutations that occur in reproductive cells (gametes) have a significant effect on evolution.

When a mutation takes place in an organism's reproductive cells, it is known as a germ line mutation, as it will be carried into the next generation. It may have one of the following effects.

- No change in phenotype – in many cases, the mutation takes place in a section of DNA with no noticeable function, or else it simply does not alter the structure of the DNA enough to have any noticeable effect.
- Small change in phenotype – sometimes a single mutation can have a small effect. Examples are increased tail size or alteration to eye shape.
- Great change in phenotype – sometimes a single mutation can cause a highly noticeable change. For example, a single mutation can lead to pesticide resistance in a specific insect or antibiotic resistance in a specific bacterium. Some single mutations can cause death.

Migration or gene flow

Sometimes, individuals from another population may migrate to join a genetically different population. So, for example, brown worms from nearby plants might have migrated to join the mostly green worm population. This would change the frequency of the brown gene in the population. Another term for migration is **gene flow**. Gene flow includes many different ways by which genes move from one place to another. Pollen and spores can blow to areas hundreds of miles away from the parent plants; seeds have been known to float from one continent to another; people travel across great distances, either just to visit or to relocate. In many cases, gene flow allows new genetic material to be introduced to populations where it did not exist before.

Genetic drift

In one generation, random events might kill off some of the individuals in a population. For example, some green worms might get crushed by an animal walking past, or might get washed away by sudden rain. The next generation would then, by chance, have a few more brown worms than the previous generation. These random changes between generations are known as **genetic drift**.

The Amish community in Pennsylvania, USA, also demonstrates genetic drift. One of the founder members had a recessive allele for a rare form of dwarfism. The Amish community is relatively isolated, and individuals tend only to marry and have children with others from within the same community. As a result, the Amish community now has an unusually high rate (about 1 in 14) of dwarfism and polydactylism (extra fingers). This is an example of the **founder effect** – a type of genetic drift in which the founder of a population had a rare allele or combination of rare alleles, which then became 'bred into' the population because of non-random mating within an isolated group. Founder effect causes a higher frequency of rare recessive conditions in the isolated population than in the general population.

Another form of genetic drift is the **bottleneck effect**. This takes place when a population reaches near extinction because of a natural disaster or similar occurrence in its habitat, for example overharvesting, drought or flood. The few surviving individuals create a genetic 'bottleneck'. Biologists believe that cheetahs are an example of the bottleneck effect. Studies on cheetahs have shown that they do not exhibit much variation in their production of enzymes. Scientists have speculated that this may be the result of mass extinctions tens of thousands of years ago, or of overhunting more recently. However, the inbreeding after the bottleneck makes cheetahs susceptible to infertility.

Figure 21.3 Dwarfism is more common amongst the Amish population than in the general population, as a result of the founder effect.

Figure 21.4 Cheetahs are particularly vulnerable to infertility, as a result of the bottleneck effect.

■ Natural selection and adaptation

Think about the example of the green and brown worms from page 241. This example demonstrates natural selection. When the plants are healthy and green, brown worms are easy for predators such as birds to see. However, the change in environmental conditions could make the vegetation go brown. This would make green worms more visible. The change in environment thus gives the brown worms a slight advantage in their chances of survival. This could help them to outnumber green worms in the next generation.

Natural selection is quite simple to understand, but it is often misunderstood. It works on the principles shown in Figure 21.5 overleaf.

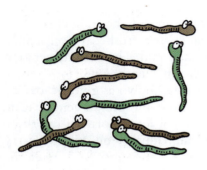

There is variation in genetically inherited traits of the individuals in a population.

Not all the individuals get to reproduce.

A trait that helps an individual to survive is more likely to be passed onto the next generation.

Figure 21.5 Natural selection in a worm population.

Key facts

- Evolution has several mechanisms, including mutations, genetic drift and gene flow. Genetic drift refers to random changes.
- Natural selection is one of the basic mechanisms of evolution. Alleles that help organisms to survive to reproductive age will become more frequent in the species genome.
- Organisms evolve to suit their environment as a result of the mechanisms of evolution, particularly natural selection.
- As the environment changes, new sets of characteristics will become more beneficial for survival.
- Natural selection refers to consistent adaptive changes.
- Natural selection may sometimes lead to new species.

Selection pressure

Antibiotic medicines target bacteria that cause infectious disease. However, this means that as soon as bacteria mutate in such a way that produces resistance to antibiotics, the medicine begins to exert **selection pressure**. Those bacteria that cannot resist the drugs die, and those that have the genes to resist the drugs survive and pass on their genes to new generations. This process happens very quickly because bacteria multiply rapidly.

Any factor that makes an organism less likely to reproduce can exert what is known as **evolutionary pressure** or selection pressure.

The malaria parasite provides another example of selection pressure. In humans, the gene which carries the mutation that causes sickle cell anaemia (sickle cell haemoglobin gene mutation) has a side effect of providing some resistance to malaria. As a result, in areas where malaria is highly prevalent, there is selection pressure that causes people with the sickle cell mutation to survive better than those that do not have it. As a result, selection pressure has caused populations in malaria zones to have a higher incidence of the gene for this condition.

Controversies about evolution

The philosopher Daniel Dennett called natural selection 'Darwin's dangerous idea'. Before Darwin, people believed that the adaptations and complexities of living organisms were the work of God's design. However, Darwin's theories showed that natural processes have shaped (and continue to shape) the features of plants and animals.

This issue still causes great controversy today.

1 Why do you think Dennett called natural selection 'Darwin's dangerous idea'?
2 In some countries, religious groups object to the study of natural selection and evolution.
  a Why do you think they do this?
  b Why do you think it is important to study natural selection and evolution?
3 'If you took all the pieces of a complicated clock and threw them together, what are the chances that you would make a clock? In the same way, it is impossible to say that our complex life forms could have originated from random natural processes.'

This statement is a common false argument against the theory of evolution. Discuss the ways that evolution differs from a random 'throwing together of pieces', and identify the mistakes that this argument makes.

The Galapagos finches

A famous example of natural selection and adaptation was the variety of finches that Darwin observed on the Galapagos Islands. Each island had its own species of finches not found anywhere else in the world. There were 14 different species in total. The finches were all roughly the same size (10–20 cm in length), brownish or black in colour. However, the most important difference between them was the way their beaks were adapted (see Figure 21.6).

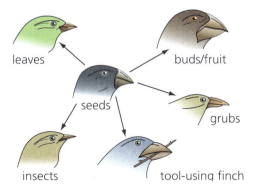

Figure 21.6 The finches' beaks have adapted to suit their diet.

The peppered moth

Another famous example of natural selection is the peppered moth. One form of the peppered moth is greyish white with dark markings, and the other form is darker with fine white markings. The second form first appeared in Manchester in the mid 1800s. Naturalists reached the following explanation:

- The white peppered moth enjoyed natural camouflage against the lichens of trees.
- Industrialization led to increased soot and pollution in the environment. This pollution killed off many of the lichens growing on tree trunks, and soot from factories also blackened the tree trunks and buildings.
- As a result, the paler form of moth would be more easily visible against the darker backgrounds of sooty buildings and trees.
- Predators such as birds could spot the moths more easily against the dark background.
- Mutations brought a new strain of peppered moth into existence, with a darker phenotype than that of the white peppered moth.
- The darker peppered moths were harder for predators to spot in environments where industry had caused pollution.
- Thus natural selection favoured the darker moths in polluted environments and the whiter moths in less polluted environments.

Figure 21.7 The peppered moth occurs in two dramatically different forms – seen here against pale and dark backgrounds.

■ Sex and mating

When two individuals reproduce sexually, they produce offspring with a combination of DNA from the two parents. This process introduces some **genetic shuffling** of the parents' genetic material.

Random mating takes place when the parents mate by chance rather than according to genotype or phenotype. When animals breed randomly, they constantly increase the chance of new combinations of genetic material taking place. The advantage is that new genetic combinations can arise by chance, but the disadvantage is that beneficial gene combinations can also get broken up. However, on the whole, random mating helps to strengthen a species.

In some cases, mating takes place in a non-random way:

- In-breeding – in very small populations, individuals may mate with partners that are closely related to them. Inbreeding increases the proportion of homozygous gene pairs in a population. This makes it more likely for recessive abnormalities to be expressed in the offspring.

- Assortative mating – sometimes individuals tend to mate with other individuals that have similar traits. For example, in humans, tall humans may tend to mate with partners that are similarly tall. This causes an increased frequency of homozygous gene pairs for the 'tallness' trait.

Check your knowledge and skills

In British cities with high levels of pollution, the darker form of the peppered moth became more prevalent. When the government put programmes in place to reduce industrial pollution in these cities, the paler form of the peppered moth began reappearing. Draw a flow diagram to illustrate the relationship between the city environment and the process of natural selection that changes the gene frequency of the peppered moth.

Unit 3

Artificial selection

When Mendel cross-bred pea plants with particular characteristics he was doing what humans have been doing for many centuries: breeding organisms with particular features. This is sometimes known as 'artificial selection' because humans replace environmental factors with their own selection choices. However, artificial selection has been taking place since long before Mendel's time.

Traditional selective breeding

Selective breeding is the process of breeding animals or plants by selecting those with particular desired genetic traits and breeding them to produce more individuals with those same traits. Traditional breeding involves sexual reproduction between organisms of the same species.

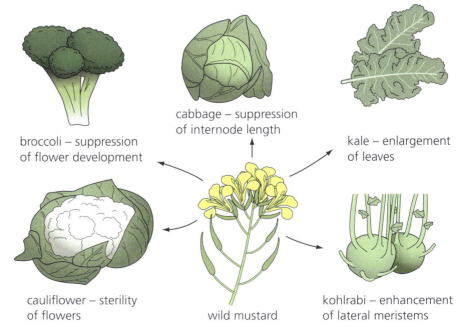

broccoli – suppression of flower development

cabbage – suppression of internode length

kale – enlargement of leaves

cauliflower – sterility of flowers

wild mustard

kohlrabi – enhancement of lateral meristems

Figure 21.8 Wild mustard (*Brassica olera*) – the original plant from which humans have bred many other cabbage-like plants.

Figure 21.9
Domesticated grains (top) are easier to harvest than wild grass grains (bottom) because the seeds hang onto the stems for longer.

Archaeologists have found the remains of cultivated wheat in Middle Eastern sites that date back more than 9000 years. The main difference between wild grass and domesticated grain is that the seeds of domesticated grain hang from the stem for much longer (see Figure 21.9). In wild grasses, the seeds drop almost immediately as they dry. Farming activities would create selection pressure for the attribute of tightly adhering seeds, because those would be the plants that the farmers would pick and replant. Some scientists, however, say that this process was still a form of natural selection as the farmers may not necessarily have spotted the individual plants with the tightly adhering seeds.

Another example of traditional selective breeding is the process that has given rise to the various forms of Brassica plant that we eat today. In order to cultivate each new crop, farmers have selected for specific attributes, as shown in Figure 21.8.

Disadvantages of selective breeding

Domestic dogs have been selectively bred for many centuries, resulting in the many different breeds of dogs we have today. Whilst highly pedigreed dogs exhibit many prized traits, they can also suffer from genetically inherited weaknesses as a result of many generations of inbreeding. In dogs, these weaknesses range from decreased lifespan to respiratory weaknesses and other inherited conditions.

Genetic engineering

For the past 10 000 years, artificial selection only took the form of selective breeding. However, recent advances in technology have allowed scientists to develop very powerful methods of artificial selection.

In laboratories, scientists can isolate specific genes and manipulate them in various ways. This process is known as gene splicing and generally involves the following steps:

- isolating the gene of interest
- inserting the gene into a vector (an organism, cell, virus or plasmid that can host the genetic material)
- transferring the vector into the organism to be modified
- transformation of the cells of the organism
- selection of the genetically modified organism for further use.

Any organism that has received a foreign gene by means of this process is known as a **transgenic organism**. For example, scientists have produced:

- silkworms that spin industrial-strength fibres, glow-in-the-dark fibres (using genes from jellyfish), or make silks that contain human proteins
- plants that have been modified to resist frost damage.

Rice with bred-in vitamin A helps prevent blindness in countries where children do not get enough vitamin A.

Tomatoes that are frost-resistant soften more slowly, allowing a longer shelf life.

Potatoes bred for their lower fat absorbency help manufacture lower-fat potato products.

Frost-resistant strawberry crops spoil less quickly.

Apples can be modified to carry genetic forms of a flu vaccine; other modifications make apples resistant to insect attacks.

Coffee has been bred with a lower caffeine content.

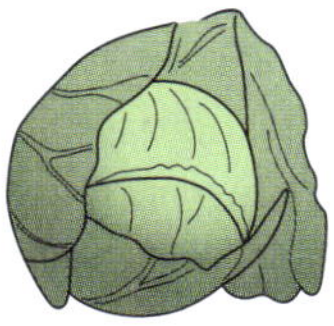

Cabbages that can resist caterpillar attacks.

Sunflowers that produce oil with lower levels of saturated fats.

Figure 21.10 Examples of plants that have been genetically modified for specific traits.

Table 21.1 compares traditional selective breeding methods with genetic engineering.

Traditional selective breeding	Genetic engineering
Slow, unpredictable process which may require many repeated cross-breedings in order to breed in the desirable traits or breed out the undesirable traits	Fast and more tightly controlled; organisms with desirable traits can be bred in one generation
Does not break up clusters of related genes	Organisms acquire one isolated gene, or several isolated genes, without the associated chains of genetic material that have evolved together with that genotype
Process involves sexual reproduction, although humans choose the organisms that reproduce	Process involves removing a piece of DNA from a cell and inserting it into a vector which is then inserted into the organism being modified; the foreign gene can behave in unpredictable ways
Only involves crosses within a single species or between related species	Can involve crosses between unrelated species, for example the frost-resisting gene from a fish has been inserted into tomatoes

Table 21.1 Traditional selective breeding versus genetic engineering.

Key facts

- We can engineer disease-resistant crops and livestock.
- We can engineer crops and livestock with particularly desirable phenotypes.
- Conventional breeding is a slow and unpredictable process.
- Genetic modification allows the introduction of specific traits in a single generation.
- Genes from different species (even from bacteria) can be crossed with other species.
- Supporters of GM foods identify many potential advantages, whilst critics identify disadvantages.

Debate about GM

Is GM the future of food production? Or is it a disaster waiting to happen? Scientists differ widely in their views of what genetic modification has to offer.

What supporters say:

- Farmers can make big profits from high-yield crops.
- GM can help improve the storage and nutritional quality of farmed food.
- The initial cost of the technology is high, as research is expensive, and that cost gets built into the price of seeds. However, because yields will improve so greatly, the cost will come down.
- GM allows the development of crops that are resistant to herbicides, so farmers can spray to kill unwanted plants while the crops survive unharmed.
- GM can develop crops that are drought-tolerant and salt-tolerant, which may be important for providing crops suitable for currently non-arable land.
- Animals can be modified to produce leaner, faster growing meat; cows can be modified for greater milk production.

What detractors say:

- Biotech companies are only developing modified seeds as a way of ensuring their own profits. Often the seeds are sterile in order to force farmers to buy new patented seeds each year. The costs will not come down as biotech companies design GM seeds to maximize their own profits.
- GM crops have not demonstrated significantly higher yields than ordinary crops.
- Little is known about the long-term effects of these crops; studies focus on short-term effects; it is possible that we may be creating selective pressure for highly resistant pests, weeds, viruses and so on.
- Pesticides introduced into GM crops can harm food webs in unforeseen ways. In Britain, a bird called the skylark was affected by the planting of GM sugar beets that were highly resistant to herbicides. This crop allowed farmers to spray extensively around the sugar beets to kill weeds. However, the weeds had previously provided the skylarks with a major source of food. In the absence of these weeds, skylarks starved if they could not find food elsewhere.
- GM crops may pose health risks to the animals that eat them. In the USA, monarch butterfly numbers fell because their caterpillars died when they fed on pollen from GM corn.
- Cross-pollination will introduce genes from GM crops into other species of plants. This may produce highly herbicide-tolerant weeds, exacerbating the problem that genetic engineering was originally trying to solve. Selection pressure could then produce 'super-weeds' and 'super-bugs'.
- There is very little scientific data about the effects of GM foods on human health. Some doctors and scientists have expressed concerns that they may trigger new allergies and diseases.

Check your knowledge and skills

1 What are the similarities and differences between natural and artificial selection?
2 Give one example of selective breeding in:
   a plants   b animals
3 What is the difference between traditional selective breeding and genetic engineering?
4 Describe the main steps in genetic engineering.

Chapter summary

Do you know?

If you are unsure of any of the facts in the list, refer to the page number in brackets.

- Charles Darwin made some important observations on his trip with the HMS Beagle, which led to the development of the theory of natural selection. (page 240)
- Organisms compete for limited resources. Those with the most beneficial traits for survival are more likely to survive and reproduce. (page 240)
- All life forms are descended from a common ancestor. (page 240)
- Evolution is the slow process of descent with modification. (page 241)
- Random changes in gene frequency are caused by mutations, migration (gene flow) and genetic drift. (page 242)
- Mutations are changes that happen at a genetic level when cells divide. (page 242)
- Migration or gene flow takes place when gametes move away from the parent organism. (page 243)
- Genetic drift refers to random changes such as natural disasters. (page 243)
- Selection pressure is a process that causes some genotypes to occur more frequently in a population. (page 244)
- Sexual reproduction introduces variations into DNA. (page 246)
- Natural selection is governed by environmental factors, whereas artificial selection is determined by human choice. (page 247)
- Artificial selection has been used for many centuries in the form of traditional selective breeding and cross-breeding. (page 247)
- Recent advances in artificial selection have led to processes that enable us to isolate specific genes and alter the DNA of new organisms by inserting or deleting genes. (page 248)

Are you able to?

If you have trouble in doing these things, refer to the page number in brackets.

- Explain the principle of 'survival of the fittest'. (page 240)
- Outline the observations that led to the theory of natural selection. (page 241)
- Define evolution. (page 241)
- Define adaptations. (page 244)
- Provide illustrations of natural selection. (page 245)
- Give examples of evolutionary adaptations in the plant and animal world. (page 246)
- Give examples of common processes of artificial selection. (page 247)
- Discuss the advantages and disadvantages of genetic engineering in plants and animals. (page 250)

22 Health and disease

Transmission and control of disease

Health is more than just the absence of disease. The World Health Organization defines health as 'a state of complete physical, mental and social well-being'. In other words, a healthy life includes wellness of the body, mind and social functions.

In this chapter you will learn about some of the diseases that affect humans, how these diseases are spread and how they can be controlled.

By the end of this chapter you should be able to:

■ Define disease and describe the ways in which diseases affect humans

■ Outline the main categories of diseases

■ Explain how infectious diseases get transmitted

■ Discuss the role of animal vectors in the transmission of disease

■ Identify natural barriers to infection: the human immune system

■ Explain how diseases are treated and controlled

Unit 1

Disease – what is it?

Diseases and medical conditions include any abnormal conditions that interrupt normal physical, mental and social functioning. Each disease or condition is associated with specific symptoms and signs.

■ Physical diseases affect one or more organs of the body, for example eczema affects the skin; glaucoma affects the eye; cancers usually start in one area of the body and then spread to others.

■ Mental illness affects the patient's thoughts and feelings. Examples include different forms of depression and personality disorders.

■ Physical and mental illnesses overlap in many ways, and it is not always possible to distinguish the two. For example, many mental illnesses are in fact caused by chemical imbalances. Many physical illnesses can lead to mental illness if they are not treated correctly. Many mental illnesses manifest in physical ways in the body – for example, headaches, stiff limbs and tiredness.

■ Infectious diseases

A **pathogen** is any biological agent that causes disease. Pathogens normally pass easily from one person to another in a variety of ways. Pathogenic diseases are thus **contagious** or communicable – they spread easily through a population. There are five groups of pathogens – bacteria, viruses, fungi, protozoa and parasites (see Figure 22.1).

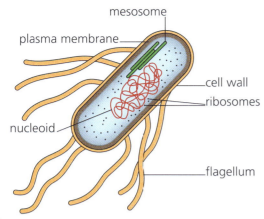

Bacteria

- Single-celled organisms that can reproduce by mitosis
- Can live outside or inside the host's body
- May be round, rod-shaped, box-shaped or spiral

Examples of diseases caused: Tuberculosis (TB), cholera, tetanus, gonorrhoea

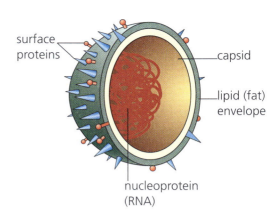

Viruses

- Tiny, multicellular organisms
- Can only function and reproduce inside living host cells
- Once inside a host living cell, they reprogramme the cell with their own DNA

Examples of diseases caused: Influenza, HIV, herpes, chicken pox

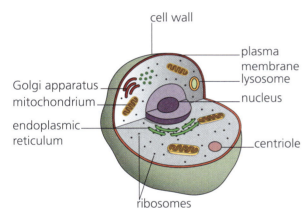

Fungi

- Multicellular, plant-like
- Lives on substrates on which it draws for nourishment

Examples of diseases caused: Candidiasis (thrush), athlete's foot, ringworm

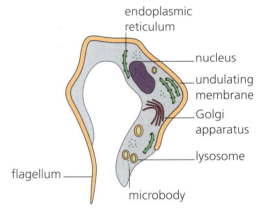

Protozoa

- Unicellular organism that lives inside the host's or carrier's body
- Usually spread through water
- Not always harmful, for example we have protozoans in our intestines that aid with digestion

Examples of diseases caused: Malaria, bilharzia, amoebic dysentery

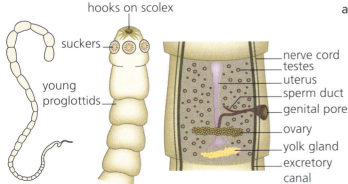

Parasites

- Complex organism
- Can live within the host organism, for example in the intestine

Examples of parasites: Tapeworms, lice, roundworms

Figure 22.1 Five types of pathogen.

Activity 22.1 Researching pathogens

For each of the types of pathogen shown above, choose a disease or illness that is caused by that type of pathogen. Research the disease, its mode of transmission, symptoms, treatment and control. Write up a detailed fact sheet for each of the five pathogens. Include the following headings on your fact sheet:

Type of pathogen:
Disease caused:
Main forms of transmission:
Symptoms:
What happens if disease is left untreated:
Available treatments:
Strategies for prevention or control:

■ Non-infectious diseases

Some diseases develop as dysfunctions in the cells, tissues or organs rather than as the result of pathogens. These diseases do not pass from one person to another. However, they may be prevalent among members of a specific population that either share a closely related gene pool or specific environmental circumstances.

- ■ Hereditary diseases – these develop as a result of genetic information in a person's DNA that causes their body to dysfunction.
- ■ Deficiencies and physiological diseases – these may arise due to a combination of factors including a person's environment and how they treat their body.

Deficiency diseases (caused by a lack of a specific nutrient)	Kwashiorkor; Malnutrition; Iron-deficiency-related anaemia	**Disease:** Kwashiorkor **Cause:** Deficiencies of protein and trace elements (iron, folic acid, iodine, selenium and vitamin C) **Symptoms:** Swelling of feet, distended abdomen, enlarged liver, thinning hair, loss of pigmentation from skin, skin irritations **Treatment:** Improving the nutrient content of the diet **Control:** Balanced diet
Hereditary diseases (caused by inherited genes)	Haemophilia; Huntington's disease; Early-onset Alzheimer's; Sickle-cell anaemia	**Disease:** Huntington's disease **Cause:** Inheritance of recessive gene **Symptoms:** Lack of co-ordination, unsteadiness, decline in mental abilities, behavioural and psychiatric problems **Treatment:** No cure available, drugs can offer some relief from symptoms **Control:** Scientists are researching gene therapy treatment for the disease
Physiological diseases (caused by organ malfunction)	Asthma; Glaucoma; Stroke; Diabetes	**Disease:** Asthma **Cause:** Combination of environmental and genetic factors **Symptoms:** Breathlessness, wheezing, gasping for breath, coughing **Treatment:** Bronchodilator inhalers, breathing techniques **Control:** Avoidance of triggers such as long-haired pets and Aspirin; medications that have a preventative effect

Table 22.1 Examples of deficiency and hereditary diseases.

Multiple sclerosis

Many diseases do not have a clear-cut set of causes. For example, doctors do not understand exactly what causes multiple sclerosis (MS). MS is a disease in which the body's immune system mistakenly recognises its own normal tissue as 'foreign' and starts to attack itself causing damage to the brain and spinal cord. This damage results in progressive loss of muscle control, vision, balance and sensation. Data suggests that the disease is caused by a combination of factors including genetics, environment and even a virus.

- Genetics – researchers believe that there may be several inherited genes that contribute to the development of MS. This is because relatives of an MS sufferer are at a greater risk of developing the disease.
- Environmental factors – MS is more common in temperate climates.
- Viruses – research has suggested that the viruses associated with measles, herpes and flu might also be associated with MS.
- Other factors – some studies have shown that hormones, particularly sex hormones, affect MS by acting as immune-response suppressors.

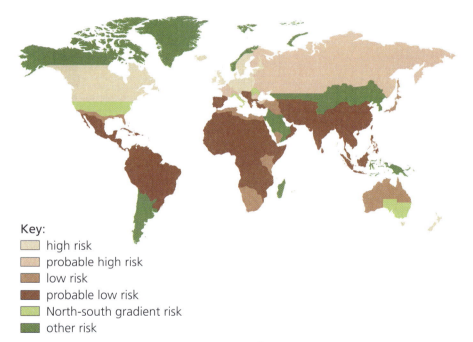

Key:
- high risk
- probable high risk
- low risk
- probable low risk
- North-south gradient risk
- other risk

Figure 22.2 This world map shows risk areas for multiple sclerosis.

Check your knowledge and skills

1 a What is a pathogen?
   b What are the differences and similarities between the following kinds of pathogens: bacteria, viruses, fungi and protozoans?
2 Give the main kinds of non-contagious diseases, and give an example of each.

Unit 2 Transmission of diseases

Different pathogens move from one organism to another using different methods.

■ **By air** – pathogens can survive in tiny droplets of water in the air around us.
■ **Direct contact** – if a person has an eye infection, the pathogens may be transferred to their hands, skin, clothing or washcloth, and then infect someone else that has had contact with these.
■ **Dust particles** – pathogens may adhere to dust particles in the atmosphere.
■ **Faeces** – when untreated sewage comes into contact with drinking water, people who drink the water may become infected with pathogens carried in the faeces.
■ **Vectors** – animals (usually insects) that carry diseases are known as vectors.

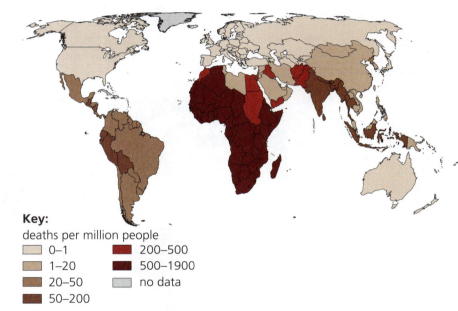

Key:
deaths per million people

☐ 0–1	■ 200–500
☐ 1–20	■ 500–1900
☐ 20–50	☐ no data
■ 50–200	

Figure 22.3 This world map shows the prevalence of vector-borne diseases.

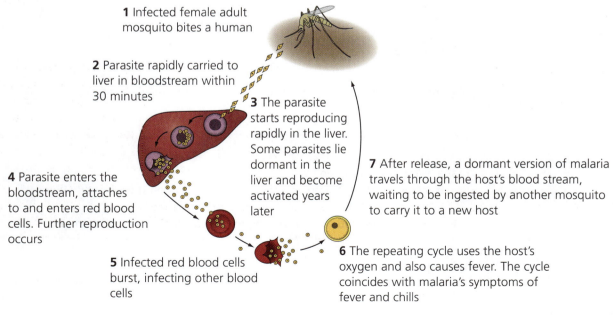

1 Infected female adult mosquito bites a human

2 Parasite rapidly carried to liver in bloodstream within 30 minutes

3 The parasite starts reproducing rapidly in the liver. Some parasites lie dormant in the liver and become activated years later

4 Parasite enters the bloodstream, attaches to and enters red blood cells. Further reproduction occurs

5 Infected red blood cells burst, infecting other blood cells

6 The repeating cycle uses the host's oxygen and also causes fever. The cycle coincides with malaria's symptoms of fever and chills

7 After release, a dormant version of malaria travels through the host's blood stream, waiting to be ingested by another mosquito to carry it to a new host

Figure 22.4 The vector, the mosquito, carries the malaria parasite.

■ Animal vectors

Malaria and mosquitoes

Vectors carry pathogens from one host to another. Examples of animal vectors include rats, mosquitoes, flies and birds.

The Black Death and fleas

The Black Death was a series of outbreaks of deadly infectious illnesses that killed between 75 million and 200 million people worldwide between 1347 and 1351. It reduced the world's population from 450 million to about 350 million.

Scientists at first thought that the Black Death was an outbreak of one disease, bubonic plague, caused by the bacterium *Yersinia pestis*, which is carried by fleas, which in turn live as parasites on rats. However, it is now known that there were several forms of the plague.

- Bubonic plague – this had a mortality rate of 30 to 75%; symptoms included earaches, painful aches in the joints, nausea and vomiting.
- Pneumonic plague – this had a mortality rate around 90%; symptoms included fever, coughing, bloody spit and phlegm.
- Septicemic plague – this had a mortality rate around 100%; symptoms included high fever and purple patches on the skin.

Modern scientists and historians are not sure whether the disease was in fact bubonic plague or haemorrhagic fever. It is believed to have started as an animal disease in Asia, and it gradually spread to fleas, rats, and eventually to humans.

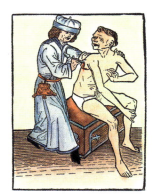

Figure 22.5 The Black Death spread across Europe in the 1300s and killed millions. Scientists believe it was spread by the rat flea.

■ Sexually transmitted diseases (STDs)

Sexual intercourse involves intense direct contact, as well as exchanges of blood and semen. This allows for easy transmission of a variety of contagious diseases.

For most STDs (see Table 22.2 overleaf), the most effective forms of control are:

- abstinence (100% effective)
- limiting sexual activity to a single partner and consistently using condoms (99% effective).

The AIDS pandemic

The **human immunodeficiency virus (HIV)** causes **acquired immune deficiency syndrome (AIDS)**. Unlike most other viruses, HIV attacks the human immune system so most carriers of HIV become ill with other diseases because they do not have a strong enough immune system to resist pathogens. HIV seeks out and destroys **T-cells**, a type of white blood cell that is vital to the immune system. AIDS is the name of the disease that develops in the late stages of HIV. A person may carry HIV for many years before developing AIDS.

Scientists have traced HIV back to a type of chimpanzee in West Africa. They believe the virus 'jumped' from chimpanzees to humans when people hunted the animals for meat and came into contact with infected blood. The virus gradually spread across the world. It was first identified in the United States in 1981.

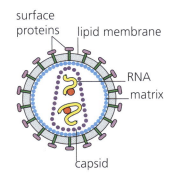

surface proteins
lipid membrane
RNA
matrix
capsid

Figure 22.6 The structure of the human immunodeficiency virus.

Disease	Pathogen	Symptoms	Treatment
Chlamydia	Bacterium *Chlamydia trachomatis*	In women – infection at neck of womb; often unnoticed In men – infection of urethra; pain whilst urinating; discharge If untreated, can lead to serious reproductive problems	Antibiotic drugs if detected in time
Gonorrhoea	Bacterium *Neisseria gonorrhoeae*	In women – often unnoticed In men – yellowish discharge from urethra; pain whilst urinating If untreated, can lead to sterility, arthritis, weakened heart, blindness	Antibiotic drugs if detected in time
Syphilis	Bacterium *Treponema pallidum*	Ulcers in genital region, followed by skin rashes and sore throat If untreated, can cause painful skin conditions, scarring in liver, blindness, paralysis, heart failure, dementia	Penicillin if detected in time; difficult to diagnose as it is often confused with other diseases
Genital herpes	Virus *Herpes simplex*	Blisters (during active phase); these heal and recur throughout sufferer's lifetime	No known cure
AIDS	Virus human immunodeficiency virus (HIV)	The disease attacks the body's immune system, so the sufferer becomes susceptible to a wide array of opportunistic infections including shingles, thrush, pneumonia, cancers and skin lesions	No known cure or vaccination

Table 22.2 STDs – the pathogens that cause them, their symptoms and treatments.

HIV cannot live for long outside the body. It is primarily found in blood, semen and vaginal fluid. The virus is transmitted in very specific ways:

- Sexual intercourse (anal, vaginal or oral) with an infected person.
- Sharing needles or syringes with an infected person.
- Exposing the foetus before birth or a baby through breastfeeding to HIV.

There is currently no vaccine or cure for HIV, although people carrying the virus can take drugs known as antiretrovirals (ARVs) to slow the replication of the virus in their bodies. This can allow people to live longer even though they are carrying the disease. However, the most effective form of control has been shown to be education about the modes of transmission and the correct use of condoms.

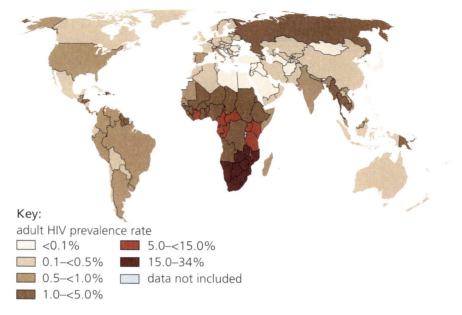

Key:
adult HIV prevalence rate
- ☐ <0.1%
- ☐ 0.1–<0.5%
- ☐ 0.5–<1.0%
- ☐ 1.0–<5.0%
- ☐ 5.0–<15.0%
- ☐ 15.0–34%
- ☐ data not included

Figure 22.7 This world map shows the prevalence of adults with HIV.

Biology at work
Chapter 22 – Biological terrorism

Biological terrorism

Biological terrorism is the deliberate use of pathogens to cause illness or death in humans, animals or plants. Countries can prepare for and respond to biological attacks:

- Detection systems can provide early warnings and identify areas contaminated by biological agents.
- Biological agents can be identified and their origins sourced.
- Data from hospitals, clinics, call centres can be analysed.

Researchers are working on experimental technologies such as:

- electronic chips with live nerve cells that can respond to bacterial nerve toxins
- fiberoptic tubes that utilize antibodies which can identify specific pathogens.

Preparation for biological attacks is complicated because there are such a large number of potential agents that can be used. Most of them are not often encountered naturally and may have long incubation periods which lead to delayed onset of disease. During this stage, they may infect large numbers of people. Furthermore, pathogens used by bioterrorists may be genetically engineered to resist known drugs, vaccines or antidotes.

Discussion

1 **a** Give three examples of biological warfare used in human history.
   b How do these examples differ from the modern-day threats of bioterrorism?
2 Responses to bioterrorism have led to new areas of study called biodefence and biosecurity. From what you have read in the article in the Chapter 22 'Biology at work' section on the CD, suggest what you think these areas involve.
3 Why do you think diseases with low mortality rates are still considered a serious threat to biosecurity?
4 How do advances in genetic engineering complicate bioterrorism defence plans?

Unit 3

The human immune system

We are exposed to pathogens all the time. Usually these do not enter our bodies, and even if they do, our **immune system** usually prevents them from spreading throughout our bodies.

■ Non-specific and specific immune responses

There are two aspects to the immune system: non-specific and specific immune responses.

Non-specific immune responses

Non-specific responses are physical, chemical and cellular defenses against microbes and other foreign bodies (see Table 22.3). These responses have the following characteristics:

- present from birth
- quick-acting
- effective against a wide range of pathogens and foreign bodies
- always function in the same way
- skin and linings of the body cavities form protective barriers against microbes
- entrances to organs have structures that protect against invasion by pathogens
- mucus membranes secrete fluids, such as saliva and mucus, to assist in trapping and repelling microbes
- internal microorganisms assist in killing pathogens in a form of symbiosis (working together for mutual benefit).

If pathogens get past the barriers created by our non-specific immune responses, the body needs to identify and remove the pathogens as effectively as possible. Specific immune responses:

- take place when a pathogen has already infected the body
- involve cells and proteins in the blood and lymph systems that attach to foreign bodies, disarm and destroy them and eliminate them from the body
- respond to specific organisms in specific ways.

Specific immune responses

Specific immune responses work by distinguishing between cells that belong to the body and cells that do not. Each cell in your body contains surface membrane proteins. These proteins combine with carbohydrates and lipids to form markers that cover the surface of every cell and which your immune system recognizes as 'self'. White blood cells (or **leukocytes**) then recognize foreign cells or viruses because they have different ('non-self') markers on their surface membranes. The leukocytes respond by attacking the 'non-self' invaders. When the immune system is working normally it recognizes and tolerate its own antigens (self-antigens). When a foreign organism (bacteria or viruses) enter the body, non-self antigens trigger an immune response.

Figure 22.8 shows which organs in the body are involved with the immune system and Figure 22.9 on page 262 summarizes the steps in an immune response.

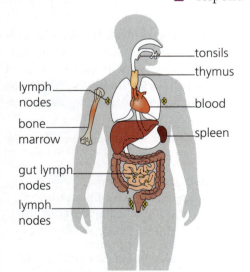

tonsils
thymus
lymph nodes
blood
bone marrow
spleen
gut lymph nodes
lymph nodes

Figure 22.8 The organs and tissues that coordinate the immune system.

Component	Function
Skin	Physical barrier that prevents microbes from entering the body
Mucous membranes	Inhibit microbes
Mucus	Traps microbes in respiratory and digestive canals
Hairs	Act as a filter for microbes and dust
Cilia	Function with mucus to trap and remove microbes and dust
Tear ducts	Secrete tears to wash away irritating substances
Saliva	Dissolves and washes away microbes from mouth and inside of mouth
Epiglottis	Prevents microbes and dust from entering trachea
Urine	Rinses urethra of microbes
Gastric juices	Destroy bacteria and toxins in stomach
Acidic pH of skin	Discourages microbial growth
Fatty acids and lysozyme in sweat	Antibacterial function
Interferon (IFN)	Chemical that protects healthy cells from viral infection
Complement	Chemical that promotes phagocytosis
Phagocytosis	Process of ingesting and deactivating foreign particles in body
Inflammatory response	Isolates and destroys microbes; stimulates tissue repair
Fever	Inhibits growth of pathogens and speeds up metabolism to help the body to fight disease

Table 22.3 Summary of non-specific immune responses.

White blood cells

There are five main types of leukocytes (white blood cells). Each plays a specific role in fighting disease.

- Neutrophils – these travel to tissues where an inflammation is developing and engulf and destroy pathogens in a process called phagocytosis.
- Eosinophils – these target larger organisms that cannot be destroyed by phagocytes.
- Basophils – these release a chemical that helps to attack pathogens.
- Monocytes – cells released from the bone marrow into the blood; when they reach the site of infection they can change into macrophages that engulf and destroy pathogens.
- Lymphocytes – there are three types: natural killer cells that can attack tumor cells or cells infected with a virus; B-cells which develop in the bone marrow; T-cells which develop in the thymus.
- B-cells and T-cells – these have an outer covering of many different molecules. These outer molecules act like complementary jigsaw pieces – they correspond to the molecular structure of a pathogen so that they can lock around it, engulf it and destroy it. The body keeps a molecular record (or 'memory') of every B- and T-cell that gets activated, so that if the same pathogen invades the body in the future, specific antibodies can be reactivated very quickly to remove it.

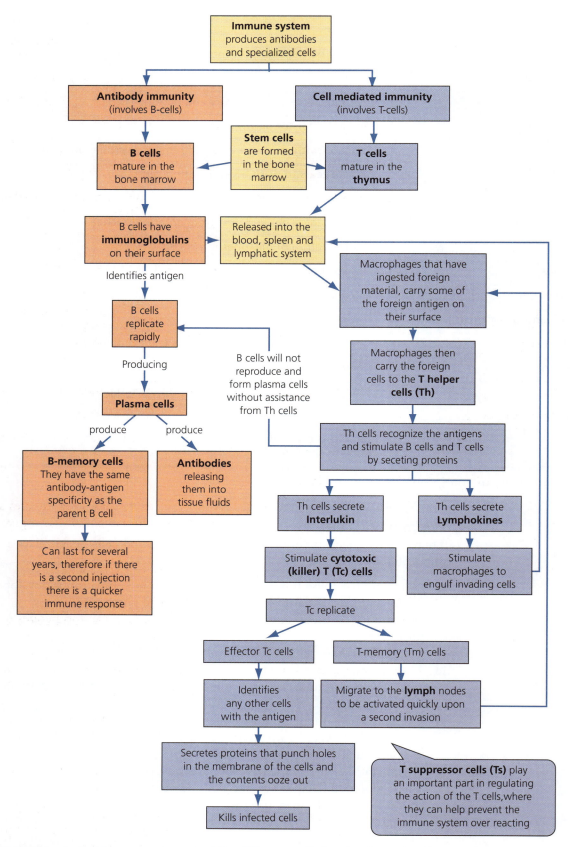

Figure 22.9 Steps in the immune response. The different cells involved are shown in Figure 22.10.

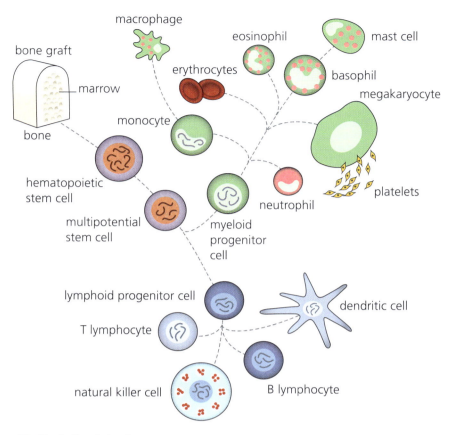

Figure 22.10 Cells of the immune system.

Autoimmune diseases

Dysfunction in the recognition system can lead to autoimmune diseases such as rheumatoid arthritis, where the body targets its own tissue. Table 22.4 gives some more examples of autoimmune diseases.

System affected in the body	Example of autoimmune disease
Nervous system	Multiple sclerosis
Digestive system	Crohn's disease; ulcerative colitis
Endocrine	Type 1 diabetes
Muscular-skeletal system	Rheumatoid arthritis
Circulatory system	Pernicious anemia
Skin	Psoriasis

Table 22.4 Autoimmune diseases.

Passive immunity

When a foetus is in the womb, it does not have an effective immune system. During this period, the mother gives antibodies from her own body to the foetus so that the baby has high levels of antibodies at birth. Breast milk also contains antibodies which help to protect the newborn's digestive system against bacterial infections until it can synthesize its own antibodies. This is known as passive immunity and usually only lasts for the first few months of the baby's life.

Naturally acquired active immunity

For the rest of your life, your body is engaged in naturally acquired active immunity – it actively responds to antigens and also prepares for and protects against future infections. Figure 22.11 shows a graph to show that B- and T-cells have a 'memory' and can destroy a pathogen that invades for the second or subsequent times very effectively.

Natural acquired active immunity takes effect when a person:

- is exposed to a live pathogen
- develops a primary immune response
- creates a biological 'memory' within their immune system
- draws on antibodies from the primary immune response in subsequent exposures to the same pathogen.

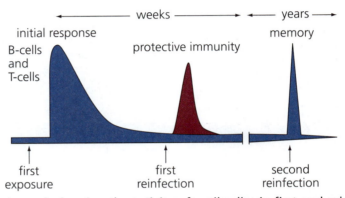

Figure 22.11 A graph showing the activity of antibodies in first and subsequent exposures to a pathogen.

Artificially acquired active immunity

This works in a similar way to naturally acquired active immunity, except that the live pathogen is artificially introduced in the form of a **vaccine** which is a substance containing modified antigens that mimic those of the live pathogens that cause a disease.

Unit 4 Treating and controlling disease

Diseases kill millions each year. Many of these diseases can be prevented and cured.

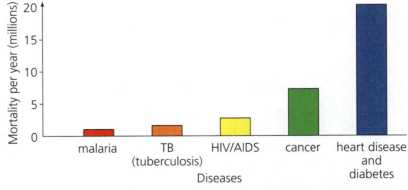

Figure 22.12 Graph showing causes and numbers of deaths worldwide.

Check your knowledge and skills

1 Explain how the flu vaccination works.

Prevention and control of infectious diseases

The transmission of infectious disease is affected by three factors: the host, the pathogen and the environment. Although it is possible to treat diseases once they have entered the body, it is more effective to prevent infection in the first place.

Host	Pathogen	Environment
Increase resistance to disease via: • personal hygiene • healthy diet • exercise • vaccinations Detect disease early via: • screenings • public health education Control vectors: • limit vector populations • quarantine contagious animals • ensure that conditions are not conducive for the breeding of vectors	Eradication of pathogens outside the body: • use of antiseptics, soaps • sterilization techniques such as heating or boiling • food safety techniques such as pasteurization Prevention of direct contact, for example: • doctors and dentists wear masks over noses and mouths • bandages to cover wounds • safety gloves for people working with food • wrappings for food • condoms to limit exchange of body fluids during sex Drugs that kill pathogens: • antibiotics • anti-fungal medications	Improvement of sanitation, including: • water purification • sewage treatment Safety procedures in the preparation and storage of food: • health and hygiene in abbatoirs, factories, dairies and other industrial food production sites • use of 'best before' and 'sell by' dates on food Improved public health programmes: • vaccination programmes • health education

Table 22.5 Ways of limiting infection from pathogens.

Controlling the vectors of disease

One way to control the transmission of pathogenic diseases is by controlling the vectors that carry it. Vectors may carry pathogens either externally (on skin, claws, fur) or internally (in the bloodstream, saliva). Understanding the life cycle of a vector can help humans to reduce the transmission of diseases. Table 22.6 overleaf shows different diseases, their pathogens and their mode of transmission.

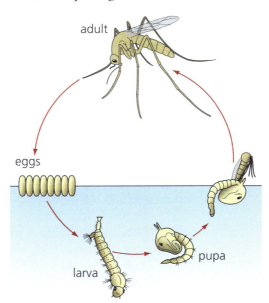

Figure 22.13 A mosquito goes through four stages in its life cycle – egg, larva, pupa and adult. By targeting specific stages in its life cycle, humans can control the spread of diseases carried by mosquitoes.

Disease and its pathogen	Vector and transmission
Chagas disease, caused by protozoan *Trypanosoma cruzi*	Triatomine bug – bites people at night and ingests their blood, then defecates near the wound. Parasites called trypomastigotes live in the faeces and enter the host through the wound or at the conjunctiva (the clear mucous membranes that cover the eyeballs under the surface of the eyelids). Inside the body, the parasites multiply
Dengue fever, caused by viruses of the genus *Flavivirus*	*Aedes aegypti* (yellow fever mosquito) or *Aedus albopictus* (forest day mosquito), which picks up and transfers the dengue virus during probing and blood feeding
Malaria, caused by protozoan parasites of the genus *Plasmodium*	*Anopheles* mosquito bites an infected person, taking a small amount of infected blood containing microscopic parasites. Around a week later, the parasites can be transferred to another human body when the mosquito feeds again (see below).
Yellow fever, caused by a virus of the genus *Flavivirus*	*Aedes aegypti* (yellow fever mosquito) and *Hamagogus* mosquitoes in Central and South America; same mode of transmission as for dengue and malaria
Rabies, caused by viruses of the *Lyssavirus* genus	Carried by any mammal; tends to be carried by bats, monkeys, raccoons, cats and dogs. If the animal bites a human, the virus enters the nervous system

Table 22.6 Vector-borne diseases.

Mosquitoes breed in stagnant water such as rain tanks, and puddles left after rain. They leave rafts of eggs which then hatch into larvae. These develop into pupae that dangle beneath the surface of the water, getting oxygen through a short tube. Many strategies for controlling mosquito populations target this stage of their development:

- pouring or spraying oil on the surface of the water to block the passage of oxygen to the pupae
- draining or drying out bodies of water, for example swamps and ponds
- introducing fish that feed on the larvae and pupae.

Other methods may repel adult mosquitoes or prevent contact with them:

- burning citronella candles (mosquitoes are repelled by the smell)
- applying insect-repellant to skin, bed linen and clothing
- wearing long-sleeved clothing that shields the skin
- ingesting quinine, a chemical that makes the blood unappealing to mosquitoes
- sleeping under mosquito nets.

There are also anti-malarial drugs such as Paludrin and Larium which act against malaria by preventing the *Plasmodium* protozoan from entering the liver and multiplying. The drawback is that they have side-effects (for example nausea) and they also mask the symptoms of malaria. If someone on anti-malaria drugs still contracts the disease it can take longer to diagnose.

■ Preventing and controlling non-infectious diseases

The graph in Figure 22.14 shows the numbers of deaths in the USA in 2000 caused by preventable or avoidable factors. Tobacco accounted for 435 000 deaths or 18.1% of total deaths in the USA. Poor diet and physical inactivity accounted for 365 000 deaths or 15.2% of total deaths.

Lifestyles in the Caribbean are fairly similar to those in the US, and obesity, heart disease, diabetes and stroke are all on the increase. These are 'lifestyle diseases' – which increase in prevalence as countries become more industrialized and people's diet and lifestyle increase their risk for developing these diseases.

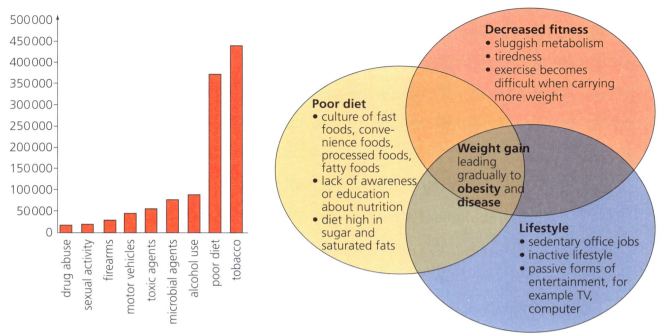

Figure 22.14 Deaths from 'lifestyle diseases' in the USA in 2000.

Figure 22.15 'Lifestyle diseases'.

Diet

Certain kinds of food in the diet can increase a person's risk of diseases such as diabetes, hypertension and heart disease.

- Simple sugars are high in calories and can lead to weight gain. Over a long period, eating too much sugar can cause dysfunctions in the liver's functioning, leading to over-production of insulin and diabetes.
- Fried and fatty foods are also very high in calories and contribute to obesity. In addition, they raise cholesterol levels. Cholesterol is a sticky type of fat that can clog the arteries and veins around the heart and contribute to heart disease.
- High salt levels can raise blood pressure and contribute to hypertension.

Exercise

An active lifestyle can:

- protect against heart disease by clearing away fatty build-up in the veins and arteries
- promote re-growth of blood vessels damaged in a heart attack
- strengthen muscles and bones
- lower blood pressure
- release endorphins, which counteract feelings of anxiety, irritation and depression associated with stress
- contribute to a sense of physical and psychological wellbeing.

Other lifestyle factors

Smoking can increase a person's risk of heart disease and cancers, particularly lung cancer. Stress is a major risk factor for many kinds of non-infectious diseases.

Figure 22.16 This graph shows the number of deaths from heart disease against income levels for men and women.

Activity 22.2 Test your fitness levels

Equipment and apparatus

- stopwatch
- bench or box or step approximately 30 cm high

Work in pairs – one person times and counts and then swap around.

Core strength

Lie in the position shown in Figure 22.17, with your knees bent at a 45 degree angle and feet flat on the floor. Rest your hands on your knees. Tighten your stomach muscles and raise your back off the floor so that your hands slide up to touch your knees. Exhale as you go up and inhale as you go down. How many sit-ups can you do in one minute? Look at the table below to see how you rank!

Figure 22.17 Starting position for sit-ups.

	male	female
Excellent	48 or more	43 or more
Good	44 to 47	37 to 43
Above average	39 to 43	33 to 36
Average	35 to 38	29 to 32
Below average	31 to 34	25 to 28
Poor	30 or fewer	24 or fewer

Endurance

Step up and down on the bench, box or step. First with one foot, then the other. Try to maintain a steady, brisk pace. Continue for three minutes. At the end of three minutes, immediately take your pulse for one minute. Look at the table of pulse rates below to see how you rank.

	male	female
Excellent	79 or less	85 or less
Good	79 to 89	85 to 98
Above average	90 to 99	99 to 108
Average	100 to 115	109 to 117
Below average	106 to 116	118 to 126
Poor	117 or higher	127 or higher

Figure 22.18 Positions for push-ups.

Upper body strength

Boys should use the standard push-up style; girls can use the bent-knee position if not strong enough for the standard position. Do as many push-ups as you can continuously until you can't do any more. Look at the table below to see how you rank.

	male	female
Excellent	56 or more	35 or more
Good	47 to 55	27 to 35
Above average	35 to 46	21 to 26
Average	19 to 34	11 to 20
Below average	11 to 18	6 to 10
Poor	10 or fewer	5 or fewer

Check your knowledge and skills

1 What is a pathogen? Give examples of two common types of pathogens and draw a simple sketch of the structure of each.
2 Distinguish between contagious and non-contagious diseases.
3 Why would the plague be less common now than it was in the 1300s?
4 List five features of the non-specific immune system and describe what they do.
5 What role do white blood cells play in the immune system?
6 Describe some of the measures that countries with a high incidence of malaria can take to prevent the spread of the disease.
7 How can someone with a history of diabetes in their family reduce their risk of developing the disease?

Chapter summary

Do you know?

If you are unsure of any of the facts in the list, refer to the page number in brackets.

- Health is a state of wellbeing in body, mind and society. (page 252)
- Disease describes any disruption to physical, mental or social health. (page 252)
- Pathogens are the biological agents that cause disease. Common pathogens include bacteria and viruses. (page 252)
- Diseases caused by pathogens are usually contagious – they spread easily through a population. (page 252)
- Diseases caused by other factors (for example genetics, deficiencies or physical malfunctions) are not contagious. (page 254)
- Vectors carry pathogens from one host to another. (page 257)
- The human body has two strategies for defending against disease: the non-specific and specific immune responses. (page 260)
- Regular exercise and a balanced diet can help to prevent disease. (page 267)

Are you able to?

If you have trouble in doing these things, refer to the page number in brackets.

- Distinguish pathogenic, deficiency, hereditary and physiological diseases, and give examples of each. (pages 253 and 254)
- Identify ways of treating and controlling each of the different types of disease. (page 254)
- Explain the role of vectors in the transmission of disease. (page 257)
- Describe the life cycle of a mosquito. (page 265)
- Describe ways to control vectors. (page 265)

23 Drugs and disease

The control and treatment of disease and the impact of drugs

Each year, between 250 000 and 500 000 people around the world die from the influenza virus. Outbreaks of disease can have very serious consequences for a population – preventing people from working and even killing large numbers of people. Over the last 150 years, advances in medicine, especially vaccinations, have allowed human beings to prevent the spread of many diseases. However, along with these advances have come many new chemicals and substances, some of which can be abused in ways that threaten human wellbeing.

In this chapter you will learn how diseases and drug abuse affect everyone.

By the end of this chapter you should be able to:

■ Define what drugs are

■ Differentiate between medicinal and other types of drugs

■ State how certain diseases affect the society and the economy of a country

■ Describe how drug abuse can affect the family and the society

■ Explain how plant and animal diseases adversely affect the economy

Unit 1 Drugs and their uses

A **drug** is any chemical substance that can alter normal actions of the body when it is absorbed. There are many ways that drugs may be given (see Figure 23.1).

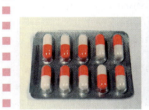
a Orally

b Anally

c Intravenously (into the veins)

d Inhalation

e Direct application

Figure 23.1 Drugs can be administered in different ways.

Medicines are widely used to prevent, treat or cure disease. Medicines contain at least one **active ingredient**. The active ingredient is the substance that gives the drug its effect. This is the reason that medicines come with many warnings and instructions on how to administer them safely. Table 23.1 lists some of the most commonly used pharmaceutical drugs and gives examples of their active ingredients.

Type of drug	What it does	Example of active ingredient
Analgesics	Relieve pain	Paracetamol
Antihistamines	Treat allergic reactions	Diphenhydramine
Antibiotics	Kill or inhibit the growth of disease-causing bacteria	Penicillin
Antidepressants	Alter the mood of people suffering from depression	Fluoxetine; sertraline
Contraceptives	Alter women's hormonal cycles in order to prevent ovulation	Progesterone
Decongestants	Medication that shrinks swollen nasal tissues to clear nasal congestion, swelling and mucus secretion	Pseudoephedrine
Vaccines	Introduce antigens into the body to stimulate an immune response	Flu antigens (flu vaccine)

Table 23.1 Common pharmaceutical drugs.

Using drugs correctly

A medical drug usually has a beneficial effect if it is taken in the right **dose**.

A dose is the correct amount of a drug that may be introduced into the body for the intended effect.

In medicine, the effect of a drug is dependent on its dose. Most doses of medicines are measured in milligrams (mg); although some are so powerful they are measured in micrograms (μg). The correct dose may vary depending on many factors:

- patient's age and weight
- mildness or intensity of symptoms
- other factors, for example other medical problems or allergies.

Over-the-counter drugs are drugs that can be bought without a prescription, either from a supermarket or pharmacy. Prescription drugs are drugs that require a written prescription from a qualified doctor. They are usually only available from a pharmacy or hospital. By law, a pharmacist is not allowed to sell these drugs to people that do not have a prescription. The prescription includes the name of the drug, its correct dosage and the name of the patient.

Biology at work
Chapter 23 – Poison is in the dose

The placebo effect

A **placebo** is a substance with no reported theraputic value but that is administered as if it did. It contains no active ingredients. A typical example of a placebo is a sugar pill. Whenever scientists do trials for a new drug, they always need to test a control group using placebos. This is because a patient who believes they are taking a drug can sometimes cause actual or perceived changes in their own condition. This phenomenon is known as the placebo effect.

Check your knowledge and skills

1 'The difference between a cure and a poison is all in the dosage.' Explain what you understand from this statement. Give two examples of substances that can poison or cure depending on the dosage.

2 Work in pairs or small groups. Go to a pharmacy and use your observations to answer these questions.

   a Make some observations about the way the pharmacy is arranged and what the arrangement of the products tells you about the scheduling of the drugs.

   b Copy and complete the table below by filling in examples of products that are available as over-the-counter medications (i.e. no prescription required).

To treat	Name of product	Description (tablets, capsules, medicine, cream)	Active ingredient
Colds and flu			
Coughing			
Headache or pain			
Insect bite or sting			

   c Identify five products that are marketed as 'natural' or 'herbal' remedies. List:
   – the name of the product
   – what it aims to treat, or what it aims to do
   – its active ingredients
   – any other ingredients
   – why you think it is considered a natural or herbal remedy rather than a drug.

3 a Explain the 'placebo effect'.
   b Why do you think it is important to set up control trials using placebos whenever a new drug is being tested?

Unit 2 Substance abuse and addiction

Some legal substances such as alcohol or drugs can be mis-used which can lead to substance abuse and addiction.

Abuse

Sometimes people use substances such as alcohol or drugs in ways that can damage their mental and physical health. This behaviour takes many different forms:

- excessive use of alcohol for its intoxicating effects
- abuse of prescription drugs (taking more than necessary, taking them when they are not prescribed, or mixing drugs)
- use of illegal drugs such as cocaine, heroin and ecstasy.

Over-indulgence on drugs or alcohol, especially when it leads to dependence and problems for the person's health and wellbeing, is known as **substance abuse**. Many drugs are illegal, partly because they are highly addictive. This does not mean that everyone who uses these drugs becomes addicted to them. However, substance abuse can lead to serious problems for users whether they are addicted or not.

Substance abuse begins in many ways and for many reasons. The substances may be legal or illegal, used regularly or occasionally. Here are just a few examples of ways in which people have started abusing substances.

- Susan was in pain and having difficulty in sleeping after a car accident. Her doctor prescribed a drug containing codeine and advised her not to take it for longer than 2 weeks. However, Susan enjoyed the deep sleep that the drug caused. She continued taking it nightly for several months. After a while, she was unable to sleep without taking the drug. In fact, as soon as she stopped taking it, she suffered nausea, vomiting, cramps, sweating and anxiety attacks.
- George's friends consume alcohol socially. However, George is unable to drink in moderation – he always wants to keep drinking until he cannot remember his actions the following day. He also finds it difficult to stop himself from drinking alcohol at home on his own. When he thinks about stopping, he has feelings of severe anxiety and helplessness.
- Dave is a regular marijuana smoker. He started smoking the drug as a teenager, with his friends. He liked the relaxed feeling that the drug gave him and it made him feel like he belonged to a cool group.
- Sarah only takes drugs at parties. She likes to take ecstasy as it gives her a strong feeling of connection with her friends. Although ecstasy is illegal, Sarah believes it is harmless as she does not take it very regularly.

Addiction

Addiction is a type of illness characterized by compulsive and uncontrollable cravings for drugs. Addicts continue to seek out and use drugs, even when they are facing extremely negative consequences for their drug use.

As you can see from the examples above, people abuse drugs in different ways and not all users are addicts. Some people become instantly addicted to a drug the first time they try it. Others may gradually develop addictions over a long period. And many may never become addicted.

Similarly, it is very difficult to predict how a person will react to a particular drug. In some extreme cases, a person may be particularly vulnerable and reactive to a particular drug and may die from overdosing the first time they try it. Others may use the same drug once or many times and never suffer any ill effects.

Some drugs are more physically addictive than others. Schedule 1 and 2 drugs are highly addictive; most users want more after the first time they try. Others are less intensely addictive. Many addicts do not realize that they have a drug problem until it begins to interfere with other areas of their life – their relationships, job, schoolwork or health. Even then, addicts frequently try to convince themselves that everything is fine and that they do not have a problem.

Tobacco

The active ingredient in cigarettes is a substance called **nicotine**. When people inhale the smoke from a lit cigarette, the nicotine has various effects on their nervous system. These include pleasure, stress relief, relaxation and a hunger-curbing effect that some people use to assist in weight control.

However, tobacco is also a highly damaging chemical substance. It is highly physically and psychologically addictive and has been identified as a leading cause of heart disease, lung cancer, emphysema, peptic ulcers and stroke.

Figure 23.2 These men are taking risks with their health by smoking.

Alcohol

In most countries it is legal to purchase and consume alcohol for persons over the age of 18. In the USA, a 'standard drink' is any alcoholic drink that contains around 14 grams of pure alcohol. A standard drink might be any of the following (different brands of alcoholic beverages contain varying amounts of alcohol):

- 284 ml beer
- 125 ml wine
- 25 ml brandy or spirits.

Because moderate drinking of alcohol is a socially accepted behaviour, alcohol addiction is sometimes difficult to notice. However, alcohol has a drug-like effect on the nervous system: it acts as a depressant (slows down the central nervous system). People who are addicted to alcohol are known as **alcoholics**.

Alcoholism is a disease. It is characterized by:

- craving – a strong desire to drink
- impaired sense of control – inability to limit or stop one's drinking
- increased tolerance – needing an increased amount of alcohol to achieve the same level of relaxation, 'buzz' or desired effect from drinking
- memory lapses or blackouts
- feelings that one's drinking pattern is out of control
- physical dependence – withdrawal symptoms (any unpleasant mental or physical effects on the body caused when a person suddenly stops using drug) may include nausea, sweating, anxiety and shakiness.

Effects of alcohol abuse

Alcohol is a brain depressant. Its short-term effects include:

- slowing down of the central nervous system (depressant effect)
- loss of motor coordination which can include slurred speech, staggering walk and impaired mental acuteness
- slowed reaction times which can cause accidents

- lowered inhibitions
- violent behaviour, in some cases
- alcohol poisoning which can lead to nausea, vomiting, seizures, loss of consciousness and in extreme cases, death.

Long-term effects of alcohol abuse include:

- psychological and physical addiction
- liver problems, including cirrhosis of the liver, liver failure
- heart disease
- drinking whilst pregnant can impair the development of the foetus and lead to foetal alcohol syndrome (serious birth defects that result from excessive alcohol use by the mother during pregnancy).

Illegal drugs

Most illegal drugs fall into one or more of the following categories.

- **Stimulants** – these quicken a region of the body such as the central nervous system, causing increased alertness. Examples include caffeine, nicotine, amphetamines. Illegal stimulants include cocaine.
- **Depressants** – these diminish or slow down activity in a specific region such as the central nervous system or brain. Examples include alcohol and tranquil-lizers. Illegal depressants include heroin and marijuana.
- **Hallucinogens** – these distort perceptions and cause hallucinations. Illegal hallucinogens include LSD and magic mushrooms.
- **Analgesics** – these lower the sensation of pain. An example is codeine. Illegal analgesics include herion.

Table 23.2 overleaf lists some illegal drugs and the effects they have.

Social and economic implications of substance abuse
Substance abuse has impacts on society at many levels.

Loss of productivity
When people are under the influence of alcohol or drugs, their faculties are impaired and they cannot carry out their normal activities. Similarly, the after-effects of substance abuse (hangovers, 'comedowns', withdrawal symptoms) can make people unable to perform. In children and teenagers, this can lead to setbacks at school, poor exam results, lowered educational and job prospects. In adults, it can lead to unemployment and financial and family problems.

Accidents
Impaired faculties can cause people to behave irresponsibly and endanger their own safety and that of others, both at home and in public. Alcohol is a key factor in thousands of household accidents and road accidents every year and is thus directly responsible for many injuries and deaths. Treatment for alcohol-related accidents costs governments millions of dollars every year.

Domestic violence
Substance abuse can cause enormous damage within families. Alcohol and drugs can cause users to behave irrationally, unpredictably and violently. Violence develops either as a direct reaction to the effects of the substances or as an indirect reaction of addiction. The families of substance abusers often find themselves the victims of neglect and abuse.

Drug	Why people use it	Effects
Marijuana (delta-9-tetrahydrocannabinol or THC) Also known as weed, ganga, dope	Pleasure, relaxation	Lessens coordination and memory; can cause psychological addiction; can cause lung disease (if smoked); leads to lowered productivity at work; long-term use can lead to fertility problems for men
Cocaine Also known as coke, crack, snow, rock	Powerful stimulant; gives sensations of pleasure, increased alertness, feeling of power and confidence	Highly physically and psychologically addictive; long-term use can damage heart, brain, lungs and kidneys; frequent snorting of cocaine through nostrils can cause damage to internal linings of nasal cavity; drug causes constriction of blood vessels which may cause heart damage or stroke, irregular heartbeats, and in some cases death; because of the illegal trade in cocaine, drug syndicates can contribute to increased crime and violence in drug-producing countries
Heroin Also known as horse, smack	Powerful analgesic that induces sensations of drowsiness, pleasure, sense of release	Highly (often instantly) physically addictive; withdrawal symptoms can be intense and include vomiting, cramps, aches, shivers, diarrhoea and sweats; addicts often die from overdoses; injections with dirty needles can be associated with other health complications including transmission of viruses and STDs, tetanus and botulism; social effects can include increased crime as addicts become desperate to support their habit, but are unable to work
Methamphetamines Also known as meth, crank, ice, speed, crystal, tik	Powerful stimulant that increases alertness, decreases appetite and gives a sensation of pleasure	Highly addictive; withdrawal symptoms include depression, abdominal cramps, increased appetite; long-term effects of frequent use include paranoia, hallucinations, weight loss, rotten teeth, damage to heart tissue
Ecstasy (MDMA)	Stimulant and hallucinogen that creates a sense of euphoria and increased energy	The active ingredient (MDMA) often gets mixed with other dangerous ingredients that cause adverse reactions; long-term effects may include depression, interrupted sleep patterns, memory loss, inability to regulate emotions
Rohypnol and GHB	Powerful sedatives; cause low blood pressure, dizziness, confusion, impaired memory	Used as a date-rape drug; victims usually receive the drug added to their drink, and then wake several hours later in a disoriented state to discover they have been abused or raped
LSD or acid and hallucinogenic mushrooms	Powerful hallucinogens	Long-term effects can include 'flashbacks' (reappearance of hallucinations in later life) and psychosis including delusional beliefs, paranoia, mood disturbances

Table 23.2 Some illegal drugs and the effects they have.

Crime

Because many drugs are illegal, the trade in these drugs is associated with syndicated crime. The organizations that produce, distribute and sell drugs use crime and violence as a method of maintaining control over their business. The bosses of drug cartels are often associated with human trafficking, pornography and other illegal businesses.

Sexually transmitted diseases

Whilst under the influence of alcohol or drugs, people tend to forget about the importance of responsible sexual behaviour. Risky sexual behaviours are more likely under the influence of substance abuse. Also, some drugs (especially meth and heroin) are associated with intravenous needle use. Sharing needles can lead to the transmission of STDs and other diseases.

Cost of social welfare and healthcare systems

As a result of all these factors, substance abuse costs governments millions in social welfare and healthcare each year. Children from abusive families may need to be placed in state care; criminals who end up in jail for drug trafficking cost the government a lot of money to maintain and rehabilitate; accidents associated with substance abuse can lead to high medical bills for both families and the government.

Check your knowledge and skills

1 For each of the following statements, discuss whether it is true or false and why.
  a 'Only illegal drugs are addictive.'
  b 'As long as you follow your doctor's prescriptions closely, you will never be at risk of abusing drugs.'
  c 'Drugs only have negative effects.'
  d 'Drugs are not only expensive for the people that buy and use them – their cost is much greater than it looks.'
2 a Why do you think some people try illegal substances in the first place?
  b Why do you think some people continue to take illegal substances?
3 Write down the meanings of the following terms and give two examples of each:
  a stimulant   b depressant   c hallucinogen   d analgesic.
4 a Is tobacco a drug? Give reasons for your answer.
  b Do you think tobacco should be illegal? Discuss this in groups.
  c Why do some people argue that marijuana should be legalized, but cocaine should not? Do you agree or disagree? Discuss in groups.

Unit 3 — The social impact of disease

Disease and ill-health have serious consequences for communities and for countries as a whole. These include the following.

■ Increased burden on the healthcare system – patients require diagnosis, treatment and medication; this requires money for hospitals and health workers; research and development of medications also requires money.

■ Loss of man hours in the workplace – patients may be unable to work, leading to a loss of skilled people from the economy.

■ Increased burden on family – often family members care for the affected person; this requires time and resources; loss of the carer's job; in the case of contagious diseases, family members may also get ill; sometimes the burden of caring for sick family members can lead to stress for the carers.

■ A drop in tourism – countries like the Caribbean rely on tourism; if disease breaks out fewer tourists will visit Caribbean countries and people's livelihoods suffer.

HIV and AIDS

Sexually transmitted diseases, particularly HIV, impact on society. There are currently around 350 000 people living with HIV in the Caribbean. The prevalence varies from one area to another, but in some regions more than one out of every 100 people are HIV-positive. This has some serious effects on society. Table 23.3 overleaf shows Caribbean statistics about HIV and AIDS.

■ The main mode of transmission is through heterosexual sex. As a result, the people most at risk of HIV and AIDS are sexually active teenagers and

	Living with HIV/AIDS		Deaths due to AIDS (2007)
Country	all people	adult (15 to 49) rate (%)	Total deaths
Bahamas	6200	3.0	Fewer than 200
Barbados	2200	1.2	Fewer than 100
Cuba	6200	0.1	Fewer than 100
Dominican Republic	62 000	1.1	3900
Haiti	120 000	2.2	7500
Jamaica	27 000	1.6	1400
Trinidad and Tobago	14 000	1.5	Fewer than 1000

Table 23.3 The number of people in Caribbean countries infected with HIV/AIDS. Data are estimates only. Source: www.avert.org

adults. This is also the age at which adults are at their most economically productive. As the transmission of HIV rises, society faces a situation where many members of the workforce are unable to work and support themselves and their families.

■ People who die of AIDS tend to leave families composed of people who are older (parents and grandparents) and younger (small children). The very old and very young are vulnerable in that they are not able to support themselves financially. They may also be vulnerable to illness if they do not have sufficient support. This can place strain on the social welfare system.

■ Treatments for HIV and AIDS are still expensive. In poorer areas, people are less able to afford their treatments.

The impact of animal and plant diseases

The agricultural industry in most Caribbean countries provides a significant part of the countries' revenues. Diseases of plants and animals can have significant consequences for humans, especially if these animals are part of a country's agriculture produce. If diseases are not controlled it may lead to economic problems. Some ways in which diseases can affect a country are listed below.

■ Plant diseases can cause a drop in yield – this would mean that fewer fruits and vegetables are produced.
■ Low yields can cause food shortages within a country.
■ A country might not have enough produce to meet demand for exports.
■ Lowered supply would lead to increased prices and might drive importing countries to seek other suppliers with better pricing.
■ There could be job losses amongst people employed in harvesting, processing and packing agricultural products.
■ The diseases BSE ('mad cow' disease) and 'bird flu' caused many farms all over the world to go out of business.

At one time Jamaica was a large exporter of papayas. Recently there has been an infestation of the ring spot virus which has led to a significant reduction in export of the fruit. Farms that were once thriving are suffering. Research is underway to find a cure for the virus or a resistant type of papaya plant.

Name of disease	Crops affected	Description
Black sigatoka	Bananas and plantains	Fungal infection causes dark brown to black streaks on leaves
Geminivirus	Different forms of the virus affect different plants: beets, beans, tomatoes, corn, peppers	Symptoms vary depending on the strain. In tomatoes, can cause curling and yellowing of leaves, stunted growth and low yield In beans, can cause mosaic effect on leaves Main vector is the white fly
Tristeza	Citrus fruits	Caused by the tristeza citrus virus. In the 1940s this disease wiped out the citrus industries of Argentina, Brazil and Uruguay. There are various strains: – seedling yellows, tree turns yellow, wilts rapidly and dies within a few years; – stem-pitting disease, fruits develop pits on their skins and plants develop pits on stems and trunks; tree may be stunted and crops poor Main vector is aphids
Ring spot virus	Papaya and cucurbits (watermelons, cucumbers, other types of melons)	Symptoms include dark rings, usually sunken, on the fruits and a yellow mosaic on the leaves. Leaves also develop a shoe-string appearance as though they have been eaten by mites. Affected plants produce a lower fruit yield; if infected very young, they may never produce fruit; Main vector is aphids

Table 23.4 Some plant and animal diseases found in the Caribbean region.

Biology at work
Chapter 23 – The papaya ring spot virus

Check your knowledge and skills

1 Mr James has a family of four including his wife, who is a stay-at-home mother, and three children who are all in school – two boys and a girl. The two boys are 17 and 15 and the girl is 6 years old. Mr James is the manager at a bank that has seen great improvement since he took up the post. Describe what effect the following could have on the family.

   a Mr James starts to drink alcohol excessively, causing him to become a violent alcoholic.

   b The 16-year-old boy starts to use marijuana.

   c Mrs James is diagnosed with terminal breast cancer.

Chapter summary

Do you know?

If you are unsure of any of the facts in the list, refer to the page number in brackets.

- In order to use a drug correctly, you must pay careful attention to the dosage. (page 271)
- The use of illegal drugs, or the use of legal drugs for any purpose besides their prescribed medical use, is known as drug abuse. (page 272)
- Drug abuse not only affects the abuser; it also affects the society and the economy. (page 275)
- Disease in humans can have far-reaching effects on the society, especially if there is an epidemic of contagious disease. (page 277)
- Disease in plants and animals can have serious impact on the economy of a country. (page 278)

Are you able to?

If you have trouble in doing these things, refer to the page number in brackets.

- Define what drugs are. (page 270)
- Differentiate between medicinal and other types of drugs. (page 271)
- State the ways that certain diseases can affect the society and the economy of a country. (page 277)
- Describe how drug abuse can affect the family and the society. (pages 275–277)
- Explain how plant and animal diseases can have an adverse affect on the economy. (page 279)

24 Life and the environment

The physical environment

Each living organism on Earth is adapted to survive in a particular **environment**.

- Some organisms have adaptations that allow them to survive underwater; marine fish have adaptations to help them survive in salty water, and fresh-water fish have adaptations to help them survive in sweet (fresh) water. Fast-flowing water contains plenty of dissolved oxygen, which is needed by some freshwater fish in order to breed.
- Animals such as seals, penguins and polar bears have adaptations such as thick fur and layers of fat (blubber) to help insulate them against the cold.
- Deserts are hot and dry; some plants and animals have adaptations to help them survive extreme heat and water scarcity. Camels can survive hot dry desert environments by storing water in their humps. Similarly, plants such as cacti have adaptations that help them conserve water.
- Tropical rainforests tend to be warm and humid, with dense populations of plants. Many plants and animals thrive in these conditions.
- Savannah (grassy plains) is a **habitat** that supports many forms of grasses and trees. Animals that live in these areas tend to be able to travel far to find food, and have adaptations that camouflage them well in their habitat.

Figure 24.1 Can you see how each of these living organisms is adapted to its environment?

The survival of an organism in an environment is dependent on a wide range of environmental factors. These factors can be divided into:

- **biotic factors**, which are influenced by other living things in the community (see Table 24.1)
- **abiotic** or **physical factors**, which are non-living factors that influence the existence of living organisms (see Table 24.2).

Biotic factor	Effect
Competition	Plants and animals compete with each other for food, light and space. This influences the number and types of organisms that live in a specific place
Predation	A predator is an animal that hunts and feeds on other organisms (known as prey). Population levels of prey are influenced by how well they are able to evade being captured; population levels of predators are in turn influenced by how much food is available, and how easily the prey can catch it
Disease	Pathogens cause disease in animals and plants. Therefore their presence or absence influences the number and types of organisms in a community
Decomposers	Decomposers facilitate the breaking down of dead plant and animal matter, and the recycling of nutrients in the ecosystem. Thus decomposers make nutrients available to new generations of organisms. Decomposers can help an ecosystem to thrive
Human activity	Activities of humans can affect the survival of other organisms. Farming, industry and civilisation have all had an effect on the survival of organisms
Grazing	The presence of large herbivores can influence the population of grasses and other plants in a field

Table 24.1 Biotic factors and their effects on living organisms.

Climate

Figure 24.2 shows a world map of vegetation zones. These zones arise due to climate differences and they support different organisms.

Key:
- tundra, ice
- coniferous forest
- broadleaf forest
- Mediterranean scrub
- grassland
- savannah
- semi-desert
- desert
- dry tropical scrub
- sub-tropical scrub
- monsoon forest
- tropical rainforest

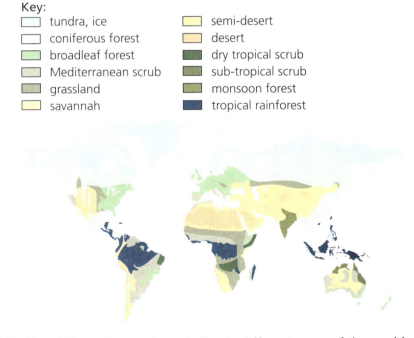

Figure 24.2 The different types of vegetation in different areas of the world influence the kinds of living organisms that survive there.

	Abiotic factors	Effect
Climate	Temperature	All organisms need particular conditions for them to thrive. Climatic conditions affect the rainfall, temperature and light intensity of an area Most organisms thrive between the temperatures of 0°C and 40°C, but some have the ability to withstand extreme temperatures. For example, penguins can survive in temperatures as low as −60°C, which makes them able to live through Arctic winters
	Light	All plants require light for photosynthesis. However, at lower light intensities some plants die, while others thrive. As an area becomes more crowded, plants compete for light. They may develop adaptations that help them to branch outward and upward to reach more light, or they may become adapted to tolerate shade
	Rainfall	Rainfall provides water that is vital for the survival of animals and plants. Some organisms, such as earthworms, thrive in places with high rainfall; others are able to withstand very little rainfall
Aquatic factors	Oxygen	Most organisms require oxygen for respiration. Organisms that live in water need to be able to respire using oxygen that is dissolved in the water. Ponds, seas and lakes tend to have more oxygen diffused near the surface than at greater depths. However, some animals have adaptations that help them to survive in deeper, less oxygenated water; Water pollution can also lower the oxygen levels in the water, as the pollution feeds bacteria that use up the dissolved oxygen. This uses up the oxygen supply, making it difficult for other animals and plants to thrive Thus in places where the oxygen levels are very low, only organisms that carry out anaerobic respiration can survive
	Salinity	Some aquatic organisms are adapted to live in salty water, whilst others are adapted for freshwater conditions. For example, fish that live in the ocean cannot live in a pond because they are not adapted to water that is not salty
Topographic factors	Slopes and plains	Some organisms are more adapted to living on level surfaces while some live on a slope
Edaphic factors (soil factors)	Particle size	Rocks break and crack over time under different climatic conditions. Very fine particles of rock gradually mix with plant matter and water and eventually form an organic mixture called soil, which is made up of particles of rock, humus, water, nutrients, air and living organisms The size of the particles of minerals determines soil quality. The finest particles form clay, which is dense. It gets very hard when dry, and is thick and sticky when wet. Larger particles form sand, which does not hold water very well. Loam is soil that is composed of a balanced mixture of sand, clay and nutrients
	Mineral content	Decomposing plant and animal matter, as well as bacteria and fungi, gradually produce a dark, rich organic matter called humus. Humus contains a supply of substances such as nitrogen and nutrients that plants can use. Nutrients and mineral ions dissolve in the water in the soil, and plants take in the nutrients and minerals as they take up water through their roots
	pH	The nutrients and organic matter in the soil determine its pH level – how acidic or alkaline the soil is. This influences what kinds of organisms can survive in the soil. Most plants survive best in soil that is slightly acidic
	Water and air content	Plants are not the only organisms that live in soil. Soil also contains organisms such as bacteria, worms, ants, termites and fungi. These organisms need oxygen for respiration and some, but not an excess, of water. Animals such as earthworms cannot live in waterlogged soil

Table 24.2 Abiotic factors and their effects on living organisms.

Soil

In some areas, the soil may be very sandy; in other areas it may be thick with clay. The ideal soil for growing plants is a loam soil – this is soil with a good balance of sand, clay and organic matter.

Soil has several layers. In order to study the composition of these layers, we take a **soil profile**. This means digging into the ground and looking at a cross-section of the soil.

Activity 24.1 Taking a soil profile

Aim: To analyse the composition of a soil sample.
Health and safety note – always wash hands after handling soil.

Equipment and apparatus

- shovel
- notebook
- pen
- camera (if possible)
- access to different areas of land

SBA skills

Observation/recording/ reporting (ORR)
Manipulation/ measurement (MM)
Analysis and interpretation (AI)

Procedure

1 Using the shovel, dig a hole about 30 cm deep. Make sure you dig straight down and clear away any loose soil so that you can see the soil layers.
2 Describe the layers of soil that you see.
3 Make notes in your notebook, and if possible take photographs.
4 Repeat the investigation in areas with different kinds of soil, for example near a river or in a forest.

Questions

1 Compare the soil profile. What similarities and differences do you notice?
2 Suggest reasons for the similarities or differences.

Activity 24.2 Analysing soil composition

Aim: To analyse the composition of a soil sample.

Equipment and apparatus

- cup measure
- large glass jar and lid
- spoon for collecting soil
- water
- soil
- sieve

SBA skills

Observation/recording/ reporting (ORR)
Manipulation/ measurement (MM)
Analysis and interpretation (AI)

Note: In water, larger particles settle more quickly as they are heavier. The settled layer of particles at the bottom of the water is known as sediment.

Procedure

1 Put a cupful of soil into the jar and top it up with three cupfuls of water. Screw the lid on tightly and shake well.
2 Let the soil settle slightly for half a minute. Then measure the thickness of the sediment at the bottom of the jar. Call this measurement A.
3 After 30 minutes, measure the sediment again. Call this second measurement B.
4 After 24 hours, take a third measurement of the sediment and call it C.
5 By subtraction, determine the thickness of the main layers (see Figure 24.3).
   $C - B$ = layer of clay
   $B - A$ = layer of silt
   A = layer of gravel and sand

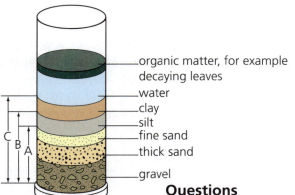

Figure 24.3 An example of a soil profile.

Labels in figure: organic matter, for example decaying leaves; water; clay; silt; fine sand; thick sand; gravel

6 Using the sieve, separate the gravel from the sand. Determine the gravel-to-sand ratio. Calculate the content (%) of each component of the soil sample.

7 Repeat this experiment with soil collected from different places (for example a garden, wood, river bank) where the soil has a different consistency or texture.

8 You can also use this technique to evaluate the composition of the soil for a pot plant. The balance of the soil can then be improved, depending on your observations.

Questions

1 What did you notice about each profile?

2 Describe the differences between the profiles of the samples from different areas.

Activity 24.3 Analysing soil moisture and permeability

Aim: To analyse the moisture and permeability of a soil sample.

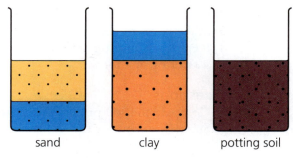

Figure 24.4 Analysis of soil permeability.

Labels: sand, clay, potting soil

The composition of a soil affects how well it holds or drains water. The amount by which a soil holds water is called its permeability. Very sandy soil does not hold water well at all (it is very permeable) so plants do not grow easily in this kind of soil. However, heavy clay soil does not drain water well, so the plants may get waterlogged. A good growing soil has a balance of water-retaining and draining properties.

Equipment and resources

- sand
- clay
- potting soil (soil and humus mix)
- three glass jars
- water

Procedure

1 Put a small handful of sand into one jar, clay in another and potting soil into the third. With your finger, press each sample against the side of the glass jar.

2 Pour some water into each jar.

3 Observe what happens.

Questions

1 How quickly do the soil particles fall to the bottom?

2 Write your observations about the different properties of clay, sand and humus.

Unit 2

Investigating an ecosystem

When we study an ecosystem, we may need to find out:

- how many of each species there is in the area
- how species are distributed
- the interrelationships of the organisms and the relationships of the organisms with the non-living part of the ecosystem.

It is impossible to count every member of every species in a particular area. This would be as time-consuming and difficult as trying to record the size, shape and colour of every grain of sand on a beach! Instead, ecologists take a **sample** from an ecosystem, using sampling methods to make sure that the samples are representative of the whole area.

The **abundance** of a species tells us the number of individuals in the ecosystem. The **density** is the number per unit of habitat: for example, number of earthworms per cubic metre of earth, or number of couchgrass weeds per square metre of lawn. Abundance and density change constantly. For example, a warm, rainy season may increase the abundance and density of weeds in an area. In turn, this may increase the population of a specific beetle that feeds on those weeds, which in turn may attract a particular bird that feeds on the beetle. The increased bird population begins to bring down the numbers of beetles, and so on.

The DAFOR scale

Scientist use the DAFOR scale to describe the abundance of species. The scale is adapted to apply to different ecosystems by assigning a value to each rating (Table 24.3).

For example, when applying the DAFOR scale to a plant species, we might use the following values.

- Dominant (D) – more than 75% cover
- Abundant (A) – 51 to 75% cover
- Frequent (F) – 26 to 50% cover
- Occasional (O) – 11 to 25% cover
- Rare (R) – 1 to 10% cover

Rating	Description
5 **D**ominant	Found throughout the system, all the time, in high numbers with a significant effect on the system
4 **A**bundant	Found all the time
3 **F**requent	Found often
2 **O**ccasional	Found sometimes, but not often
1 **R**are	Found occasionally
0 **A**bsent	Not found at all

Table 24.3 The DAFOR scale.

Sampling methods

Some of the most commonly used sampling methods for studying organisms in an ecosystem are described below.

Quadrats

A **quadrat** is a square frame of a given area, such as $0.25\,m^2$ or $1\,m^2$. It is usually made of wood or wire. The frame may be divided into smaller squares forming a grid. It is used to sample species present in a habitat.

The quadrat is usually thrown at random over the selected area several times (usually ten times). After each throw the number of each species of plant and slow-moving animal is recorded. Usually a sketch of the area accompanies the report. Various statistics can be produced from the data collected. These include the percentage coverage for each species in the area, the distribution of the species, their density and their frequency.

The researcher places the quadrat over an area of the habitat to be studied, and then attempts to record all the organisms found within the area of the quadrat. The quadrat is used to:

- compare samples from areas of consistent size and shape
- study the distribution of species in a large area
- generate average numbers of each species in a large area
- observe the combinations of species that can be found in an area.

The size of quadrat that is chosen depends to a large extent on the type of survey being conducted.

Not all quadrats are $1\,m^2$ in area. Some may be smaller in size, or rectangular (or even circular) in shape. However, by using the standard size of quadrat, the investigation findings will usually be comparable to other published research. There are also other types of sampling units used to study different areas.

- Aquatic microorganisms or water chemicals – the equivalent form of sampling would be to collect water samples in standard-sized bottles or containers.
- Parasites on fish – each individual fish would form the sampling unit.
- Aphids that live on a specific flower – a flower might be the sampling unit.

Figure 24.5 A quadrat is a frame used to sample a species in a given habitat.

In cases where the units are not the same size (for example leaves, fish, flowers) the information needs to be standardised by using a weighted mean, which takes into account the different sizes of the sampling units.

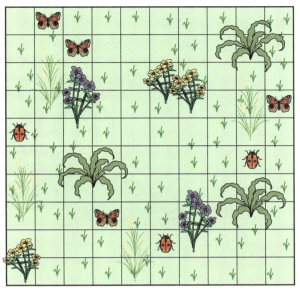

Place:	Name of site	
Date:		
Time:	Quadrat 1	Quadrat 2
Number of different kinds of flowering plants		
Names of flowering plants		
Number of different types of grasses		
Number of grasses		
Number of different kinds of minibeasts		
Names of minibeasts		

Figure 24.6 Distribution of plant species in a quadrat and a table to record data.

Line transects

A **line transect** is a line, usually made of a piece of string several meters long, with markings at specific intervals along the length of the line. Line transects are usually used to analyse an area where there is a change in habitat, for example from coast line to inland or down a slope. This system shows the way different species occur in zones along the environmental gradient.

The line or strip is placed across the area selected for study. Then the investigator walks along the line and records the organisms found at each marked interval. This method is mainly used for plants and slow-moving animals.

The data are normally displayed in the form of a diagram, using symbols for different species, which are drawn to scale. This is a useful way of being able to clearly visualise what changes are taking place along the line. It shows patterns in the type of species that tend to reside in particular areas.

When using a line transect, it is important to use an appropriate interval. If the interval is too great, then many of the organisms present may be overlooked. If the interval is too small, the transect method may become too time-consuming and may yield more data than the researcher needs. This can also make it difficult to identify patterns as the transect method becomes cluttered with too much detail.

a

2	4	6	8	10	12	14	16	18	20
	A	B	C	D	D	D	D	E	E

b

Figure 24.7 a A typical line transect of a coastal area; **b** a coastline and inland area.

Capture–recapture

The **capture–recapture** method is used to estimate the size of a population of a species of animal in an area. First, a sample is trapped or caught. The animals are counted, marked in some way (for example by tagging or with paint marks), then they are released back into their habitat. At a later point (for example a few days or a week later), a second sample is trapped and counted. A record is made of the ratio of the second sample that was caught previously.

To estimate the population size with this method, use this equation:

$$\frac{(\text{number of organisms captured in the first sample}) \times (\text{number of organisms captured in the second sample})}{(\text{number of marked organisms in the second sample})}$$

For example, the number of grasshoppers in a field is to be measured. A randomly chosen square meter of ground is marked off. Ten grasshoppers are captured and, taking care not to harm them, are marked using a coloured marker and then released. A day later, 15 grasshoppers, of which three are marked, are captured in the same area.

$$10 \times \frac{15}{3} = 50$$

Capturing methods

The methods used to capture animals vary depending on the animal. Figure 24.8 shows the different methods used.

Activity 24.4 Investigating percentage and density of weeds in a lawn

Aim: To investigate the percentage and density of weeds in a lawn.

Equipment and apparatus
- quadrat $0.25\,m^2$
- notebook
- pencil

Procedure
1 Select a section of lawn at your school.
2 Randomly throw your quadrat on to the lawn.
3 Wherever the quadrat lands, count the number of grass blades seen and the number of weeds.
4 Throw the quadrat a total of ten times. Do the count after each throw.
5 Record the data in a suitable table.
6 Calculate the percentage and density of weeds in the lawn.

Activity 24.5 Investigating types of organisms at a coastal area

Equipment and apparatus
- line transect (10 m in length)
- stakes
- notebook
- pencil

Procedure
1 Select a coastal area at low tide and in calm conditions.
2 Stake one end of the line transect about 1 metre into the water, and stretch it up towards the land as far as it will go.
3 Walk along the transect and record the organisms at each interval.
4 Present your data in a table and in a graph.

Check your knowledge and skills
1 What is the DAFOR scale and how is it used?
2 Describe two suitable methods for capturing each of the following:
   a moths   b crabs.
3 What do you understand by sampling?
4 Describe a suitable example in which you would use each of the following sampling methods:
   a a line transect   b a quadrat.

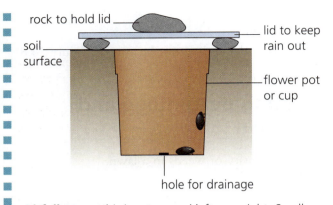

Pitfall trap – this is set up and left overnight. Small bugs and insects fall into the jar as they move over it. Disadvantages – prey and predators may be caught together.

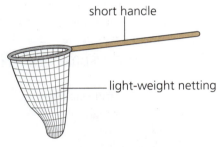

Insect net – the net is swept over an insect and the handle is rotated to close the opening so that the insect cannot escape.

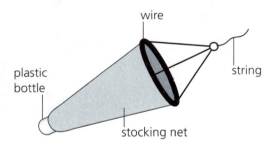

Plankton net – the opening allows organisims into the funnel; water flows through the stocking net and organisms fall into the bottom tube.

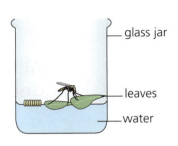

Jar – this jar is set up with conditions that a female insect likes for laying her eggs, for example water and leaves in a jar will attract mosquitoes. Disadvantage – an unwanted insect may be attracted.

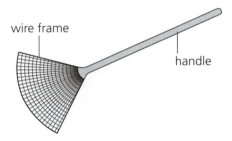

Sweep net – the net is swept across long grass. Small animals get caught in the bottom of the net.

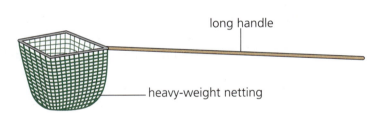

Fishing net – this is similar to a butterfly net, but usually has stronger netting as fish are heavier.

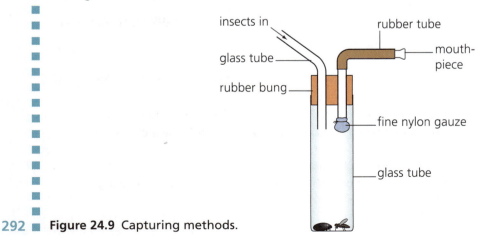

Pooter – this allows insects, such as ants or termites, to be captured by encouraging then into the in-going tube, and sucking them through the other tube. When not sucking up insects, the ends of the tubes need to be covered to prevent the insects from escaping.

Figure 24.9 Capturing methods.

Human population growth

Unit 3

A very important area of ecology is to assess the growth of populations in an ecosystem. When a habitat provides favourable conditions for living and reproducing, a **population** tends to increase, sometimes multiplying many times over. The **population growth rate** (**PGR**) is the increase in a country's population over a given period, usually a year. This rate is usually expressed as a percentage of the population at the start of that period.

There are four main factors that affect population size:

- immigration (organisms arriving to join the population)
- natality (births)
- emigration (organisms leaving the population to settle elsewhere)
- mortality (deaths).

Immigration and natality increase a population's size, while emigration and mortality decrease a population's size. These four factors are influenced by many other limiting factors. Figure 24.9 shows population growth rates for different countries. Table 24.4 shows the main terms used when studying populations.

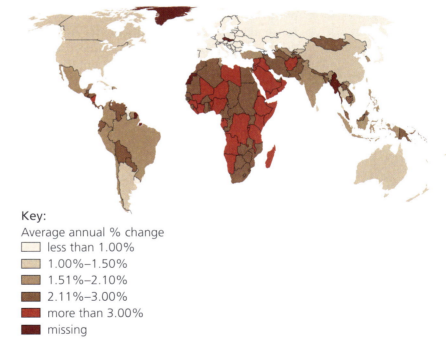

Key:
Average annual % change
- less than 1.00%
- 1.00%–1.50%
- 1.51%–2.10%
- 2.11%–3.00%
- more than 3.00%
- missing

Figure 24.9 This map shows the population growth rates for different countries around the world over the period 1990–95.

Check your knowledge and skills

1 Developing countries tend to have higher population growth rates than developed countries. Discuss the possible reasons for this. Think about the following factors in your discussion:
- education
- women's rights
- health services
- cultural differences
- vaccinations
- food shortages.
- contraception

Term	Definition
Population	All the organisms within a given area belonging to the same species
Population size	Number of individuals in a population
Population density	Number of individuals of a certain species per unit area or volume
Population distribution	Pattern of dispersal of a population within that area
Limiting factor	Factors that determine whether an organism can survive in an area, for example availability of water, space, food

Table 24.4 Terms used when studying populations.

A **population curve** shows the pattern of population growth over time. In general, a population begins with a **lag phase**. This is a period where the population remains stable, or grows very slowly. If circumstances are favourable for reproduction, the population gradually grows, until the gradient of the curve changes. At this point, it is not only the population that increases; the rate of population growth increases too. A population continues multiplying until a factor controls or curbs its growth. At the height of its growth, it reaches a rate called **exponential growth** where the increase in the total population occurs faster with each successive time interval. Exponential growth creates a positive feedback system. The bigger the population gets, the more rapidly it grows.

Any species that is increasing at an exponential rate faces a situation where it will outstrip available resources such as space, air, water, food. Eventually one of these factors will cause the population growth rate to decelerate until it eventually reaches equilibrium. A population which is at equilibrium has about equal numbers of births and deaths, so the population numbers remain stable. Logistic growth is a term used to describe growth that lags, increases, reaches a plateau and then gradually decreases (in other words, a population curve usually follows a logistic growth pattern).

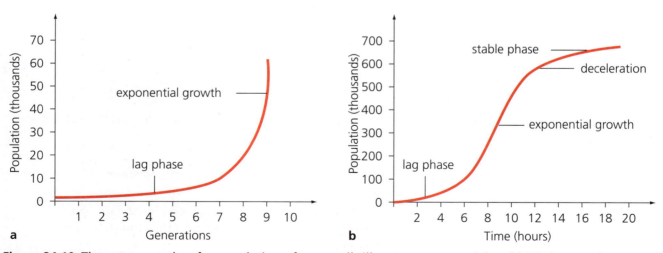

Figure 24.10 These two graphs of a population of yeast cells illustrate exponential and logistic growth. Exponential growth, shown in **a**, creates a J-shaped curve. Logistic growth, shown in **b**, follows a sigmoid (S-shaped) curve.

Check your knowledge and skills

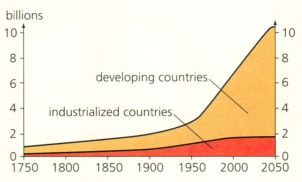

Figure 24.11 World population changes and predictions 1750 to 2050.

1 What do you notice about the world population before 1950?
2 What does the graph show after 1950?
3 The 1900s were a period of many discoveries and developments in the fields of medicine and sanitation. How do you think this affected the world's population?
4 Do you think the world's population can continue growing at this rate into the future? Give reasons for your answer.
5 Distinguish between biotic and abiotic environmental factors, and list five of each.
6 Look at each of the soil samples shown on the right and suggest what could be done to improve the quality of each.
7 Describe three methods of collecting insects without harming them.
8 Describe the capture–recapture method of sampling and give the formula you would use to calculate a population size using this method.
9 Describe, using graphs, the usual shape of a population curve.

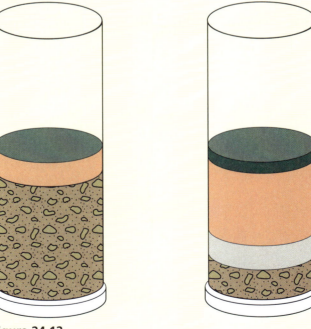

Figure 24.12

Chapter summary

Do you know?

If you are unsure of any of the facts in the list, refer to the page number in brackets.

- A habitat is a place where a particular organism is found. (page 281)
- An environment is made up of physical and biotic factors. (page 283)
- Soil provides water, nutrients and oxygen to living organisms, particularly plants. (page 285)
- Air provides various important gases that help to support and sustain life on Earth. (page 286)
- The abundance of a species is the number of individuals in the ecosystem. The density is the number per unit of habitat. (page 287)
- A population is a group of individuals of the same species within a particular habitat. (page 287)

- Sampling methods give us ways of estimating population figures. (page 288)

Are you able to?

If you have trouble in doing these things, refer to the page number in brackets.

- Explain what the physical environment is. (page 281)
- Describe the impact of the physical environment on organisms. (page 284)
- Categorise the main types of physical factors. (page 284)
- Plan and carry out an investigation of a typical ecosystem. (page 291)
- Identify and determine the appropriate tools for investigating an ecosystem. (page 292)
- Discuss population and growth within an ecosystem. (page 293)

25 Humans and their environment

Human activity and its environmental impact

The world's population is growing. Each person needs food, water and shelter. These needs, as well as daily activities, place constant demands on natural resources.

Human survival is also determined by many of the same factors outlined in Chapter 24. However, humans are threatening their own prospects for survival by overusing and damaging the Earth's natural resources.

By the end of this chapter you should be able to:

■ List the types of resources and put them into the two main categories

■ Describe human effect on the environment

■ Explain possible solutions to some environmental problems

1 Resources and their limits

The environment provides **resources** for sustaining life. Natural resources include:

■ Air – natural processes purify and recycle the air we breathe.
■ Water – the water cycle moves water from one area to another and cleans it.
■ Food – recycled plant and animal matter needs to break down to provide nutrients for new plants to grow.
■ Medicines – many medicines are made from extracts of plants and animals.
■ Minerals – these substances are essential to life.
■ Fuel – most of our technologies rely on fuels that come from natural resources, such as the **fossil fuels**, coal, gas and oil.
■ Fibres for clothing and manufacturing – cotton, wool and other cloths are made from natural fibres.
■ Building materials – buildings and homes are constructed from various materials found in the environment.
■ Land – wherever humans live and work, they use space that would otherwise be occupied by plants and other animals.

Many natural resources are **finite resources**, which means that there is a limited quantity of the resource available. Once they get used up, they cannot be replaced.

■ Water

The Earth has a finite quantity of fresh water available in aquifers (underground channels), surface waters (rivers and lakes) and in the atmosphere. The water in the oceans is plentiful but has high levels of salinity. Current technologies do not yet offer us efficient ways of converting seawater to drinkable fresh water.

The United Nations and other organisations use the term 'water crisis' to describe the limited water resources we have available to meet human demands. Some environmental agencies claim there is no water crisis. However, those who claim that we do have a water crisis point out the following.

- Nearly 900 million people do not have adequate access to safe drinking water.
- Approximately a third of the world's human population does not have access to water for sanitation and waste disposal.
- Overuse of groundwater has led to lower yields of agricultural crops.
- Pollution of water supplies is damaging biodiversity.
- Lack of water contributes to civil conflicts and warfare.
- Lack of clean, safe water supplies contributes to the spread of disease.
- Contamination of drinking water with untreated sewage threatens the safe water supply in developing countries.

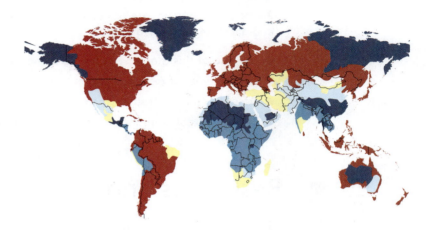

☐ **Physical water scarcity:** water resource development is approaching or has exceeded sustainable limits. More than 75% of river flow is extracted for agriculture

☐ **Approaching physical water scarcity:** more than 60% of river flow is extracted. These areas will experience physical water scarcity in the near future

☐ **Economic water scarcity:** limited access to water, even though natural local supplies are available to meet human demands. Less than 25% of water extracted for human needs

■ **Little or no water scarcity:** abundant water resources relative to use, with less than 25% of water extracted for human purposes

■ **not estimated**

Figure 25.1 Areas around the world suffering from reduced water resources.

■ Minerals

Minerals include substances such as gold, silver, platinum, zinc, lead, oil, copper, aluminium and iron. There are around 4000 known minerals, of which about 100 are very commonly found. The mineral resources of a country help to determine the value of the economy. Mineral resources in the Caribbean include:

- bauxite (Dominican Republic, Haiti, Guyana, Jamaica)
- nickel and nickel ore (Cuba, Dominican Republic)
- oil (Barbados, Cuba, Trinidad)
- asphalt (Trinidad).

■ Fossil fuels

Most modern technology requires **fossil fuels** such as coal, gas and oil. These are all finite resources. Increasing population creates greater demand for these resources. We cannot continue using these fuels at an ever-increasing rate, as they will simply run out. For this reason, it is increasingly important for humans to:

- ■ conserve energy
- ■ come up with energy-efficient technologies for transport, manufacturing and communications
- ■ find alternative sources of energy.

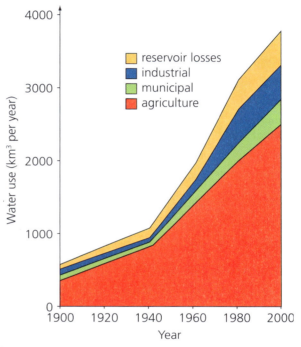

Figure 25.2 World water use over the period 1900 to 2000.

Unit 2 Environmental damage

Humans use natural resources far more intensively than any other living organisms on Earth, and our activities have a significant impact on the environment.

Building and construction

Unlike most animals, we create shelters such as houses. We also use buildings for factories and other businesses. The construction of buildings has many impacts on the environment. Land has to be cleared to provide space for buildings. This means that plant-life is destroyed and animals lose their natural habitat. The demand for building materials, such as wood and stone, also requires activities such as tree felling, stone quarrying and mining. These all damage the environment.

Sewage and waste

We all produce organic waste in the form of sewage. This needs to be treated and disposed of properly, as it carries disease and toxins. When untreated sewage flows into oceans it kills wildlife.

Humans also produce large quantities of waste products. For example, unwanted packaging, used nappies, rubble — all these help to fill-up landfill sites.

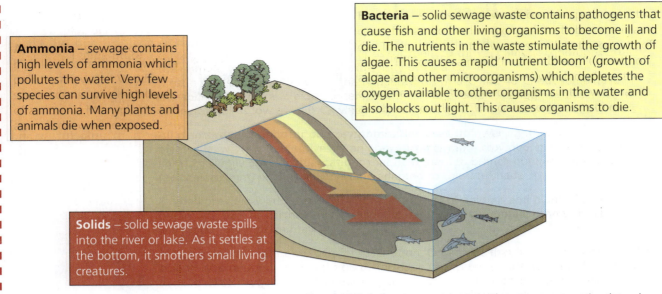

Ammonia – sewage contains high levels of ammonia which pollutes the water. Very few species can survive high levels of ammonia. Many plants and animals die when exposed.

Bacteria – solid sewage waste contains pathogens that cause fish and other living organisms to become ill and die. The nutrients in the waste stimulate the growth of algae. This causes a rapid 'nutrient bloom' (growth of algae and other microorganisms) which depletes the oxygen available to other organisms in the water and also blocks out light. This causes organisms to die.

Solids – solid sewage waste spills into the river or lake. As it settles at the bottom, it smothers small living creatures.

Potential dangers – nutrients in the manure can stimulate algae growth. Decaying algae can deplete dissolved oxygen and cause more fish kills. Manure can also harbour dangerous bacteria, such as *giardia* and *cryptosporidium*. These pathogens are a risk for seafood, seafood eaters, and swimmers.

Figure 25.3 How untreated sewage can poison water environments.

Farming

Industries that supply food for humans damage the environment in various ways. Humans clear land for crop farming and also for grazing livestock. Farming methods also have impacts on the environment, depending on the techniques used.

■ Pesticides – these are designed to kill specific pests but in doing so they interrupt food webs by knocking out the insect or animal that feeds another organism in the food chain. Sometimes, pesticides also poison other animals, especially if they run into the water supply; they can also be toxic for humans.

■ Fertilizers – these are intended to boost the nutrient levels in the soil to help crops grow. However, these dissolve into the water that runs off from farmland. Once the fertilizers get into the water cycle, they can over-stimulate growth of algae in lakes and rivers. This can block out light that allows other animals to survive, leading to deaths of aquatic animals.

Unsustainable fishing practices

In many areas of the world, unregulated fishing industries have damaged the populations of fish. The depletion of fish stocks is a major threat to food security – especially in countries where people rely on the seas for their food.

In some areas in Asia, fishermen use cyanide fishing – a method of poisoning fish in order to catch them. They squirt cyanide, a powerful poison, into coral holes and crevices where the reef fish hide. The poison stuns the fish and makes them easy to catch. However, it also poisons the living coral and many of the animals that live around reefs. Less than half of the fish caught using this method actually survive long enough to be sold to aquariums and restaurants.

Another technique is 'blast fishing', in which fishermen use dynamite to create explosions underwater. These kill the target fish, but also destroy all other plant and animal life in the area, destroy reefs and create debris that pollutes the oceans.

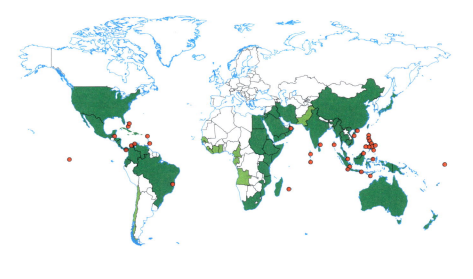

Key:

■ reported case of 'blast fishing'

■ countries with direct responsibility for coral reefs

■ countries with less developed coral communities

Figure 25.4 This map shows areas with reported incidents of 'blast fishing'.

The most successful alternative to overfishing is aquaculture – a method of farming fish sustainably.

The mind map shown in Figure 25.5 summarises some of the many human activities that impact adversely on the environment.

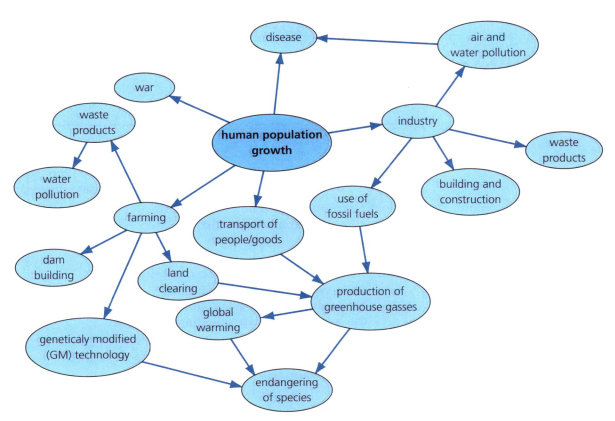

Figure 25.5 Some of the human activities that damage our environment.

The greenhouse effect

The sun is Earth's only source of external heat. During the day, rays of light from the sun heat the Earth's surface. At night, when the Earth is in darkness, the Earth continues to radiate the heat it received during the daylight hours. The Earth is surrounded by a thick layer of gases called the **atmosphere**. Gases in the atmosphere trap the heat energy from the sun. This is known as the **greenhouse effect**. This effect keeps the Earth's average temperature fairly stable, at around 15°C.

Some **greenhouse gases** occur naturally in the atmosphere. Over the last 200 years, human activities have increased the quantities of these greenhouse gases in the atmosphere. Table 25.1 on page 303, lists the most commonly occurring greenhouse gases.

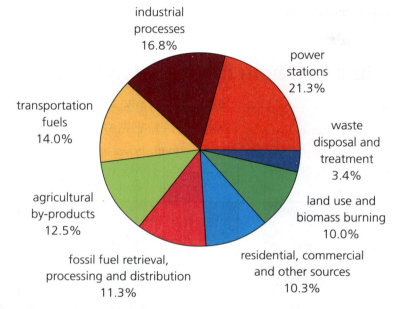

Figure 25.6 How each sector contributes to the annual emissions of greenhouse gases.

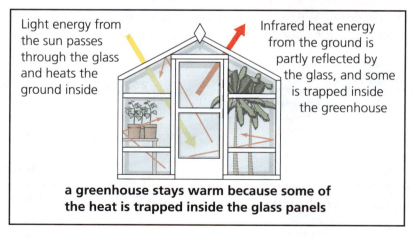

a greenhouse stays warm because some of the heat is trapped inside the glass panels

Figure 25.7 The greenhouse effect.

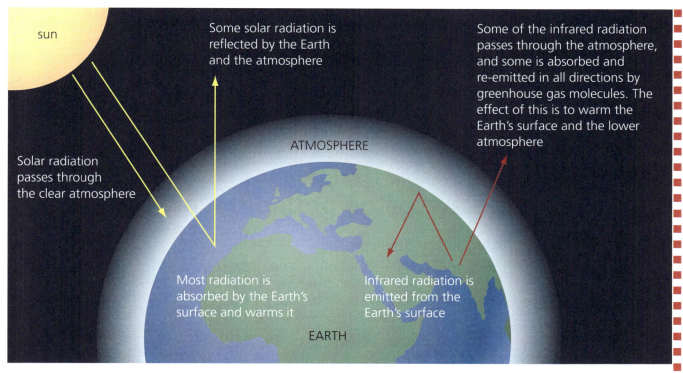

Figure 25.8 The effect of the greenhouse effect on Earth.

Greenhouse gas	Natural occurrence	How human activity has increased levels of this gas
Carbon dioxide (CO_2)	Plants and animals release it during respiration Natural product of forest fires Trees and other plants remove it from the atmosphere to use it for photosynthesis	Burning fossil fuels (coal, oil and gas) for energy and electricity Clearing forests, which results in build-up of CO_2 as there are fewer trees to remove it
Methane	Forms naturally in bogs and marshes when dead plant and animal matter decays Also a by-product of burning	Paddy farming of rice; Cattle farming (cows release enormous amounts of methane as they digest their food; their manure also releases methane) Pipeline losses Emissions from landfills Sewage systems
Nitrous oxide	Occurs naturally in the oceans and as a product of lightning activity	Fertilizers increase NO_2 concentrations
Water vapour	Water evaporates into the atmosphere as part of the water cycle	No significant ways that human activity increases water vapour levels
Chlorofluorocarbons (CFCs)	Do not occur naturally	Man-made substances that contain chlorine, fluorine and carbon They were invented in the 1930s for industrial uses, for example, fridge coolants and aerosols
Ozone (O_3)	High-level ozone occurs naturally about 10 to 50 km above the Earth's surface Does not occur naturally in significant concentrations closer to Earth's surface	Product of complex gaseous reactions with emissions from cars and other industrial activity Ozone at Earth's surface is a pollutant that interferes with photosynthesis Stunts plant growth and irritates eyes

Table 25.1 Common greenhouse gases.

Global warming

Without the greenhouse effect, the Earth would be too cold for humans to survive. However, if greenhouse gases increase significantly, the temperature on Earth is likely to rise. According to current scientific predictions, the temperature of the Earth's surface may increase by 3 °C by the end of the 21st century.

Rising sea levels

As temperatures rise, the polar ice sheets and glaciers in Greenland and Antarctica are melting. This increases the volume of water in the sea, causing rising sea levels. Over the past 100 years, sea levels have been rising by 1 to 2 mm per year. Scientists predict that the sea level could rise by another half a metre in the next 100 years. This will threaten low-lying coastal areas such as Holland and Bangladesh. It will also threaten low-lying islands in the Pacific and Indian Oceans.

Impact on agriculture

Global warming will have significant effects on agriculture. Many areas, especially in the southern hemisphere, will suffer from droughts and food shortages. It is possible that new crops will be able to grow in areas that were previously too cold to support agriculture. However, warmer conditions can precipitate an increase in pests and diseases that affect crops.

Infectious diseases

Warmer conditions also encourage pathogens. Rising sea levels and increased levels of water vapour in the atmosphere can help the spread of waterborne diseases.

Loss of biodiversity

The Earth supports a vast range of species of plants, animals and microorganisms which live within ecosystems, including deserts, rainforests, tropical and sub-tropical climate zones and coral reefs. The range of species and ecosystems is called the **biodiversity** (short for biological diversity). Global warming threatens biodiversity as many species have adapted over many generations to survive within a specific range of climatic conditions. As the sea temperatures rise, many species of corals and fish are dying as they cannot survive the higher temperatures. Also, as fish stocks move from their normal habitats as a result of rising sea temperatures, larger predators lose their food supplies.

Extreme weather events

The world's oceans are also getting warmer. In some areas, oceans are approaching temperatures of 27 °C or warmer during the summer season. Also, increased ocean temperatures lead to higher evaporation levels. This can assist in the formation of hurricanes and cyclones. Warmer oceans are also changing rainfall patterns and contributing to an increase in extreme weather events, such as storms and floods. In 2004, a spate of severe hurricanes devastated many Caribbean islands as well as areas of the United States. Many people were injured or killed and cities were badly damaged. Scientists have called these kinds of weather events super-storms and relate them to the effects of global warming.

Environmental solutions

All over the world, environmental organizations and individuals are working towards reducing the effects of human damage to the environment. However, because of the global scale of the problem, governments are recognising the importance of international agreements to ensure sustainable solutions for the environment.

■ International responses

The Millennium Ecosystem Assessment

The Millennium Ecosystem Assessment (MEA) is a report that measures the damage that human activities cause to the environment. The assessment was launched by the United Nations in 2001. It took five years to put together, drew together the work of more than 1300 ecologists and scientists from 95 different countries, and cost 24 million US dollars.

The MEA found that the past 50 years have brought significant damage to the world's ecosystems. This damage is accelerating as the human population grows, together with its need for resources. The report also found that international policies may be able to remedy the current rate of environmental degradation.

Some of the specific findings of the report include:

- 60% of the world's ecosystems have been seriously damaged.
- Approximately 25% of the Earth's land surface is now cultivated.
- We are currently using 40–50% of our available fresh water, and running out of fresh water supplies.
- More than 25% of the ocean's fish populations are over-harvested.
- Species extinctions are currently hundreds of times higher than ever before.

These dramatic findings brought to the world's attention the seriousness of the environmental issues facing our planet.

Reducing emissions

The main international agreement that sets up targets for reducing carbon emissions worldwide is the Kyoto Protocol. This treaty has been signed by more than 160 countries worldwide and accounts for more than 55% of the world's greenhouse emissions. The only countries that have refused to sign the agreement are the USA and Kazakhstan. However, the USA is currently the world's largest producer of greenhouse gases. It is expected to sign a negotiated protocol at the United Nations Climate Change Conference in Copenhagen in 2009.

Many businesses are also recognising the significance of climate change and are contributing towards efforts to reduce carbon emissions.

■ Individual responsibilities

No individual person can stop global warming. But there are many choices we can make as consumers and as citizens to help contribute towards a more sustainable future.

A person's **carbon footprint** is the amount of carbon dioxide (CO_2) that they release into the atmosphere as a result of their activities.

A tonne of carbon is released every time one person:

- travels 5000 miles in an aeroplane
- travels 2500 miles in a medium-sized car
- cuts down and burns a tree that is 40 ft high and 1 ft in diameter.

Did you know?
Carbon dioxide is the main greenhouse gas that is contributing to global warming. Almost everything we do adds a little more carbon dioxide to the atmosphere – using hot water, turning on the kettle, travelling in planes.

Smaller amounts of carbon are constantly released when we use energy in our daily activities.

Helpful hints for reducing your carbon footprint

Here are some suggestions for actions you can take to reduce your personal contribution to carbon emissions. There are many ways to do this. The main principle is to conserve energy wherever possible: for example, by reducing the amount of heat and waste you produce. Here are some tips.

- Use less fossil fuels – cars and other vehicles use fossil fuels. By using public transport, car pools and shared rides, you can reduce levels of fossil fuels being used in your town or city. If you walk or ride a bike instead of travelling in a vehicle powered by fossil fuels, you save one pound of carbon for each mile you travel.
- Fly less – if people in business use telephone calls and video conference technology instead of flying to meetings, they save time, money and carbon emissions. A tonne of carbon is released for every 5000 miles you travel in an aeroplane.
- Use energy-efficient lightbulbs – incandescent bulbs use more power and produce more heat than compact fluorescent bulbs. You can save 45 kg of carbon for each incandescent bulb replaced with a compact fluorescent bulb.
- Recycle and use recycled products – less power is used making products from recycled materials than is used making them from new materials. Recycling paper also reduces deforestation, thus saving trees that can remove carbon from the atmosphere.
- Inflate the tyres on the car – if your family has a car, keep the tyres properly pumped up. This helps the car to run more efficiently, burn less fuel and emit less carbon.
- Plant indigenous trees – trees use CO_2 from the atmosphere and produce oxygen. A tree can remove between 315 to 3175 kg of carbon from the atmosphere over its lifetime. A tree that shades your home can help to reduce the cost of running an air conditioner, saving up to 2000 more pounds of carbon over its lifetime.
- Turn off heaters and coolers – heaters and air-conditioners use a lot of power. Do without, or if you can't, install a timer switch so that you only use these devices for short periods during the day. Similarly, a time switch on your hot water cylinder can save a lot of energy. Even moving the thermostat down by 2 degrees in winter and up by 2 degrees in summer can save around 900 kg of carbon per year.
- Use renewable energy sources – find out whether your area offers electricity generated by windmills, solar panels or other clean technologies. Support these sources of energy.
- Buy local produce – importing fresh foods adds 'carbon miles' to the food you eat. Buying local produce helps to support the local economy and discourages wasteful transportation of food.
- Unplug appliances – kitchen appliances, televisions, VCRs, computers, phone chargers and MP3 chargers all use power when they are plugged in, even when they are switched off.

Figure 25.9 Wind has the potential to supply more than 30% of the world's electricity.

Figure 25.10 Solar panels capture some of the light energy from the sun and convert it to heat energy.

Figure 25.11 The tides, waves and currents of the world's oceans carry wave energy.

■ New technologies

With the human population constantly growing, we cannot use finite resources at the current rates of consumption. We need to find ways, not only to conserve resources, but also to identify new technologies that will meet our needs sustainably into the future.

Wind power

Advantages:

- widely available
- clean (does not pollute the environment)
- does not produce greenhouse gases.

Disadvantages:

- turbines create noise pollution
- animals can get caught and killed in the blades
- some people consider wind farms unsightly.

Factors that need improving or refining:

- wind does not blow all the time, so it is necessary to find ways of dealing with unpredictable or inconsistent supply
- at present, there is no way to move the power from the place where it is generated.

Solar energy

The sun provides enough energy to supply the Earth with all its power. However, we do not yet have the technology to capture this energy effectively.

Ocean power

In the future, engineers may be able to harness wave energy. In 2008, British engineers tested underwater buoys that harness wave energy from 50 metres below the ocean's surface.

Waste heat

Power stations generate energy, but about 40% of the energy generated is given off into the atmosphere as waste heat which contributes to global warming. Some researchers have suggested that by installing small domestic generators into homes we can use waste heat for heating domestic water.

Electric cars

Motor designers are working towards new technologies to create cars that do not require fossil fuels. An example is the electric car.

Eco-friendly architecture

Architects are designing houses that incorporate features to reduce their carbon emissions by more than 80%. Thousands of houses in Europe have been built to include:

- high levels of insulation
- solar panels for domestic heating
- internal ventilation and heat recovery systems.

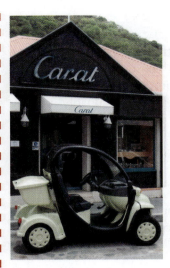

Figure 25.12 A few motor companies have already developed cars that run on electricity.

Second-generation biofuels

For a while, scientists experimented with the idea of 'bio-fuels'. These are fuels that can be grown as crops. However, it soon became clear that growing crops for fuel requires land clearing, encourages deforestation and reduces the available space for food crops, thus contributing to food shortages. At present, vast areas of the rainforests are still being cleared to make way for palm oil plantations, causing severe loss of biodiversity. Recent research has shown that it is possible to break down waste wood into liquid fuels. More research is required before this becomes a reality.

Carbon capture and storage

Coal is the Earth's cheapest and most widely available fossil fuel. We currently have enough coal to supply world power requirements for hundreds of years. However, burning coal produces very high CO_2 emissions which are contributing to global warming. Coal technologists have come up with a possible solution. If new technologies can find ways to capture and store the CO_2 emissions safely, the carbon may be used as an energy source.

Carbon capture and storage (CCS) systems have been used since 1996, when oil companies in the North Sea began pumping CO_2 back into aquifers beneath the seabed. Similar schemes are being used at other sites (see Figure 25.13). However these projects do not yet clear significant amounts of the carbon emissions produced by human activities (24 billion tonnes annually). Viable CCS technologies are not expected to be available before about 2030, at the earliest.

Biochar

Biochar is a charcoal produced by burning agricultural waste in an airless environment. It is extremely stable and can be stored for hundreds of years without releasing carbon into the atmosphere. It also increases soil fertility.

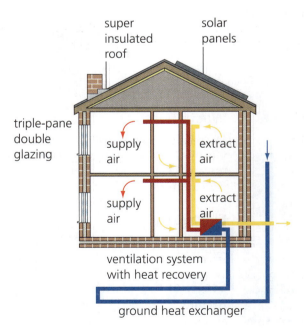

Figure 25.13 The PassivHaus movement in Germany has resulted in the design of highly eco-friendly houses in Europe.

Key:
- ☐ sedimentary rocks with high potential for storage
- ☐ stationary emitters
- ☐ stationary emitters

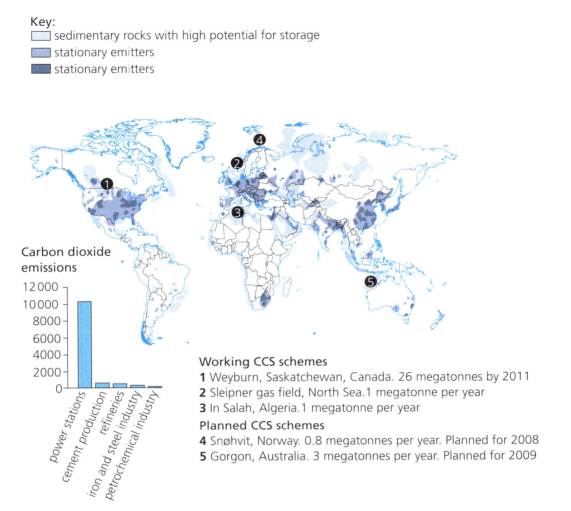

Carbon dioxide emissions

Working CCS schemes
1 Weyburn, Saskatchewan, Canada. 26 megatonnes by 2011
2 Sleipner gas field, North Sea. 1 megatonne per year
3 In Salah, Algeria. 1 megatonne per year

Planned CCS schemes
4 Snøhvit, Norway. 0.8 megatonnes per year. Planned for 2008
5 Gorgon, Australia. 3 megatonnes per year. Planned for 2009

Figure 25.14 40% of the world's carbon dioxide emissions are produced by coal-burning industrial plants and power stations. Most are not close to sites to bury their carbon dioxide, so a network of pipelines would have to be laid.

Research

1 Choose one of the new technologies discussed above and research it further. Write a report about developments in this technology.

Check your knowledge and skills

1 List five renewable resources and five non-renewable resources that you use daily.
2 Identify three finite resources which are likely to run out in your lifetime. Suggest which will run out fastest, and why.
3 How will global warming affect life in your country? Discuss at least five ways that life in the Caribbean could change as a result of global warming.
4 List ten ways you can reduce your personal contribution to carbon emissions.
5 Find out what is meant by 'offsetting' carbon emissions.

Chapter summary

Do you know?

If you are unsure of any of the facts in the list, refer to the page number in brackets.

- The world's population is growing at an unprecedented rate. (page 297)
- Our natural resources are finite. (page 297)
- Human activities have a serious impact on the environment. (page 299)
- The greenhouse effect traps heat in the Earth's atmosphere. (page 302)
- One of the greatest effects of human activities is the emission of carbon and other greenhouse gases into the atmosphere. (page 302)
- With high levels of greenhouse gases in the atmosphere, the Earth's temperature can rise significantly, killing off many of the organisms on the planet and threatening human life in many ways. (page 304)
- Reducing carbon emissions is the main way we can slow down global warming. (page 306)

Are you able to?

If you have trouble in doing these things, refer to the page number in brackets.

- Identify natural resources. (page 297)
- List finite resources. (pages 297 to 299)
- Discuss the ways human activities affect the environment. (page 299)
- Explain the main causes of global warming. (page 302)
- Describe the greenhouse effect. (page 303)
- Identify greenhouse gases. (page 303)
- Describe strategies for reducing carbon emissions. (page 306)

Index

Chapter references in **bold** refer to the 'Biology at Work' sections on the accompanying CD-ROM

abiotic components of the environment 9, 283, 284
abscisic acid (ABA) 175
absorption 73, 81–2
abstinence from sexual intercourse 222, **Chapter 19**
abundance of species 287
accommodation (eye) 163
active immunity 264
active ingredients of drugs 270
active sites 53
active transport 34
addiction 273–5
adrenal glands 166, 167
adrenaline 166
adrenocorticotrophic hormone (ACTH) 198
aerobic respiration 88, 89–90, 92
African dust **Chapter 3**
AIDS (acquired immunodeficiency syndrome) 257–8, 277–8
alcohol abuse 274–5
alimentary canal 73, 74
alleles 228, 229, 232
alveoli 99
amino acids 50–1
amylase 78, 81
anaemia 44, 85
anaerobic respiration 88, 90–2
animal cells 24, 25–6
   effect of solutions 38
antacids **Chapter 7**
antagonistic muscles 177
anther 205, 206
antibody immunity 262
antidiuretic hormone (ADH) 147–8, 166, 198
anti-malarial drugs 266
aorta 109, 110, 111
aphids 128, **Chapter 11**
appendicular skeleton 171, 172
aquatic ecosystems 11, 284
arteries 107, 108
artificial selection 247–50
asexual reproduction 201–5
assimilation 73
assortative mating 147
asthma 254
astigmatism 165
ATP (adenosine triphosphate) 34, 88
atria 110
autoimmune diseases 263
autotrophs (producers) 11, 12, 59
auxins 175, 194–6
axial skeleton 171, 172
axons 156, 157

bacteria 253
   anaerobic respiration 92
balanced diet 83
ball-and-socket joints 176–7
bar graphs 141
barrier contraception methods 220–1, 222
base-pairing 228
basophils 261
B-cells 261
Benedict's solution 46
beriberi 43
bicuspid valve 111
bile 74, 81
binary fission 182
biochar 308
biodiversity 304
biofuels 308
biogas 92
biological terrorism 259, **Chapter 22**
biomass, pyramids of 15

biotic components of the environment 9, 283
Biuret test 51
Black Death 257
black sigatoka 279
blind spot 162
blood 105, 113–14
blood donation **Chapter 10**
blood types 237
blood vessels 105, 107–8, 109
body mass index (BMI) **Chapter 12**
bone 171, 174
bone marrow 174
bone marrow implants **Chapter 16**
bottleneck effect 243
Bowman's capsule 138, 139
brain 160–1
   neuroplasticity **Chapter 15**
brain cells 30
breathing 97, 98–101
   difference from respiration 88
bromothymol blue 100
bronchi 98
bronchioles 99
bronchitis 103
budding 201
buildings, and the environment 299, 306

calcitonin 166
calcium, use in the body 44, 174
Calvin Cycle 69
cancer 103
capillaries 107, 108
   in lungs 99
capillary action, xylem 119
carbohydrates 44–6, 52
   testing for 46–8
carbon capture and storage (CCS) 308, 309
carbon cycle 19, 88
carbon dioxide
   excretion 135, 136
   as a greenhouse gas 303, 305, 306
   levels in the atmosphere 66
   levels in exhaled breath 99–100
   use in photosynthesis 59–60, 61, 64
carbon footprint 305–6
carbon monoxide 103
cardiac muscle 110
cardiac sphincter 73, 79
carnivores 11
carpel 206
cartilage 171
cataracts 165
cell cycle 181
cell division 181
cell-mediated immunity 262
cell membrane 24, 27
   maximization of diffusion 36
   permeability 33
cells 22
   comparing plant and animal cells 25–6
   function of different parts 27
   specialized 29–30
   structure 24
   study using a microscope 23
cellulose 45, 70
cell wall 27
cementum 76
central nervous system (CNS) 155, 160
centrioles 185
centromere 185, 211
cerebellum 160
cerebrum (cerebral hemispheres) 160
cervical vertebrae 173
Chagas disease 266
chemical digestion 74, 78–81

chlamydia 258
chlorofluorocarbons (CFCs) 303
chlorophyll 59, 64–5
chloroplasts 24, 27, 59, 67
chromatids 185, 211
chromosomes 28–9, 201, 227–8
chyme 79
cilia 98
circulatory system 105
classifying organisms 7–8
climate 283
climate change 304
cocaine 276
co-dominance 237
colds **Chapter 14**
collecting duct, kidney 138, 140
colour-blindness 236
commensalism 17
communities 11
companion cells, sieve tubes 127
competition 283
concentration gradients 34
   root hairs 118
concept mapping 161
condoms 220, 222
conjunctiva 162
conjunctivitis 164
connective tissue 31
consumers (heterotrophs) 11, 12–13
continuous variation 229–30
contraceptive methods 218–22
controls 63
cornea 162
corpus luteum 215
cotyledons 131, 187
courtship 216
cranial reflexes 158
cytokinesis 185
cytoplasm 24

DAFOR scale 287–8
Darwin, Charles 240, 245
death, causes of 264
decomposers 11, 12–13, 283
deficiency diseases 43–4, 85, 254
denaturing of enzymes 54
dendrites 156, 157
dengue fever 266
density of species 287
dentine 76
deodorant 136
deoxygenated blood 111
dependent variables 56
descent, principle of 241
designer babies **Chapter 21**
de-starching a plant 61
development 6, 181
   human 196–7
diabetes 84
diaphragm (muscle) 100
diaphragms (contraceptive) 220, 221, 222
diastole 112
dicotyledons 187
diet 82–4, 267
diffusion 34, 35–6
   gas-exchange surfaces 96
   in unicellular organisms 105, 136
digestion 73, 74
   chemical 78–81
   mechanical 77
diploid number 29, 201
disaccharides 44, 46
discontinuous variation 229, 231
diseases 252, 283
   social impact 277–8

Index

see also autoimmune diseases; infectious
   diseases, non-infectious diseases; sexually
   transmitted diseases
distal convoluted tubule   138, 140
DNA (deoxyribonucleic acid)   28, 227–8
dogs, inbreeding   248
dominant genes   225, 226, 228, 231–2
   co-dominance   237
   incomplete dominance   234–5
dose of drugs   271
drawing diagrams   25
drugs   270–1
   illegal   275, 276
   substance abuse and addiction   272–7
dry mass   190
duodenum   73, 80

E10 petrol   91
ecological pyramids   15–16
ecosystems   11, 287–8
Ecstasy (MDMA)   276
EEG (electroencephalography) machines   160
effectors   146
egestion   73
eggs
   food storage   132
   human, maturation   211
electric cars   307
electron microscopes   23
embryo   217
emphysema   103
emulsification of lipids   74, 81
emulsion test   50
enamel   76
endocrine system   155, 166–7
endorphins   198
endoskeletons   170
endosperm   131
energy   87–8
   pyramids of   16
energy content of food   82–3
energy requirements   83
energy transfer, food chains   12–13
environment   9–10, 281–2
   sustainable solutions   304–8
environmental damage   299–301
enzymes   53–4
   in digestion   74, 78–9, 81
   effect of pH   55
   investigation of properties   56
eosinophils   261
epidermal tissue   31
epididymis   213
epiglottis   98
epithelial tissue   31
ethanol, formation by fermentation   90
etiolation   188
eukaryotes   182
evolution   241
   Charles Darwin's theory   240
   controversies   245
   natural selection   243–6
   random mechanisms of change   242–3
exams
   answering questions   147
   essay questions   114
   terms used in questions   93
   'use of knowledge' questions   133
excretion   6
   in animals   135–6
   kidney   137–40
   in plants   134–5
exercise   267
   effect on pulse and breathing rates   93
exocrine glands   155
exoskeletons   170
extracellular digestion   72
eye   162–4
eye defects   164–5

faeces   82

fair tests   63
fallopian tubes   214
farming
   environmental damage   300
   organic   **Chapter 6**
far sightedness   165
fats   48
fatty acids   49
feeding relationships   11
fermentation   90
fertilization
   flowering plants   207
   humans   217
fertilizers   300
fevers   **Chapter 14**
finite resources   296
fishing, unsustainable practices   300–1
fitness levels   268–9
flaccid cells   39–41
flowers, structure   205–6
foetus   217
follicle development   215
follicle-stimulating hormone (FSH)   166, 198,
   215
food, energy content   82–3
food chains   11–12
   energy transfer   12–13
food samples, testing for nutrients   52
food storage   129
   animals   132
   plants   130–2
food webs   13–14
fossil fuels   299, 306
founder effect   243
fovea   162
fragmentation   202
fructose   44
fruits   208
fungi   253
   sporification   203

Galapagos finches   245
gall bladder   73
gametes   210, 232
gas-exchange surfaces   95–7
gastric juice   79
geminivirus   279
gene flow   243
genes   28, 181, 227, 228
   dominant and recessive   225, 226, 231–2
genetic crosses   232–3
genetic diversity   229
genetic drift   243, 244
genetic engineering   248–50
   designer babies   **Chapter 21**
genetic shuffling   246–7
genetic variation   229–31
genital herpes   258
genotype   226, 228, 232
geotropism   175
germination   187–9
gibberelins (GAs)   175
gills   96
ginger ale   91
glands, endocrine and exocrine   155, 197
glaucoma   165
global warming   304
glomerulus   138–9
glucagon   145, 166
glucose   44, 45
   blood levels   144–5
   production in photosynthesis   69, 70
glycerol   49
glycogen   45, 88, 132
goblet cells   30
goitre   44, 85
gonads   166, 198, 210
gonorrhoea   258
graphs   56
grazing   283
grease-spot test   50

greenhouse effect   66, 302–4
greenhouse gases   302–4
growth   6, 180–1
   effect of auxins   194–6
   effect of human hormones   196–8
   measurement   190–2
   in plants   186–9
growth curves   193–4
growth hormone   166, 198
guard cells   30, 66, 67, 120–1

habitat   9, 281
haemodialysis   **Chapter 13**
haemoglobin   113
hallucinogens   275, 276
haploid number   29, 201
heart   105, 110–13
heart disease   112, 267, 268
height measurement   190
hepatic portal vein   81
herbivores   11
hereditary diseases   254
heredity   227
heroin   276
heterotrophs (consumers)   11, 12–13, 72–3
hibernation   132
hinge joints   177
HIV (human immunodeficiency virus)   257–9,
   277–8
holozoic nutrition   73
homeostasis   143–4
   negative feedback   144–6
   osmoregulation   147–9
   temperature regulation   149–50
homeotherms   151
homologous pairs, chromosomes   29, 201, 211
honeybees, reproduction   202–3
hormonal contraception methods   221, 222
hormones   155, 166–7
human genome project (HGP)   **Chapter 20**
humans, impact on the environment   283,
   299–301
Huntington's disease   254
Hydra, reproduction   201–2
hydrochloric acid, production by stomach   79
hypertonic solutions, effect on cells   38
hypertension   84
hypothalamus   197
hypotheses   125
hypotonic solutions, effect on cells   38

immune system   260–2
   cell types   263
immunity
   active   264
   passive   263
inbreeding   246–7, 248
incomplete dominance   234–5
independent variables   56
indoleacetic acid (IAA)   195–6
infectious diseases   252
   prevention and control   265–6
   transmission   256–7
ingestion   73
inorganic compounds   42
insect-pollination, adaptations   206, 207
insulin   145, 166
intercostal muscles   100
internal environment   143–4
interphase   185
intrauterine devices (IUDs)   221–2
invertebrates, sensitivity   153–4
involuntary actions   158
iodine, use in the body   44
iris   162
iron, use in the body   44
isotonic solutions, effect on cells   38

joints   176–7

karyokinesis   182

312

kidney 137–40
kidney disease **Chapter 13**
kinetochores 185
kwashiorkor 85, 254
Kyoto Protocol 304

labelling diagrams 25
lacteals 81
lactic acid 92
lamina, leaves 65
large intestine 73, 82
leaves
    adaptations to function 66
    passage of water 120
    structure 65–7, 117
    testing for starch 58–9
length measurement 190, 192
lens 163
lenticels 120, 121
lifestyle diseases 266–7
ligaments 171, 176, 177
light 284
    effect on photosynthesis 61–3
    effect on seed germination 188
lignin 120
limewater 100
limiting factors, photosynthesis 61
lipase 78, 81
lipids 48–9, 52
    testing for 49–50
liver 73
    detoxification of harmful substances 136
    storage of food 132
living things, characteristics 6–7
'lock and key' hypothesis 53–4
loop of Henle 138, 140
LSD 276
lumbar vertebrae 173
lungs 98, 99
    diffusion 36
luteinising hormone 166, 198, 215
lymph 109
lymphatic system 109
lymphocytes 113, 261
lysis 38

magnesium, use by plants 44
magnification 97
malaria 256, 266
malnutrition 84
maltase 78
maltose 78
marasmus 85
marijuana 276, **Chapter 23**
mass-flow hypothesis 127
mass measurement 190, 192
mating 246–7
mechanical (physical) digestion 74, 77
medulla oblongata 160
meiosis 210, 211, 212
melanocyte-stimulating hormone (MSH) 198
Mendel, Gregor 225
menopause 215, 216
menstrual cycle 215
meristems 186
mesophyll 31, 67
metabolism 53, 83, 134
methamphetamines 276
methane 92, 304
microbiology **Chapter 3**
micropyle 187
microscopes 23
mid-brain 160
migration 243
milk teeth 75
Millennium Ecosystem Assessment (MEA) 305
mineral resources 297
minerals 43–4
    deficiency diseases 85
    use by plants 70–1
mitochondria 24, 27, 89

mitosis 181, 182–5
monocytes 261
monosaccharides 44, 45, 46
'morning after' pill 221
mosquitoes, as vectors of disease 266
motor neurones 30, 156, 157
mouth, role in digestion 78
movement 6
    animals 176–8
    plants 175
mucus membranes 98
multicellular organisms 22
multiple sclerosis (MS) 255
muscle cells 30
muscles 31, 177–8
mutations 242
mutualism 18
myelin sheath 156, 157

natural family planning (NFP) 218, 222
natural selection 244–6
negative feedback 144–6, 215
nephrons 137, 138–40
nerve tissue 31
nervous system 155
neurones 156–7
neuroplasticity **Chapter 15**
neurotransmitters 159
neutrophils 261
nicotine 103, 274
nitrogen, use by plants 44, 71
nitrogen cycle 20
nitrogen-fixing 20
nitrogenous wastes 135, 136
nitrous oxide 303
non-infectious diseases 254–5
    prevention and control 266–7
non-reducing sugars, testing for 47, 48
non-specific immune responses 260, 261
nucleotides 227–8
nucleus 24, 27, 28
number, pyramids of 15
nutrient cycling 19–20
nutrition 6
    heterotrophic 72–3
    plants 69–71

obesity 84, 132, 267
observations 8–9
ocean power 307
oesophagus 73, 77, 78–9
oestrogen 166, 198, 214, 215
oils 48
omnivores 11
organelles 22–3, 24, 27
organic molecules 42
osmoregulation 147, 147–9
osmosis 34, 37–8, 40
    behaviour of cells in solutions 38–9
ovaries 166, 167, 198, 210
ovulation 215
oxidation 89
oxygen
    in aquatic environments 284
    effect on seed germination 188
    production during photosynthesis 68
    use during respiration 89–90
oxygenated blood 111
oxygen debt 92
oxytocin 166, 198
ozone 303

pacemaker 112
pair bonding 216
pancreas 167
    hormone production 166
    role in digestion 73, 80
pancreatic juice 80
papaya ring spot virus 278, 279, **Chapter 23**
parasitism 18, 72–3, 253
parathyroid glands 167

parenchyma 31
parthenogenesis 202–3
passive immunity 263
passive transport 34
    diffusion 35–6
    osmosis 37–9
pathogens 252–4
pea plants, inherited traits 225–6
peppered moth 146
pepsin 78, 79
peptidase 78
peripheral nervous system (PNS) 155–6
peristalsis 74, 77
permanent teeth 75
permeability, cell membranes 33
pesticides 300
pH, effect on enzymes 54, 55, 79
phagocytes 113, 114
phenotype 226, 228, 232
phloem 69, 116, 117, 127
phosphorus, use by living things 44, 71, 174
photolysis 68
photosynthesis 58, 59–60
    factors affecting the rate 60–1
    investigation of 61–4
    light-dependent stage 68
    light-independent stage 69
    oxygen production 68
    role of leaves 65–7
phototropism 175
physiological diseases 254
pineal gland 167
pistil 205
pituitary gland 166, 167, 197, 198
placebo effect 271
placenta 167, 217–18
plague 257
plant cells 24, 25–6
    effect of solutions 39
    microscopic observation 26
plant diseases 278–9
plant roots, mitosis 183–4
plants
    adaptations 122–3, 130, 149, 206–7
    gas exchange 97
    tissue types 31
plant skeletons 170
plaque 76
plasma 113
plasmolysis 39
platelets 113, 114
plumule 186
poisons **Chapter 23**
pollination 206–7
polysaccharides 45, 46
populations 10
positive feedback 146
potassium, use by plants 44
potometer 124
predation 283
pregnancy 217–18
pre-implantation genetic diagnosis
    (PGD) **Chapter 21**
prenatal development 196
primary growth 186
producers (autotrophs) 11, 12
progesterone 166, 215
prokaryotes 182
prolactin (luteotropic hormone) 166, 198
prostate gland 213, 216
protein energy malnutrition (PEM) 84
proteins 50, 50–1, 52
    testing for 51
protozoa 253
proximal convoluted tubule 138, 139
pulmonary arteries 110, 111
pulmonary veins 110, 111
Punnett squares 226–7
pupils 162, 163–4
    reflex actions 158
pyramids, ecological 15–16

quadrats 288
qualitative and quantitative observations 8

rabies 266
radicle 186
rainfall 284
random mating 246
Ranvier, nodes of 156, 157
receptors 146
recessive genes 225, 226, 228, 231–2
records 8–9, 39
red blood cells (erythrocytes) 30, 113, 114
reducing sugars, testing for 46, 48
reflex actions 157–8
regeneration 202
relay neurones 156, 157
rennin 79
reproduction 6
    see also asexual reproduction; sexual
        reproduction
reproductive systems
    female 213–14
    male 211, 213
resources 296–8
respiration 6, 87–8
    aerobic 89–90, 92
    anaerobic 90–2
respiratory system 98–9
respirometer 89–90
retina 162, 163
rhizomes 204
rhythm method of contraception 218, 219,
    222
ribosomes 27
rickets 43, 85
root hair cells 30, 70–1
    water uptake 118
root pressure 118–19

salinity, aquatic environments 284
salivary glands 73
sampling methods 287, 288
saprophytes 72
saturated fats 49
sclera 162
scurvy 43, 85
seawater, desalination Chapter 4
secondary growth 186
secondary sexual characteristics 213, 214
secretion 136
seeds
    dispersal 208–9
    dormancy 186
    food storage 131
    structure 186–7
selection pressure 244
selective breeding 247–8, 249
selectively permeable membranes 33
semi-lunar valves 112
seminal vesicles 213, 216
seminiferous tubules 213
sense organs 156
sensitivity 6, 152
    in invertebrates 153–4
    in plants 152–3
sensory neurones 156, 157
sepals 206
septum of the heart 111
set points 145
sewage, environmental damage 299, 300
sex cells 30, 210
sex determination 235–6
sex-linked traits 236
sexual intercourse 216–17
sexually transmitted diseases (STDs) 257–8,
    276, 277
sexual reproduction 200, 204, 205
    in humans 210, 211–18
    in plants 205–9
shivering 150
short sightedness 164

sieve-tubes 127
sino-atrial node 112
skeletons 169–70
    human 171–4
skin 150
small intestine 73, 80–1
smoking
    effect on health 102–3, 267, 274
    no-smoking laws Chapter 9
soil 284–6
    artificial Chapter 24
solar energy 307
solutions 37
    effect on cells 38–9
specialized cells 29–30
specific immune responses 260–1
specificity of enzymes 54
sperm 211, 213, 216–17, 218
spinal cord 160, 173
spinal reflexes 158
spindle, mitosis 185
sporification 203
spring scales 192
stamen 205, 206
starch 45, 70, 132
    testing for 47, 48, 58–9
stigma 205, 206
stolons 204
stomach 73, 79
stomata 65, 67, 120–1
storage molecules 70
style 205, 206
substance abuse 272–3
substrates 53
sucrose 44
sugar, effect on teeth 77
sulphur, use by plants 71
sunlight, use in photosynthesis 59–60
supportive tissue, plants 31
support systems 169–70
survival of the fittest 240–1
sweating 150
symbiosis 17–18
synapses 159
synovial joints 176–7
syphilis 258
systole 112

tar, effect on the body 103
target cells 167
taxonomy 7
T-cells 261
teeth 75–7
temperature 284
    effect on enzymes 54, 144
    effect on photosynthesis 61
    effect on seed germination 188
temperature regulation 146, 149–50
tendons 171, 177
terrestrial ecosystems 11
testa 187
test crosses 233
testes 166, 167, 198, 210, 211, 213
testosterone 166, 198, 213
thoracic vertebrae 173
thylakoid membrane 67
thymus 167
thyroid gland 166, 167, 197
thyroid-stimulating hormone (TSH) 166, 198
thyroxine 166, 197
tissue fluid 108, 109
tissues 31
tooth decay 76–7
trachea 98
transgenic organisms 248
translocation 127
transmission of diseases 256–7
transpiration 60, 118, 120–1
    influencing factors 122
    investigations 125–6
transpiration stream 119

transport across cell membranes 34
    diffusion 35–6
    osmosis 37–9
transport systems
    animals 105–6, 114
    plants 69, 116–20, 127
triglycerides 49
tristeza 279
trophic levels 12
tropisms 153, 175
trypsin 78, 81
tubal ligation 219, 222
turgid cells 39–41, 117, 170

ultra-filtration 137, 139
unicellular organisms 22
    diffusion 105, 136
    support 169
unsaturated fats 49
urea 135
ureter 137, 140
urethra 140
urinary system 137
uterus 214
    changes over menstrual cycle 215

vaccines 264
vacuoles 24, 27, 134
vagina 213–14
vascular tissue, plants 31, 116, 117
vasectomy 219, 222
vector-borne disease 128, 256–7, 265–6
vegetarian diets 84, 133
vegetative propagation 204
veins 107, 108
vena cava 109, 110, 111
ventricles 110
vertebral column 171–4
villi, small intestine 81
viruses 22, 253
    aphids as vectors 128
vitamins 43, Chapter 5
    deficiency diseases 85
vitreous humour 162

waste heat utilisation 307
waste products, environmental damage
    299
water
    effect on seed germination 188
    excretion 135, 136
    importance to living things 43
    use in photosynthesis 59–60, 63
water concentration, effects of changes 144
water loss, prevention by plants 122
water resources 296–7, 299
    desalination of seawater Chapter 4
water transport, plants 117–20, 123–4
water vapour 303
weight 84
wet mass 190
white blood cells (leukocytes) 30, 113, 114,
    260, 261
wind-pollination, adaptations 206, 207
wind power 307
wind speed, effect on photosynthesis 61
withdrawal, as method of contraception 218,
    222

'X' chromosomes 235–6
xerophthalmia 43, 85
xerophytes 122, 149
xylem 30, 69, 116, 117, 170
    structure 120
    water movement 118–19

'Y' chromosomes 235–6
yeast, anaerobic respiration 90–1
yellow fever 266

zygote 217